CARBON DIOXIDE AND GLOBAL CHANGE:

EARTH IN TRANSITION

Sherwood B. Idso

ISBN 0-9623489-1-0

TABLE OF CONTENTS

PREFACE

There should be little doubt in the mind of anyone that the times in which we live are indeed unique; for in a development never before experienced in the history of life on Earth, a single species has emerged from the backdrop of the biosphere to gain such power and prominence that it possesses the capacity to totally change the face of the planet and the circumstances of all that inhabit it. The species, of course, is man; and the phenomenon, his unbridled passion for processing the planetary resources upon which all life depends for sustenance and for returning the remains to the environment as the altered waste products of his supra-metabolic activity.

Foremost among the host of societal effluents thus foisted upon the fabric of nature is carbon dioxide or CO_2, expired directly by all animal life, but produced by man most prodigiously through the burning of fossil fuels and the felling of forests. So significant have these latter activities become over the past few centuries, in fact, that man is currently responsible for increasing the concentration of this trace atmospheric constituent by a full 30%, with half of the increase occurring in just the past four decades.

One of the great concerns which many people share about this alteration of the atmosphere is that it may lead to an unprecedented warming of the Earth via the infamous greenhouse effect: that it will melt the polar ice caps, raise sea levels, flood coastal lowlands, and wreck all manner of havoc upon agriculture and the planet's natural ecosystems, possibly even to the point of driving many species to extinction. Others, however, pointing to the great biological benefits of atmospheric CO_2 enrichment in enhancing vegetative productivity and plant water use efficiency, have heralded the rapidly rising CO_2 content of Earth's atmosphere as a blessing in disguise, a much-to-be-desired phenomenon which could ultimately lead to a dramatic "greening of the Earth" and provide the means we may one day desperately need to adequately feed the ever-increasing numbers of our kind. As a result of this dichotomy, in the words of R.S. Scorer,[1] "politicians do not know what or whom to believe -- those who say that a major sector of the environment is in danger; those who say there is no global danger; or those who say it will take 20 years to discover the extent of the man-made contribution to the problem."

As an aid to resolving this dilemma, I have attempted to put it in proper perspective by producing the volume for which these words serve as preface. It begins with a presentation of the broad context of the problem and is followed by a brief rehearsal of the many dire predictions which result from state-of-the-art climate model projections. Next, revealing a bit of my own personal prejudice, it critically reviews the status of current climate modeling work and describes a number of recent developments in the empirical approach to climate change. The first half of the book then concludes with a review of current research efforts directed to detecting the first intimations of the predicted climate catastrophe.

The second half of the book is biologically oriented. It begins with a comprehensive review of what we know about the direct effects of atmospheric CO_2 enrichment on plant physiological processes and their likely modification by a number of environmental constraints. Here too there is a search for the first signs of the predicted "greening of the Earth." Then comes a brief look at the vastly neglected field of CO_2 effects (or non-effects) on animals and a comprehensive picture of where the world may be headed as a result of mankind's "fertilization" of the atmosphere with CO_2.

Although I have tried to be objective throughout, and to fairly tell both sides of this intriguing story, it is perhaps unavoidable that my own views on the topic should come to the fore. Hence, to better enable the reader to form his own opinion of the matter, I have carefully documented nearly every statement of fact. This documentation is by numerical superscript reference to book and journal citations assembled at the back of the book. In addition, nearly every key word of both the text and the reference citations is included in an appended subject index which is followed by an author index. It is hoped that this material will also be helpful to scientists seeking to identify key areas of activity and the ways in which their particular specialties fit into the broad spectrum of CO_2/global change research.

In concluding these words of introduction, I note that many voices are calling for immediate action on a number of national and international policy initiatives designed to deal with the phenomenon of CO_2-induced global change. Some are warranted. Some are not. Still others may actually be detrimental to our future well-being. Hence, what we must clearly have, before we embark upon what could well be the greatest human endeavor of all time, is a solid foundation of knowledge upon which to base our efforts. Though the problems we face as the currently dominant species of planet Earth are as imminent as they are potentially catastrophic, we cannot begin to effectively broach them until we truly understand them. Hopefully, this book will contribute in some small way to this important worldwide effort: our species' one last attempt to comprehend the complexity of Earth's climate system and its synergistic coupling with the biosphere, that we may forever preserve the stability, diversity and vitality of Earth's biota in the face of the burgeoning mass of humanity which threatens to summarily destroy the incomparable and irreplaceable results of eons of creative activity.

Sherwood B. Idso

4 July 1989

Tempe, Arizona

Reference:

1. Scorer, R.S. (1988) The atmosphere -- Product and host of life: Its influence on human purposes. Presidential address to the Royal Meteorological
   Society, London, June 1988. <u>Weather</u> 43, 419-429.

1 DEFINING THE PROBLEM

.1 Earth in Transition

Earth is an evolving planet; there is nothing about it which is static. In our own lifetimes, for example, we regularly experience seasonal extremes of heat and cold,[1] years of drought,[2,3] and periods of flooding.[3,4] We witness the subtle creep of desert sands,[5,6] heralding the spread of aridity,[7,8] and the ominous dieback of forests,[9,10] mutely attesting to the changing character of the landscape.[11,12]

On longer time scales, far beyond the ken of man,[13] even the continents shift;[14-16] while mountains rise,[17] but to be reduced to rubble.[18] And through it all, life either adapts or is swept from the scene.[19-21]

Global change. It is an intrinsic property of the system. Earth's orbital parameters vary ever so slightly, and ice ages come and go.[22,23] Sea levels fluctuate,[24] land surfaces expand and contract,[25] and the biosphere alternately flourishes and suffers massive extinctions.[26,27]

But the relationship of life to inanimate nature is not always one of submissive response to the immutable demands of the physical environment. The forces which shape the face of the land and set bounds to the sea, producing the many and varied niches which species radiate to fill,[28,29] are themselves often products of life processes. Indeed, the two are sometimes so intimately intertwined that cause and effect in the great panorama of global change are difficult to distinguish, leading some to speculate that the planet itself is alive.[30-32]

Be that as it may,[33-36] the breath of life permeates Earth's atmosphere.[37] Oxygen, for example, is a by-product of plant photosynthesis;[38] while a host of additional gases owe their origin to a number of other biotic entities and enterprises.[39] Carbon dioxide, however, comes primarily from the Earth itself, released in the cataclysmic eruptions of great volcanoes and by outgassing at mid-ocean ridges and the restless margins of the huge plates which form the basis of our planet's shifting crust.[40-42] Nevertheless, modulated over time by weathering and sedimentation processes, which are influenced by numerous biological phenomena,[43,44] the CO_2 content of the atmosphere is also largely determined by the properties and products of Earth's biota.[45]

Again, this relationship too is reciprocal, as CO_2 in its turn exerts a profound influence upon the biosphere: both directly, through its dual effects on plant photosynthesis and transpiration,[46-48] and indirectly, through its multi-faceted effects on world climate.[49-52] And herein lie the seeds of what is proving to be one of science's greatest dilemmas.[53,54]

Before formulating the problem explicitly, however, it is necessary t
explore these two complex aspects of the CO_2-biosphere interaction i
somewhat more detail.

1.2 The Brightening of the Sun

Although currently a rather minor constituent of the atmosphere, being
present at a mean concentration of only 350 parts per million (ppm),[55] CO_2
has long played a major role in the development and sustenance of life or
Earth. Consider, for example, what is generally referred to as the parado:
of the faint early sun.[56,57]

According to accepted theories of stellar evolution, the energy output or
the sun has been steadily increasing since the birth of the solar system,
when its luminosity is believed to have been about 25 to 30% less than what
it is now.[58-61] All else being equal, this reduced solar energy flux shoule
have consigned the world's waters to a state of eternal ice and precluded the
development of life.[62] However, we know from the geologic and fossil records
that both liquid water and life have been persistent features of the planet
for at least the past 3.8 billion years.[63-69] And hence the paradox.

So, one might logically inquire, what kept Earth's climate mild enough to
maintain a viable biosphere throughout this period?

Practically all scientists conversant with the problem feel that the
answer to this question is the CO_2 "greenhouse effect." According to this
theory, the CO_2 content of Earth's early atmosphere was much greater than it
is today,[61] perhaps by as much as several orders of magnitude.[38,70-72] And
at such high CO_2 concentrations, the greater infrared opacity of the
atmosphere would have significantly restricted the loss of surface-emitted
thermal radiation to space, thereby elevating the near-surface air
temperature of the planet several degrees above what it would have been under
CO_2 concentrations characteristic of today.[73]

As time progressed and the energy output of the sun increased, however,
less greenhouse warming would have been needed to keep the climate suitable
for life. Thus, there must have been a process for removing CO_2 from the
atmosphere which intensified as the near-surface air temperature rose.

Probably the leading candidate mechanism to answer this requirement is the
geochemical carbonate-silicate cycle, the primary characteristics of which
have been well known for several years.[74-78] Starting with an increase in
near-surface air temperature, the process is intensified by the consequent
increase in oceanic evaporation, which leads to an increase in global rainfall
and continental runoff. And as atmospheric CO_2 dissolves in rainwater to form
carbonic acid,[79] and as soil-air CO_2 arising from the decomposition of organic

atter produces even more,[80] more CO_2 is thus removed from the air, both irectly and indirectly, thereby reducing the strength of the CO_2-induced reenhouse warming and counterbalancing the temperature increase caused by the rightening of the sun.[81]

The cycle continues with the weathering of rocks containing calcium-ilicate minerals. The carbonic acid in the rainwater reacts chemically with hese rocks to release calcium and bicarbonate ions which make their way to he ocean via groundwater, streams and rivers. In the sea, these ions are ncorporated into the shells and skeletons of various marine organisms as alcium carbonate. The captured carbon is then effectively immobilized for undreds of thousands of years, as the organisms die and fall to the ocean ottom, where the carbonate sediments which they create are gradually ransported by the spreading of the sea floor to the margins of the ontinents. There, the sediments are subducted beneath the land, only to ltimately be transformed once again into new silicate rocks. This process f carbonate metamorphism produces gaseous CO_2 as a by-product, which then ompletes the cycle by making its way back to the atmosphere via volcanos and oda springs;[80] and it is the balance between these two processes of CO_2 upply and removal which controls the CO_2 content of the atmosphere over time cales on the order of hundreds of thousands of years.[82]

Estimates of the temperature dependence of this atmospheric CO_2 depletion rocess have resulted in two major representations of the phenomenon.[83,84] oth approaches suggest that as temperature rises, the rate of CO_2 equestering is intensified by an amount which lowers the atmospheric CO_2 ontent by the degree required to reduce its greenhouse effect by just a ittle less than what is needed to compensate for the increase in temperature aused by the increase in solar luminosity. Consequently, although the arming of the Earth due to the brightening of the sun cannot be forestalled orever by this process, it can be slowed considerably.

In thus removing CO_2 from the atmosphere, marine organisms play a major ole in keeping the Earth fit for life. Even in the absence of the marine iota, however, the carbonate-silicate cycle would presumably still operate; or as the concentration of calcium and bicarbonate ions in the ocean rises n response to weathering, there must ultimately come a point at which alcium carbonate forms and precipitates even without the intervention of ife.[44]

Over time scales on the order of 500,000 years or more, the purely hysical aspects of the carbonate-silicate cycle account for about 80% of the arbon exchanged by the solid Earth and the atmosphere.[44] Over smaller ntervals of time, however, biological processes generally dominate.[85]

The evidence most often cited as substantiating this claim is the high partial pressure of CO_2 in soil pore space, where its concentration may be 1 to 40 times greater than what it is in the atmosphere.[86,87] These high concentrations of CO_2 in the soil are maintained by the oxidation of organic detritus; and the rate of oxidation approximately doubles with each 10 ° rise in temperature.[88] Hence, as near-surface layers of soil and air warm, the soil biota produce more CO_2 to react with soil water in the weathering of silicate minerals, which, as previously described, ultimately lowers the CO_2 content of the atmosphere and counteracts the tendency for warming.

The weathering of silicates and the hydrologic cycle also provide other nutrients to the sea in addition to calcium and bicarbonate ions, notably phosphorus[89,90] and nitrogen.[91,92] And as the influx of these nutrients stimulates the marine biota to still greater primary production, the atmospheric CO_2 content is even further reduced. Indeed, if these biological processes were not presently at work, the properties of the purely physical part of the carbonate-silicate cycle suggest that the CO_2 content of the atmosphere would be several times greater than what it is today.[89-92] Hence it has been reasonably argued that life is in fact operating to keep the planet's physical state, and especially its temperature, suitable for the biosphere's continued existence.[88,93]

According to this view, then, which is generally referred to as the "Gaia Hypothesis," life on Earth has created its own opportunities, so to speak, and especially over the past 400 million years or so, when vascular land plants have exerted a rather decisive influence on the weathering rates of rocks and the magnitude of riverine organic matter transferred to marine sediments.[94] But there are some unsettling indications that, from a geological perspective, time may be running out for the planet's biota.[83]

1.3 Death Knell of the Biosphere?

From a probable value on the order of 70,000 ppm at 3.5 billion years ago,[71] the CO_2 concentration of Earth's atmosphere has dropped[95] to a current mean value of only 350 ppm;[55] and as this drop represents approximately 99.5% of the total decrease possible, it is clear that we are very close to the lower limit of effective temperature regulation by CO_2 in the face of a brightening sun. As Lovelock and Whitfield[88] have described the situation, "in human terms the crisis is still infinitely distant, but in terms of the life span of the biosphere, rich with familiar metazoans, we might forecast an end to the long spell of cool and favourable climate."

Although from the perspective of man it may thus seem that there yet remains a good deal of time before we need worry about this postulated eventuality, its potential occurrence may well be much closer than we might care to acknowledge. For one thing, the CO_2 compensation point for plant photosynthesis, that is, the atmospheric CO_2 concentration below which plants can no longer photosynthesize enough to maintain essential life processes, generally lies between 50 and 100 ppm for most C_3 or Calvin-Benson type plants, which make up the bulk of Earth's vegetation.[96] In addition, analyses of the CO_2 concentrations of air bubbles trapped in ancient ice indicate that atmospheric CO_2 values as low as 180 ppm may have been reached during the coldest part of the last ice age.[97-101] Hence, as we are poised to begin another such glaciation any time now,[102] according to the periodicities exhibited by past glaciations,[103-106] we could well see many plants struggling to survive in mere thousands of years, as opposed to the 100 million years suggested by Lovelock and Whitfield,[88] all else being equal.

Rarely, however, does all else remain unaffected in complex environmental transformations; and guided by God, Gaia or just plain Global Serendipity, a new player has recently appeared on the scene which may yet save the day for the biosphere.[107] On the other hand, he could also prove its ultimate downfall.[108] The player? Man.

1.4 Man and Carbon Dioxide

Fig. 1.1 shows the course of population growth for our species over the past 2,000 years. Prior to about 1400 AD, there was apparently very little change in the number of people inhabiting the planet. Subsequent to that time, however, it is clear that mankind has been proliferating at an ever-increasing rate. And since man has such a penchant for the burning of fossil fuels such as coal, gas and oil,[112] which produces CO_2 as an ubiquitous by-product,[113] as well as for the felling of forests,[114,115] which also releases CO_2 to the air,[116-118] there is reason to believe that the recent population explosion of our species may be significantly impacting the CO_2 content of the atmosphere.[119,120]

In order to evaluate this hypothesis, it is necessary to know the course of atmospheric CO_2 concentration over the period for which we have a reasonably good representation of global population growth; and from Fig. 1.1 it is clear that that period stretches from about 1600 to the present. Hence, Fig. 1.2 depicts what is known about atmospheric CO_2 over this period of Earth's history.

As can be seen from this relationship, the concentration of CO_2 in the atmosphere has risen very analogously to the pattern of world population growth over the past four centuries. Indeed, from Fig. 1.1 and 1.2, and some additional population and CO_2 data contained in Table 1.1, it is possible to

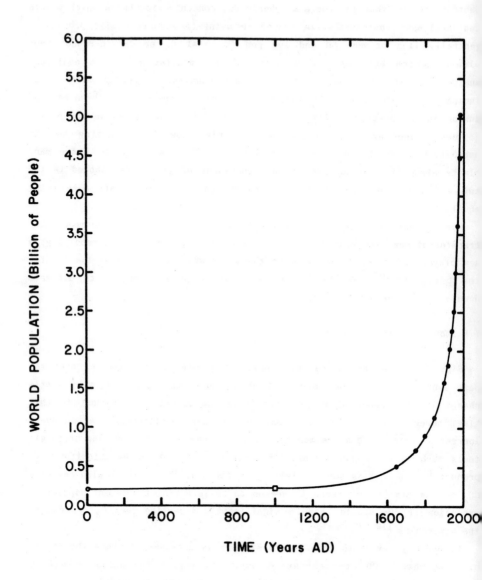

Fig. 1.1   World population vs. time, based upon data from Mills[109] (□), Westing[110] (○) and Bogue[111] (●).

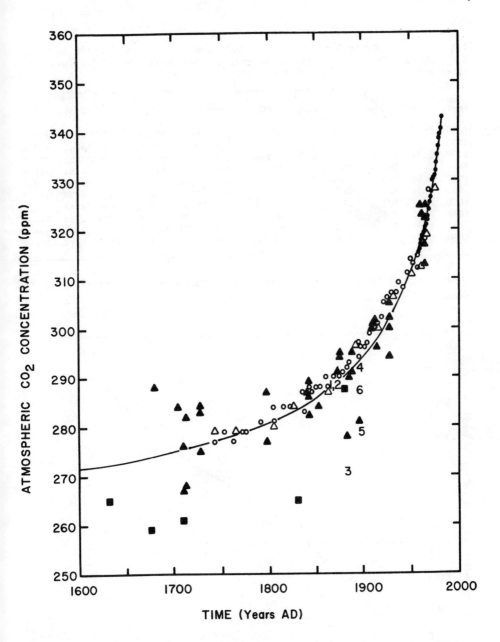

Fig. 1.2   Mean atmospheric CO_2 concentration vs. time, based upon Antarctic ice core data from Siple Station ($\triangle$ [121] and o [122]), Station D57 ($\blacksquare$ [123]), and Law Dome ($\blacktriangle$ [124]), modern air concentration measurements at Mauna Loa, Hawaii ($\bullet$ [55]), and some nineteenth century air concentration measurements at a number of sites around the world (1,[125] 2,[126] 3,[127] 4,[128] 5,[129] 6[130]), as appropriately adjusted by Fraser et al.[131]

Table 1.1   World population in billions of people,[111,132] and mean atmospheric CO_2 concentration in ppm, as determined by the relationship constructed to fit the data of Fig. 1.2.

Year	People	CO_2	Year	People	CO_2	Year	People	CO_2	Year	People	CO_2
1650	0.51	273.0	1959	2.96	315.9	1969	3.61	324.1	1979	4.38	336.7
1750	0.71	277.5	1960	3.01	317.0	1970	3.68	325.5	1980	4.45	338.6
1800	0.91	281.0	1961	3.08	317.8	1971	3.76	326.4	1981	4.53	339.3
1850	1.13	286.0	1962	3.14	318.5	1972	3.84	327.7	1982	4.61	340.6
1900	1.59	294.3	1963	3.20	319.2	1973	3.92	330.1	1983	4.68	342.8
1920	1.81	299.7	1964	3.26	319.5	1974	4.00	330.5	1984	4.76	344.3
1930	2.02	303.0	1965	3.32	320.5	1975	4.08	331.0	1985	4.84	345.6
1940	2.25	306.7	1966	3.40	320.7	1976	4.15	332.0	1986	4.94	346.8
1950	2.51	311.0	1967	3.47	321.9	1977	4.23	333.6	1987	5.03	348.6
1958	2.90	315.2	1968	3.54	322.6	1978	4.30	334.9	1988	5.11	351.2

construct a meaningful plot of the latter parameter as a function of the former, as shown in Fig. 1.3.

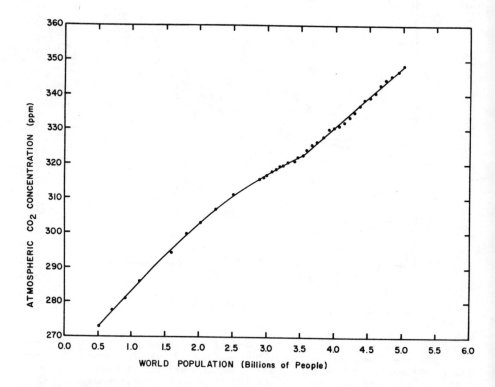

Fig. 1.3 Mean atmospheric CO_2 concentration vs. world population, as derived from the data of Table 1.1.

Clearly, this figure is highly suggestive of a strong impact of man on Earth's atmospheric CO_2 content. In fact, it suggests that the burgeoning population of humanity has recently outstripped all other biological and geophysical effects of the past several million years in terms of returning CO_2 to the air.[133]

What are the ramifications of this phenomenon?

1.5 Carbon Dioxide and Global Change

The great magnitude of man's contribution to the rapidly rising CO_2 content of Earth's atmosphere suggests that humanity, if it does not soon self-destruct,[134,135] may well play a significant role in forestalling the demise of the biosphere predicted by Lovelock and Whitfield.[88] It additionally suggests that mankind may play a significant role in subverting any imminent glaciation. Unfortunately, it also suggests that humankind could bring on a "super-interglacial," different from anything the Earth has experienced for millions of years and something which civilization may be hard put to endure.[136]

On the plus side of this balance sheet, the increasing CO_2 content of Earth's atmosphere is like a breath of fresh air for the planet's close-to-suffocating vegetation,[83,88,107] as nearly all plants utilize CO_2 as one of the primary raw materials essential to the process of photosynthesis. Literally hundreds of laboratory and field experiments have established this fact.[137] In the mean, they suggest that a 330 to 660 ppm doubling of the air's CO_2 content will increase plant productivity by about a third,[138,139] while reducing plant transpiration by approximately the same amount.[140] Hence, for a 330 ppm increase in atmospheric CO_2 concentration, plant water use efficiency, or the amount of dry matter produced per unit of water transpired, essentially doubles.[141,142] And for a 660 ppm increase it triples.[143] Consequently, the whole face of the planet will likely be radically transformed[144,145] -- rejuvenated,[146] as it were -- as the atmospheric CO_2 content reverses its long history of decline and returns, in significant measure, to conditions much closer to those characteristic of the Earth at the time when the basic properties of plant processes were originally established.[133] Indeed, it is the thesis of many that this "greening of the Earth" is already in progress (see Sec. 9).

On the minus side of the ledger, however, is the ominous threat of climate change. Will the increase in the CO_2 content of Earth's atmosphere caused by the activities of man bring on a super-interglacial[136] with all of its attendant ill effects?[147,148] Are the polar ice caps destined to melt, raising sea levels and inundating coastal lowlands?[149-151] Will tropical hurricanes intensify with warmer sea surface temperatures?[152-154] Will CO_2-induced cooling of the stratosphere exacerbate the destruction of Earth's protective ozone layer?[155,156] And will insufficient summer soil moisture,[157-159] along with increases in plant diseases[160-162] and insect pests,[163,164] destroy the agricultural productivity of large areas of the world's croplands?[165-170]

How about the planet's fragile water resources?[171,172] Will they be reduced below their ability to sustain even the world's current population?[173-176] What about fisheries,[177-179] forestry,[180-187] and human health?[188] Indeed, what about the extinction of entire species?[189-192] Will all succumb to the dire effects often predicted to result from the greenhouse effect of increasing atmospheric CO_2 caused by the proliferation of humanity?[193-199]

These are the truly burning questions confronting man today, as he emerges from the background of the biosphere to become the chief force in the continuing evolution of the planet. For better or for worse, mankind's mere presence is proving to be a phenomenon of paramount importance for the entire assemblage of the world's biota. And this is the dilemma we face: to discover our true role in the great drama of global change, and especially with respect to our flooding of the Earth's atmosphere with CO_2, an event unprecedented for both its rapidity and magnitude. Typically referred to as mankind's great "geophysical experiment,"[200] it has been described[136] as something so vast and so sweeping that "were it brought before any responsible council for approval, it would be firmly rejected."

Will man's production of CO_2 thus prove the salvation of life on our planet?[201] Or is the rising CO_2 content of the air the very essence of the problem?[202] This is the environmental issue of our time[203-206] and the primary question which science must confront, as humanity marches inexorably forward to meet and make its future.

2 PREDICTING MAN-INDUCED CLIMATIC CHANGE

2.1 The Role of Models

How can we reasonably estimate how the climate will change in response to the increase in the CO_2 content of Earth's atmosphere caused by the activities of man? Obviously, the processes which determine a planet's climatic state are much too numerous and complex to be physically reproduced in any laboratory. As a result, scientists seeking to answer this question have generally resorted to mathematical simulations of how the Earth's climate system is supposed to work.[1,2] In fact, it has been claimed by some[3] that "numerical simulations of changes in global climate due to increasing concentrations of greenhouse gases offer perhaps the only means of anticipating quantitatively how these gases may affect our planet."

Although clearly too narrow in its focus,[4,5] this statement is, nonetheless, an accurate portrayal of the current thinking of many scientists, as a 1985 state-of-the-art review of the subject[6] concludes that "the only applicable method for projecting future climates is the construction of mathematical models based on the full set of fundamental physical principles governing the climate system." Hence, it behooves us to briefly review the field of climate modeling. Exactly what is it? How does it work? And what does it predict for Earth's future?

2.2 Modeling Earth's Temperature Response to CO_2

The object of climate modeling is to apply fundamental physical principles, such as the conservation of mass, momentum and energy, to the various Earth sub-systems (atmosphere, oceans, sea ice, land surfaces and glaciers, for example) which act in concert to determine the globe's climatic state.[7] In increasing order of complexity, such schemes produce what are variously referred to as energy balance models,[8] radiative-convective models,[9] statistical-dynamical models[10] and general circulation models.[11,12] Although differing in many respects, all of these approaches to climate simulation seek to reduce the many processes which combine to produce Earth's climate to a set of simultaneous equations which can be solved by numerical means with the aid of a computer.[13-15]

In spite of the great complexity of the most advanced of these models, plus the vast amounts of computer time required to run them to a state of equilibrium,[16] the past dozen or so years have seen a surprisingly large number of such studies conducted, where the objective has been to determine the ultimate climatic consequences of a 300 to 600 ppm doubling of the

Earth's atmospheric CO_2 concentration. In general, energy balance and radiative-convective models have produced mean global warmings ranging from 1.3 to 3.3 $^{\circ}C$ for this scenario, while statistical-dynamical and general circulation models appear to be converging on a mean global warming of about 4 $^{\circ}C$.[17]

This warming, of course, is not expected to be uniform, as the relative distributions of land and ocean,[18-21] snow cover and sea ice,[22-25] and actively-transpiring vegetation and barren ground[26-29] all exert important secondary influences upon the primary radiative perturbation caused by the CO_2 increase. Also important is the latitudinal distribution of water vapor,[30,31] which varies in phase with the mean zonal air temperature.[5] Hence, as a result of all of these interacting factors, most general circulation models of the atmosphere predict a doubled CO_2 equilibrium warming of only 1 or 2 $^{\circ}C$ near the equator, but a warming of several times that amount in Earth's polar regions,[14,32,33] a relative distribution which is also suggested by certain empirical observations.[5,34]

2.3 Hydrological Predictions

Because of the amplification of CO_2-induced warming at high latitudes, many scientists are concerned about the stability of the polar ice caps.[35-39] They are worried that the predicted air temperature rise will release the tremendous reservoirs of water currently held there, with the devastating effect of flooding coastal lowlands.[40-43] Even without any net melting of ice, there is the additional threat of sea level rise due solely to the thermal expansion of the world's oceans upon warming.[44-47] And if the worst estimates turn out to be correct, it has been calculated that the value of flooded real estate in the United States alone would approach one trillion 1980 U.S. dollars.[7]

Another aspect of the hydrologic cycle which may be significantly altered by man's perturbation of the atmospheric CO_2 content is the world's rainfall pattern.[48] Although on a global basis precipitation should be increased on the order of 10% for a 300 to 600 ppm doubling of the atmospheric CO_2 concentration,[49] there have been several model predictions of significant regional reductions leading to deleterious decreases in summer soil moisture over extensive regions of middle and high latitudes of the Northern Hemisphere.[50-55] Combined with an increased demand for surface evaporation due to increased temperatures, dire consequences have thus been predicted for agriculture.[56-59]

Closely linked to the problem of soil water availability are concerns about changes in streamflow and river volume,[60,61] where significant CO_2-

induced reductions have also been predicted.[62] As CO_2 is an effective antitranspirant,[63] however, there have been some predictions of CO_2-induced streamflow increases as well.[64,65] In fact, some climate model studies[25,66-68] have even refuted the predictions of regional soil moisture decreases.[69,70] As a result, newer assessments of hydrological issues have emphasized the likelihood of increased variability in these phenomena as a result of increasing atmospheric CO_2,[71] predicting both droughts and floods at the same time.[72,73]

2.4 Effects of Trace Gases

Augmenting the climatic changes anticipated as a result of man's flooding of the air with CO_2 are the similar impacts of a number of other anthropogenically-produced gases.[74-77] Foremost among this menagerie of minor but important components of the atmosphere are methane (CH_4), nitrous oxide (N_2O) and a host of chlorofluorocarbons (CFCs). Again, the reality of the concentration increases of these other "greenhouse gases" is supported by both direct measurements over the past decade or so[78-81] and by analyses of air bubbles trapped in ancient polar ice.[82-88]

Although the individual climatic impact of each of these trace gases is much less than that of CO_2, their combined influence is about equivalent.[89-92] Hence, the current best estimate of the most advanced general circulation models of the atmosphere in use today is that, by the time the CO_2 content of Earth's atmosphere has reached 600 ppm, the expected equilibrium warming of the planet should be about 8 °C relative to its mean temperature at a CO_2 concentration of 300 ppm, with half of the warming coming from the greenhouse effect of CO_2 and the other half coming from the combined effects of all of the other greenhouse gases produced by man.

2.5 A Time of Reckoning

When is the predicted warming expected to occur? As long ago as 1980, some scientists were claiming that it should have already been evident, although they could not detect it at that time.[93] Neither has anybody been able to conclusively identify it subsequently,[94-99] although several researchers have claimed that the apparent 0.5 °C global warming of the past century[100-104] is at least consistent with the model predictions,[102,105-109] as are certain observed changes in rainfall patterns.[110,111] Hence, in an attempt to increase the possibilities for early detection and confirmation of enhanced greenhouse warming due to the production of greenhouse gases by the activities of man, a considerable effort has been made to model the transient response of Earth's climate to an increase in atmospheric CO_2.[112-114]

Due primarily to the great thermal inertia of the world's oceans,[115,116] the lag time between the point of occurrence of a CO_2 doubling (to 600 ppm) and the actual attainment of the predicted CO_2/trace gas equilibrium warming has been estimated to range from a few decades to as long as a century.[25,117-121] Consequently, as the "equivalent" CO_2 concentration (which incorporates the calculated climatic effects of all other greenhouse gases) has risen about 40% above its preindustrial level,[122] it is the belief of most scientists who have studied the subject that an unequivocal expression of CO_2/trace gas warming should be manifest within the next decade or two,[99,123,124] if the predictions of the climate models are correct.[118] And it is this big "if" to which we will turn our attention in Sec. 3.

2.6 The CO_2-Ozone Connection

One final environmental consequence of atmospheric CO_2 enrichment which has been hypothesized to occur on the basis of model calculations is a dramatic reduction of stratospheric ozone concentrations. With the surprising discovery of the now-famous Antarctic "ozone hole" reported by Farman et al.[125] in 1985, tremendous attention has been paid to this potentially serious problem,[126] which could lead to a dangerous increase in the receipt of ultraviolet radiation (particularly UV-B radiation in the 280-320 nm range) at the Earth's surface, not the least of the deleterious consequences[127-132] of which may be a significant increase in the incidence of human skin cancers, cataracts, and a number of other maladies.[133-136]

The possible relationship of this problem to the phenomenon of atmospheric CO_2 enrichment arises from the fact that as general circulation models of the atmosphere predict CO_2-induced temperatures increases at the Earth's surface, they call for simultaneous decreases in stratospheric temperatures.[66,118] And with even a slight drop in air temperature above the tropopause, there is a vastly increased potential for the formation of polar stratospheric clouds,[137,138] which are believed to provide extremely effective sites for the heterogeneous chemistry which results in ozone depletion.[139-141] Very briefly, it is believed that the nitric acid trihydrate and/or water ice particles of these clouds provide effective catalytic surfaces which greatly enhance the chemical reactions that convert normally inactive HCl and $ClONO_2$ into ClO, which is readily dissociated by solar ultraviolet radiation with the appearance of the spring sun.[142,143] Each one of the highly reactive chlorine atoms thereby released is believed to have the capacity to attack and destroy some 100,000 ozone molecules before it is removed from the atmosphere in rainfall as hydrochloric acid.[144] Consequently, if the many linked models of is total and tortuous process are all correct, the rising CO_2 content of rth's atmosphere could well be a substantial contributing factor to the tentially serious weakening of Earth's protective shield against harmful UV-radiation, via its role in decreasing stratospheric temperatures and ereby enhancing the formation of polar stratospheric clouds.[145]

3 EVALUATING THE MODELS

3.1 General Model Inadequacies

How much faith should one place in the predictions of today's mos[t] advanced climate models? With so much riding on the outcome, this is [a] question with which we must deal forthrightly; for there are those who hav[e] urged for some time now[1-4] that we attempt to slow the continuing rise i[n] atmospheric CO_2 content by a number of different means, even though th[e] problem is but perceived "through a glass darkly."[5]

A good starting point for this evaluation is the assessment of th[e] situation provided by Ruth Reck[6] at the 1981 LaJolla Institute workshop o[n] "The Responsible Interpretation of Atmospheric Models and Related Data." Reck first pointed out that "confidence in the predictions of atmospheri[c] models is limited, (1) by the uncertainty in our knowledge of the existin[g] atmospheric system as it is defined by input data, and (2) by the ability o[f] the model to describe all pertinent atmospheric processes." She the[n] astutely observed that to overcome these limitations, "it is necessary tha[t] the model sensitivity to parameter perturbation be equivalent to tha[t] observed in the real world" and that "this requires that the model includ[e] all necessary physical mechanisms as well as pertinent coupling scheme[s] between parameters."

How did the climate models of the late 1970s and early 1980s fare in thi[s] regard? In a paper published in 1979, Choudhury and Kukla[7] noted that the CO_2/climate problem had so many facets to it that it was just too complicate[d] to be reliably solved by the simple annually-averaged general circulatio[n] models of the atmosphere then in use, and that "predictions derived from the[m] should be considered highly tentative at best." Other climate modeler[s] pointed out that "the predicted global warming for a given CO_2 increase [was] based on rudimentary abilities to model a complex system with many non-linear processes."[8] And as late as 1985, it had still to be admitted that even "the most comprehensive global climate models greatly oversimplify or misrepresent key climate processes,"[9] a long-standing problem which had earlier caused Kellogg[10] to suggest that "no climate model, however sophisticated, can hope to include all the factors and interactions of the planetary system."

Striving to overcome these limitations, climate modelers have continued to ply their trade, ever increasing in sophistication and depth and detail of treatment. Concurrently, however, certain empiricists have severely criticized their work,[11-18] which has led to the development of an intense and protracted controversy.[19-34]

So how does the situation stand today? For one thing, as the models have gotten better, so also have people's appreciation for their limitations increased, including that of the models' creators.[35]

Consider, for example, the judgement of two world-class climate modelers-- M.E. Schlesinger and J.F.B. Mitchell -- as expressed in a comprehensive state-of-the-art analysis[36] published in May of 1987. After an exhaustive study of several of the most respected general circulation models (GCMs) of the atmosphere in use at that time, they concluded that "there is an inherent limitation in our ability to validate the accuracy of GCM perturbation simulations, which thereby affects our confidence in the accuracy of the GCM simulations of CO_2-induced climate change."

In expanding on this statement, they emphasized that the models "frequently employ treatments of dubious merit, including prescribing the oceanic heat flux, ignoring the oceanic heat flux, and using incorrect values for the solar constant," pointing out that "such approximations indicate that the models are physically incomplete and/or have errors in the included physics." Noting further that "CO_2-induced climate changes simulated by different GCMs show many quantitative and even qualitative differences," some of which derive from such simple things as the choice of vertical finite difference approximations,[37] they thus concluded that "we know that not all of these simulations can be correct, and perhaps all could be wrong."

Much the same conclusion emerges from the 1988 model intercomparison of Grotch,[38] where the predictions of the various models not only did not always agree well with each other but in some instances even failed to adequately reproduce the proper annual cycle of climate over the world, as noted by others as well,[37-40] and which Weickmann and Chervin[41] suggest the models should be capable of doing "with a certain amount of fidelity." Such is also the tale of the recent model intercomparison of Gutowski et al.,[42] where some of the models' regional control-climate energy fluxes actually changed in different directions upon doubling the atmospheric CO_2 content. In fact, in a comparison of GCM predictions with actual real-world observations, Weare[43] found the mean difference between modeled and measured net solar radiation at the Earth's surface to be approximately 25%, the mean difference between modeled and measured net thermal radiation at the Earth's surface to be approximately 50%, and the model-generated El Nino anomalies of planetary albedo and outgoing thermal radiation at the top of the atmosphere to differ from those observed in nature by fully 400%! Very similar results have also been obtained by a number of other investigators;[44-47] and Neeman et al.[48] have demonstrated that even if a model can accurately simulate the observed parameters of the present climate, this ability is not, in and of itself, a guarantee of the correct simulation of climate sensitivity, which is an absolute prerequisite for climate change experiments.

Even more revealing than these observations are the recent findings of Boer and Lazare.[49] By merely utilizing different horizontal grid resolutions in a standard GCM, they obtained differences in the predicted base state of the model climate "which rival or exceed those found in the kinds of strong external forcing experiments often performed with such models," i.e., CO_2 doubling, indicative of the fact that "the numerical solutions to the governing equations depend importantly, and in nonobvious ways, on resolution and parameterization," such that "parameterizations which are acceptable at low resolution ... may no longer be suitable at higher resolution, so that increased resolution degrades the simulation rather than improving it," as is also evident in the work of others.[50-52] Consequently, it is not surprising that in updating the study of Schlesinger and Mitchell, Cess and Potter[53] found that "more recent GCM simulations have actually broadened [the] range of apparent disagreement."

As but one example of this phenomenon, recent model predictions by the Goddard Institute for Space Studies suggest that one of the regions where unambiguous CO_2-induced warming should first appear is the ocean area around Antarctica;[54] while a contemporary model study of the Geophysical Fluid Dynamics Laboratory suggests that there will be no CO_2-induced warming in this region, "and even a slight cooling."[55] Hence, it is very clear, in the words of Ramanathan,[56] that our "theoretical understanding of the [Earth's] climate system is by no means complete;" and as intimated by Kellogg,[10] perhaps it never will be. Yet some climate modelers still argue "that atmospheric heat transport is actually better estimated from theoretical circulation models than from field observations"![57]

3.2 Poorly Parameterized Processes

What are some of the areas in which we lack understanding of climatic elements and events? Although many could be listed, such as meridional and vertical dynamical heat fluxes,[58] heat transport by ocean currents,[59] ocean convective dynamics,[60,61] deep water formation and circulation,[62-68] oceanic biogeochemical cycling,[69,70] sea ice dynamics,[71,72] the widespread occurrence of ice crystal haze and its role in the radiation balance and air chemistry of Earth's polar regions,[73,74] and large-scale periodic phenomena such as the Southern Oscillation,[75] tropical deforestation,[76,77] and land surface hydrological regime and energy partitioning,[78-83] we will deal here, for illustrative purposes, with but one process. Nevertheless, it is one of the most important phenomena of significance to climate, and it pertains to something to which nearly everyone relates almost daily: clouds.

Clouds cover about half of the Earth's surface at any given time,[84] and hey reflect away a goodly portion of the incoming solar radiation from the sun, providing thereby an effective cooling mechanism for the planet.[85] Clouds also, however, act as barriers to the loss of surface-emitted thermal radiation to space and thereby tend to preserve the near-surface warmth of and and sea.[86] A logical question to ask, therefore, is which effect dominates?

In a detailed analysis of this question, Slingo and Slingo[87] concluded that cloud feedback is "one of the greatest sources of uncertainty in current predictions of the effects of increasing CO_2 on climate," and that in spite of all of the advancements of the past decade, "there remains a concern that the cloud forcing in the model[s] may still be unrealistic." In fact, Del Genio and Yao[88] bluntly state that "cloud processes are so poorly understood and crudely represented in climate models that even the sign of the total cloud feedback is in doubt."

In spite of this serious deficiency in our knowledge base, most state-of-the-art climate models include interactive cloud sub-routines; and these models suggest that the net effect of cloud feedbacks (changes in coverage, altitude and type) is to enhance the CO_2 greenhouse effect.[89-93] But empirical observations seem to indicate otherwise. As long ago as 1986, for example, preliminary satellite data from NASA's Earth Radiation Budget Experiment suggested that "clouds appear to cool earth's climate, possibly offsetting the atmospheric greenhouse effect,"[94] and that conclusion now appears to be confirmed.[95] (See also Sec. 5.14.) In addition, Nullet and Ekern[96] have found that cloud cover definitely acts as a negative feedback to surface air temperature fluctuations for the Hawaiian Islands area of the tropical north Pacific, which has been demonstrated to be representative of the oceanic climate in general.[97]

The significance of this finding is illuminated by the computer study of Webster and Stephens,[98] which indicates that a mere ten percent increase in the amount of low-level cloudiness would be sufficient to completely cancel the warming effect of a doubling of the atmospheric CO_2 content, and the more recent analysis of Ramanathan,[56] which suggests that a cloud-induced increase in the planetary albedo of only one percent is sufficient to offset the warming predicted to accompany a CO_2 doubling. Hence, as a result of these observations, plus the many deficiencies associated with several aspects of cloud parameterizations,[36,93,99,100] Ramanathan[56] has wisely concluded that "since clouds are treated primitively in GCMs, it is premature to make reliable inferences from GCM studies." And if this one deficiency makes GCM predictions unreliable, how much more should one be skeptical of their results in light of their many other poorly parameterized processes?

Consider, for example, the further conclusions of Schlesinger and

Mitchell.[36] Noting that present day GCMs "do not resolve all of the physica[
processes that may be of importance to climate and climate change and whic[
span the 14 orders of magnitude from the planetary scale (10^7 m) to the clou[
microphysical scale (10^{-6} m)," plus the fact that "contemporary computer[
permit the resolution of physical processes over only 2 orders of magnitude,
they point out that "even a thousandfold increase in computer speed, which i[
not projected to occur within this century, would allow the resolution of onl[
1 more order of magnitude!"

"Clearly," they thus conclude, "even the GCMs are, and will continue t[
be, critically dependent on their treatments and parameterizations of th[
physical processes that occur on the unresolved subgrid scales," just as the[
are also substantially dependent on such seemingly mundane criteria a[
upper[101] and lower[102] model boundary locations and initial conditions.[10[
Yet, even these deficiencies are but the tip of the iceberg.

3.3 Generally Omitted Processes

Consider the effect of increasing air temperature on cloud liquid wate[
content,[104,105] which is still missing from many GCMs. Since a warmer
troposphere would contain more water vapor and, hence, more condensate, the
clouds in a warmer world would likely have greater liquid water contents than
they do now.[106-110] And since the infrared opacities of low- and mid-leve[
clouds saturate very quickly,[111] this increase in cloud liquid water content
produces a negative or stabilizing feedback upon the initial CO_2-induced
warming, without any change in cloud coverage, altitude or type. That is,
since the heat conserving "greenhouse" properties of these clouds are already
near their maximum, while their reflectances for solar radiation still have
the capacity to considerably increase, the net effect of simultaneous changes
in both cloud properties due to an increase in cloud liquid water content is
to produce an impetus for cooling.[112]

"This negative feedback can be substantial," wrote Somerville and Remer[113]
in demonstrating the significance of the phenomenon. In fact, they found that
when observational estimates[114] of the temperature dependence of cloud liquid
water content were employed in the radiative-convective climate model which
they utilized, "the surface temperature change caused by doubling CO_2 [was]
reduced by about one half," as has been suggested by a more complex GCM as
well.[112]

It should also be noted, however, that for optically thin high cirrus
clouds, just the opposite effect has been predicted.[115-117] Consequently, as
indicated by Roeckner,[116] the ultimate net sign of the feedback is "crucially
dependent on the spatial distribution of cloud amount and cloud optical

operties," such that, in the words of Schlesinger,[115] "many additional
delling and observational studies will be required to determine the sign of
e actual cloud optical depth feedback. let alone its magnitude."
vertheless, when probable changes in all types of clouds are considered
multaneously, the net feedback appears to be negative.[118]

A closely related phenomenon is the stimulation of convective downdrafts
ich characteristically occur with increasing cloud liquid water content.[119]
is phenomenon is generally predicted to increase all types of clouds[88] and
ould thus provide still another impetus for cooling.[95]   Yet modeling of
his interaction, which has significant implications for Hadley
rculation,[120] Walker circulation,[121] and the 40-50 day oscillation of the
mosphere,[122-124] is still something of a "primitive art,"[88] while mesoscale
ganization, another important component of tropical convection,[125,126] has
t to even be addressed within the context of GCMs.[88]

As a second example of a potentially important climatic process which is
issing from most GCMs, let us consider another phenomenon related to
ouds.[127,128]   This more complex chain of events is actually composed of a
mber of processes which have been conceptually linked by Charlson et al.[129]
orking backwards, these investigators first noted that an increase in the
lbedos of marine stratus and altostratus clouds would have a powerful
ooling influence on Earth's climate, and that a significant albedo increase
 this nature may be brought about by a relatively moderate increase in the
tmospheric population of cloud condensation nuclei (CCN).[130]   And as the
ceanic CCN population is believed to be largely produced from gas-to-
article conversions of oceanic dimethylsulfide (DMS) emissions,[131,132] which
re a by-product of the activity of marine phytoplankton in the ocean's
hotic zone,[133,134] Charlson and his colleagues theorized that as
emperatures rise in response to some initial impetus for warming, the
roductivity of marine phytoplankton would likewise rise, resulting in a
reater production of oceanic DMS and its release to the atmosphere, where
reater gas-to-particle conversions would increase the air's CCN population
nd, ultimately, the albedos of marine stratus and altostratus clouds, via a
arrowing of the cloud droplet spectrum and a decrease in the mean radius of
he cloud droplets, both of which effects would then tend to counteract the
nitial impetus for warming.[135]

The validity of this unique theory of climate regulation is supported by
any different types of evidence.   At the bottom rung of the conceptual
adder, for example, is the demonstrated propensity for oceanic phytoplankton
o increase their productivity with increasing temperature,[136-138] a
henomenon which is further supported by observational assessments of global
atitudinal productivity distributions.[139,140]   Secondly, it has been

experimentally demonstrated that as marine phytoplankton photosynthesize they produce a substance called dimethylsulfonio propionate,[141] which : believed to be instrumental in balancing the high osmotic pressure c seawater[142] and which remains within them while they live, but whic disperses throughout the surface waters of the oceans when they either die[14 or are eaten by zooplankton.[144] Then, a portion of it slowly decomposes t produce DMS,[145] which is the most abundant volatile compound of sulfur i seawater.[146-148]

Now a part of the DMS in the surface waters of the world's oceans diffuse into the atmosphere, where it is rapidly oxidized by OH and NO_3 radicals an converted into sulfuric and methanesulfonic acid particles,[149-157] which ar carried aloft to ultimately function as cloud condensation nuclei,[134,158-16 some of which produce new clouds and some of which dramatically increase th albedos of pre-existent clouds,[129,161-164] in harmony with a number o theoretical and observational studies.[165-170]

The end result is a sizeable cooling effect. In the words of Lovelock,[17 "it is comparable in magnitude with that of the carbon dioxide greenhouse, bu in opposition to it." Together, these two mechanisms would thus appear t constitute a global "Gaian thermostat," which effectively works to stabiliz the temperature of the planet and prohibit the radical greenhouse warmin predicted by current climate models.

Land plants too may act in a somewhat analogous fashion.[172-174] In thei detailed study of the Smithsonian Institution's long-term measurements of th spectral distribution of solar radiation at Mount Montezuma, Chile, and Tabl Mountain, California, for example, Roosen and Angione[175] discovered tha biosols, or aerosols produced by plants, can effectively attenuate the flux o solar radiation as it passes through the atmosphere. Finding that th botanical component of the seasonal modulation of this flux was about 0.5% a these mountain stations, and noting that since the responsible aerosols ar generated and concentrated near the Earth's surface, making the atmospheri attenuation of solar radiation due to land plants much larger at lowe altitudes, they concluded that "physical theories may not be sufficient t explain climate change, and considerations of biological processes may b required," a position also recently espoused by Meszaros.[176]

The potential importance of this phenomenon is further highlighted b Ramanathan's[56] contention that "an increase in the planetary albedo of jus 0.5% is sufficient to halve the greenhouse effect of CO_2 doubling," plus th many observations which suggest that the terrestrial biosphere is alread responding to the aerial fertilization effect of the atmosphere's rising CO content and may therefore be producing more biosols.[177]

Another way by which terrestrial biosols and DMS-derived CCN may act t temper any tendency for warming is through their effects on precipitation, especially drizzle over the world's oceans. Under normal circumstances, according to Albrecht,[178] the primary mechanism by which liquid water is removed from low-level marine clouds is drizzle. Consequently, he argues that in areas where CCN concentrations are relatively low, an increase in aerosol concentration would decrease the mean cloud droplet size, thereby decreasing the amount of liquid water removed from the atmosphere by drizzle and increasing the total global coverage of low-level cloud, in accordance with the observations of a number of experimental investigations.[179-181] This chain of events, of course, tends to cool the planet, as noted earlier in this subsection, by increasing the mean global albedo more than it increases the greenhouse properties of these clouds. In addition, drizzle from stratus clouds tends to stabilize the atmospheric boundary layer,[182,183] due to cooling of the subcloud layer as a portion of the drizzle evaporates; and since the resultant stratification of the boundary layer reduces the transport of water from the ocean surface to the cloud layer, it contributes to a reduction of cloud cover. Hence, this second impetus for warming (a vapor source limitation to increased cloud coverage and albedo) may also be thwarted by enhanced CCN production.

Then there are the effects which humanity itself may exert on the cloud characteristics of the world. Due to the increase in atmospheric particulate concentration caused by the air polluting activities of man,[184] clouds may either become darker, increasing the temperature of the Earth,[185] or lighter, cooling the planet.[186] In a detailed study of this topic, Twomey et al.[187] concluded that the brightening effect "is the dominant one for global climate and that the climatic effect is quite comparable to that of increased carbon dioxide, and acts in the opposite direction." Indeed, even though elemental carbon can comprise a large fraction of the fine particle mass of urban aerosols,[188] and could thereby make clouds darker,[189] radiative transfer calculations indicate that this effect is overwhelmed by the brightening effect of the increased number of additional water droplets.[189]

The reality of this phenomenon is further substantiated by a number of experimental studies which have confirmed that pollution increases the atmospheric concentration of cloud-nucleating particles,[190-193] as well as the observation that the atmosphere over the north Atlantic has become measurably more polluted over the past half-century or so.[194] Since the Pacific and Southern Oceans are still more or less pristine, however, Twomey et al.[187] calculate that these other oceans could sustain an albedo increase of one-half to one percent per decade for the next 50 to 100 years, which would readily counterbalance any CO_2-induced impetus for warming.

3.4 The Bottom Line

In view of their basic inadequacies, poor representations of certain important climatic processes, and outright omissions of many others, it is clear that even the best and most complex climate models presently in use, as well as those likely to be developed within the next few decades, are, and will continue to be, incapable of properly answering the important questions man must soon resolve relative to the climatic consequences of the rapidly rising CO_2 content of Earth's atmosphere. Indeed, these models can not as yet even properly simulate the major glacial-interglacial oscillations of the Quaternary,[195,196] which has led to a protracted search for a "missing mechanism" for the translation of weak orbital forcing into global equivalent insolation changes, which has in turn resulted in a number of speculations[197-200] but no satisfying explanation. Consequently, Neeman et al.[195] have concluded that "it is even conceivable that the major glacial-interglacial transitions are a consequence of nonorbital forcing, such as genuine solar constant changes or equivalent solar constant changes induced by purely internal variations in the global radiation budget, resulting from changes in the biosphere, cloud distribution, or concentration of atmospheric constituents" -- mechanisms such as those discussed in the preceding two subsections and which are also supported by mounting evidence for short-term global climatic oscillations of large magnitude.[201-210]

So complex is the situation, in fact, that even the climate models themselves are beginning to predict more than one stable equilibrium state for a given set of boundary conditions[211,212] something about which oceanographers have speculated for some time now.[213-215] Hence, it is totally unrealistic to expect that any GCM will be able to properly simulate the real-world climate system for perhaps even several decades, as is also implied by the much less stringent considerations of Schlesinger and Mitchell.[36]

But there is another way to broach the issue; and that is to look to observable characteristics of nature herself for our answers. In the words of Twomey et al.,[216] written in response to a criticism[217] of their utilization of this approach,[187] "we cannot accept the proposition that, just because measurements may be difficult and their interpretation complicated ..., we should abandon measurements and trust to 'extensive model calculations' to draw conclusions about the real atmosphere .... Rather than indulge in endless modelling, it is more instructive to approach the problem from another perspective." And it is this perspective of "observing the [system] itself rather than relying on mathematical models" -- for which J. Reid was recently awarded the "coveted" Albatross Award[218] -- to which we will next turn our attention; for as Schlesinger[219] has summed up the situation with respect to GCM predictions, we "have every right to be very, very skeptical of the results." Indeed, says G.L. Stephens,[220] "there is hardly a single aspect in dealing with climate change that can yet be described with confidence."

4   THE EMPIRICAL APPROACH TO CLIMATE CHANGE

4.1   General Philosophy

In seeking to determine the climatic effects of rising atmospheric CO_2 concentrations, in the words of Wallace Broecker,[1] "there are no easy solutions, and we must gear up for the long, hard job of working out how Earth's climate operates."

Obviously, climate models will play an important role in certain aspects of this program; however, all models are based upon, and largely draw their strength from, what has been independently learned by other means, chiefly of an empirical nature.[2]   Hence, the real basis for most of what we know about Earth's climate system is, and always will be, rooted in the sometimes mundane but ever-important activities of observation and measurement.[3]

The significance of this general philosophy for the problem of climate change has twice been emphatically stressed by the U.S. National Research Council,[4,5] two committees of which have stated in no uncertain terms that "observed surface temperatures of Mars, Earth, and Venus confirm the existence, nature, and magnitude of the greenhouse effect."   Consequently, we will begin our empirical investigation of the CO_2 greenhouse effect on this largest of possible scales, with a comparative planetology analysis of surface temperatures and atmospheric CO_2 partial pressures on these three adjacent members of the solar system.

4.2   Mars, Earth and Venus

As a result of spacecraft missions to our planetary neighbors, we know that the greenhouse warming of Venus is approximately 500 °C,[6-8] while that of Mars is only 5 or 6 °C.[8,9]   Likewise, we know that the large surface air temperature elevation on Venus is maintained by a massive 93-bar atmosphere, of which about 96% is CO_2,[8] and that the much smaller greenhouse warming of Mars is also maintained by an atmosphere of almost pure CO_2.   In the case of Mars, however, the atmosphere is considerably more tenuous, oscillating over the Martian year between surface pressures of only 0.007 and 0.010 bar.[10]

The two points defined by these data are plotted on the log-log scale of Fig. 4.1. Also shown there are a number of points which describe the Earth's climatic state at 500-million-year intervals extending back to 3.5×10^9 years Before Present (BP), as calculated[11] from considerations pertaining to the paradox of the faint early sun, which was described in Sec. 1.2.   These calculations are based upon the standard assumption of a 25% reduction in solar luminosity at 4.5×10^9 yr BP and a linear increase over time leading

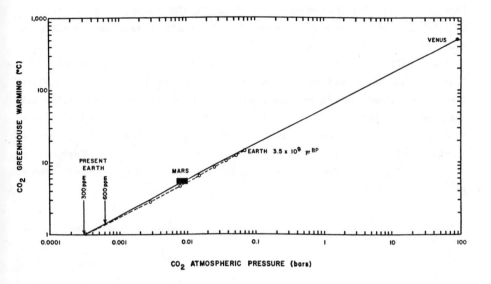

Fig. 4.1 A comparative planetary climatology relationship for Mars, Earth and Venus based on the greenhouse warmings of Mars and Venus which are produced by their atmospheric partial pressures of CO_2 (solid line). Also shown is the nearly identical relationship which derives from standard considerations related to the Earth's paleoclimatic record and the paradox of the faint early sun (dashed line). Adapted from Idso.[11]

to the present value of the solar "constant." The calculations also employ the same CO_2 variation with time as that utilized by Lovelock and Whitfield[12] in their analysis of the problem. Finally, the indicated warmings are those required to just compensate for the effects of reduced solar luminosity, based upon the evidence cited in Sec. 1.2 for the Earth having been at least as warm as it is today over most of the past 3.8 billion years.

In viewing the total assemblage of these empirically-derived data points, it is clear that the only relationship which will adequately describe them in this format is a linear one, specifically, a straight line drawn through the two points in which we have the most confidence (those pertaining to Mars and Venus, due to their being derived from actual measurements). Based upon the resulting solid-line relationship, it would appear that a 300 to 600 ppm doubling of the Earth's atmospheric CO_2 content could only increase the planet's mean surface air temperature by about 0.4 $^\circ$C (to a CO_2-induced warming of 1.4°C at 600 ppm from a CO_2-induced warming of 1.0°C at 300 ppm). In addition, there is no other simple line which can be made to connect these two points and produce any greater warming. Hence, the observed surface temperatures of Mars, Earth, and Venus would indeed appear to confirm "the existence, nature, and magnitude" of the greenhouse effect, as stated by the U.S. National Research Council; but they yield a result for contemporary

Earth, i.e., a predicted warming for a 300 to 600 ppm doubling of the air's CO_2 content, which is a full order of magnitude less than that produced by essentially all state-of-the-art GCMs.

When applied to the problem of much larger CO_2 pressures in the very early histories of Earth and Mars, however, most energy balance and radiative-convective climate models produce results which are basically compatible with the empirical relationship of Fig. 4.1. For example, the several letters plotted on Fig. 4.2 represent results of this type obtained for early Earth by Kasting and Ackerman[13](A), Kasting et al.[8](K), and Durham and Chamberlain[14](D), and for early Mars by Pollack[9](P), Cess et al.[15](C), Toon et al.[16](T), and Hoffert et al.[17](H).

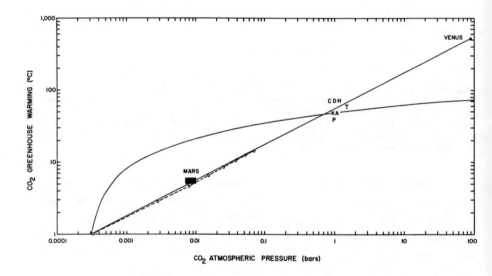

Fig. 4.2   Same as Fig. 4.1, but including results of climate model calculations of Kasting and Ackerman[13] (A), Kasting et al.[8] (K) and Durham and Chamberlain[14] (D) for early Earth, and Pollack[9] (P), Cess et al.[15] (C), Toon et al.[16] (T) and Hoffert et al.[17] (H) for early Mars.

How does one explain this inconsistency, i.e., the fact that the models and the empirical approach to climate change represented by the relationship of Fig. 4.1 produce nearly equivalent greenhouse warmings for the imposition of a 1-bar pressure of atmospheric CO_2 but differ from each other by a full order of magnitude when predicting the consequences of a mere 300 to 600 ppm doubling of the atmospheric CO_2 content on contemporary Earth?

Some insight into this dilemma is provided by the relationship proposed by the U.S. National Research Council[5] to reproduce the basic results of the more complex GCMs over a range of CO_2 partial pressures:

$$\Delta T = (T_D/\ln 2) \ \ln(P_A/P_{A,R}) \tag{4.1}$$

where P_A is the atmospheric partial pressure of CO_2, $P_{A,R}$ is the reference or base value of P_A (taken here to be 300 ppm), and T_D is the equilibrium warming of the mean global air temperature at the planetary surface for a 300 to 600 ppm doubling of the atmospheric CO_2 content (taken here to be 4°C, as per recent model results).

Now the solid curve of Fig 4.2 is a plot of this equation for the stated values of $P_{A,R}$ and T_D; and as is readily discerned, it agrees with the empirical relationship at only two points: the initial $P_{A,R}$ value of 300 ppm (which is a forced fit) and a CO_2 partial pressure of 0.7 bar. Hence, the general agreement between the various climate model predictions plotted in Fig. 4.2 and the empirical relationship developed in Fig. 4.1 is due to nothing more nor less than the fortuitous choice of the specific value of P_A used in the plotted model projections. And it is clear from the form of the plot of equation 4.1 that that value, and the initial $P_{A,R}$ value of 300 ppm, are the only two CO_2 partial pressures for which the theoretical predictions are in agreement with real world, i.e., terrestrial planetary, observations.

4.3 Water Vapor Feedback

Another interesting aspect of the empirical relationship of Fig. 4.1(2) is that total greenhouse warming is specified solely in terms of atmospheric CO_2 content, whereas the CO_2-induced warming predicted for Earth by current GCMs is due chiefly to a positive water vapor feedback effect.[18] It is noteworthy, however, that the measured greenhouse warmings of Fig. 4.1(2) for Mars and Venus do have whatever effects may be due to water vapor feedback already imbedded in them, as the measured greenhouse warmings of those planets represent the combined climatic consequences of all of the greenhouse gases in their respective atmospheres. Consequently, since water vapor is practically non-existent on Mars,[10] intermediate on Earth,[6,9] and large on Venus (in an absolute sense),[6,9] -- varying in much the same way that cloud amounts vary among the three planets -- yet, as the data of Fig. 4.1(2) define but one single and simple relationship without any adjustments for these differences, it follows that the great CO_2-induced water vapor feedback effect that the GCMs predict for the Earth is most probably a model artifact and not a characteristic of the real world.[19]

But how could this be? Several possible explanations have been provided by Hugh Ellsaesser.[20] In addition, Lindzen et al.[21] have demonstrated that replacing the conventional convective adjustment in a radiative-convective climate model with a physically-based cumulus-type parameterization of

30

convection reduces the model sensitivity to doubled CO_2 by up to 80% in tropical regions. Consequently, there is considerable evidence to suggest that the large water vapor feedback phenomenon which appears in today's state-of-the-art GCMs may well be much reduced, or even non-existant, in the real world, as is implied by the data of Fig. 4.1(2).

4.4 The Nature of Greenhouse Warming

Consider next the basic mechanics of the greenhouse phenomenon. The fact that higher surface temperatures are maintained by an enhanced receipt of thermal radiation at the planetary surface (originating with the greenhouse gas) implies that the surface air temperature rise produced by an increase in atmospheric CO_2 should be proportional to the increase in thermal radiation which it produces at the planet's surface. Now for a 300 to 600 ppm doubling of the Earth's atmospheric CO_2 concentration, this latter radiative enhancement has been estimated to be about 4 Wm^{-2}.[5] In addition, the current mean surface air temperature of the globe is about 15 ^oC;[22] and since its current mean downward-directed emittance is approximately 0.89,[22] it follows that the current mean flux of thermal radiation to the Earth's surface is about 347 Wm^{-2}.[23,24] Hence, a 300 to 600 ppm doubling of the air's CO_2 content increases the flux of thermal radiation to the surface of the planet by approximately (4 Wm^{-2}/347 Wm^{-2}) x 100% or 1.15%.

How does this result compare with the implications of Fig. 4.1(2)? First of all, we note that the current mean greenhouse warming of the Earth is about 34 ^oC.[22] Thus, since the warming predicted to accompany a 300 to 600 ppm doubling of the air's CO_2 content is determined from Fig. 4.1(2) to be approximately 0.4 ^oC, the corresponding percentage increase in the total greenhouse warming of the planet is seen to be (0.4 ^oC/34 ^oC) x 100% or 1.18%, which is essentially identical to the percentage increase in the flux of thermal radiation received at the Earth's surface due to such a doubling of the air's CO_2 content. Consequently, the greenhouse warming predicted by Fig. 4.1(2) would appear to be precisely what should be expected from the basic nature of the greenhouse phenomenon and the known radiative properties of CO_2.

4.5 A Limit to Greenhouse Warming?

Consider next a hypothetical planet with a 1-bar atmosphere of pure CO_2 which absorbs all of the solar radiation incident upon it. From Fig. 4.1(2), the greenhouse warming of this planet is seen to be about 55 ^oC. However, based upon the results of the preceding subsection, if its planetary albedo s increased from 0.00 to 0.30, its greenhouse warming must decrease to (1.00 0.30) x 55 $^{\circ}$C or 38.5 $^{\circ}$C.

In a sense, such a planet is much like Earth, which also has a mean global lbedo of 0.30.[25] In the real world, however, CO_2 is but a minor atmospheric onstituent, and its greenhouse properties are largely overshadowed by those f clouds and water vapor. Nevertheless, these other atmospheric components roduce the same type of effect as CO_2 in returning thermal radiation to the arth's surface. Hence, Earth may not be so different from our hypothetical lanet after all. And carrying the comparison to its logical conclusion, this nalogy suggests that since the current real-world greenhouse warming is pproximately 34 $^{\circ}$C,[22] and since the maximum greenhouse warming of our omparable hypothetical planet is 38.5 $^{\circ}$C, Earth may well be constrained to arm no more than about 4.5 $^{\circ}$C due to further increases in the greenhouse roperties of its atmosphere, all else being equal, i.e., total atmospheric ressure, planetary albedo, etc.

.6   Micrometeorological Evidence for a 4.5 $^{\circ}$C Pseudo-Maximum Greenhouse Warming

Is there any empirical evidence to suggest that there may indeed be an pper limit to Earth's near-surface air temperature and that that limit -- a "pseudo-maximum," in that it does not allow for any change in global albedo-- ay be only about 4.5 $^{\circ}$C warmer than the current mean global value?

First of all, there are the readily measured radiative properties of the tmosphere. Due to the powerful continuum absorption of water vapor at high oncentrations,[26] the lower atmosphere over the tropical oceans is already ssentially opaque to thermal radiation.[20]   Like a bucket which is filled to the brim and can get no fuller, the atmosphere there is already producing the aximum greenhouse warming possible; and further increases in atmospheric CO_2 oncentration can produce no additional impetus for warming. Indeed, Matthews and Poore[27] have concluded from studies of certain palaeoclimatic indicators that the surface temperatures of the tropical oceans have maintained their present values all the way back through the Cretaceous, which is the period of warmest terrestrial climate yet documented.[28]

As an example of how this phenomenon operates, consider the data of Fig. 4.3, which depicts the post-midnight difference between the foliage temperature of an alfalfa field at Phoenix, AZ and the temperature of the overlying air under a variety of climatic conditions, ranging from completely clear skies to full cloud cover, over an humidity range which produced clear-sky atmospheric emittances (e_a, calculated from the equation of Idso[29]) ranging from 0.73 to 0.96.[30]

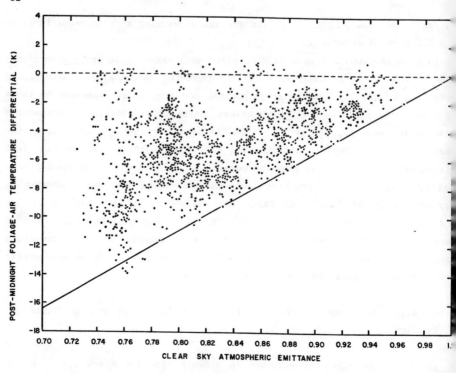

Fig. 4.3 Measured foliage-air temperature differential vs. calculated clear sky atmospheric emittance for the four half-hour periods immediately followin midnight at various times throughout a two-year period for a field of alfalf at Phoenix, Arizona under all types of sky conditions ranging from clear t completely overcast. Adapted from Idso.[30]

Since the assemblages of leaves which were viewed by the infrare thermometer which measured the foliage temperatures have so little therma inertia, each data point of this figure represents a state of quasi equilibrium. That is, although both foliage and air temperatures (T_f and T_a vary with time, the response of the foliage is so rapid that at any give instant the relationship of T_f to T_a and the emittance of the atmosphere i such that if the state of the atmosphere was suddenly frozen in time, T would not change substantially from its value at that particular instant Consequently, the vertical distance between the upper and lower boundaries o the data distribution of Fig. 4.3, at any specified value of e_a, represent the maximum greenhouse warming which could occur for <u>any</u> surface, give enough time, as a result of an increase in e_a from its initial clear-sk value to the maximum e_a value of unity (again assuming no change in surfac albedo, of course).

Now there are two ways by which T_f may rise relative to T_a within th context of Fig. 4.3. In the first case, if the sky is clear and the humidit ises, e_a will increase; and T_f will rise relative to T_a according to the diagonal solid-line relationship describing the lower boundary of the data. In the second case, e_a could remain constant as a cover of clouds developed (which would increase the true emittance of the atmosphere while leaving the calculated clear-sky value, e_a, unchanged), allowing T_f to rise relative to T_a along a vertical path emanating from the initial e_a value characteristic of the original clear-sky condition. (Of course, any number of paths between these two extreme examples are also possible.)

Note further in this regard that as clear-sky e_a increases, and the initial T_f-T_a starting point for the imposition of cloud effects draws ever closer to zero, the upper limiting value of T_f-T_a remains the same. This phenomenon is a direct manifestation of the quasi-equilibrium property of the system mentioned previously. That is, the upper T_f-T_a limit cannot be due to lack of sufficient time for equilibrium to be achieved; for cloud-induced T_f-T_a excursions at a constant clear-sky e_a value of 0.95, for example, have just as much time available for their completion as do cloud-induced T_f-T_a excursions at a constant clear-sky e_a value of 0.75, yet the upper limiting T_f-T_a value in both cases, as just noted, is the same. Clearly, if it were possible for T_f to warm further, there is more than ample time for it to do so at the higher clear-sky e_a value; but it does not. Consequently, Fig. 4.3 provides a good representation of the maximum surface warming which can be induced by "filling the bucket" of atmospheric emittance for the variety of initial conditions characterized by the clear-sky e_a range there depicted.

Utilizing Fig. 4.3, then, it is possible to determine how much the surface temperature of the Earth would rise if its atmosphere were somehow transformed into a perfect blackbody radiator and its emittance increased to unity (again with no change in surface albedo); and the first step of this procedure is to calculate the current mean global value of e_a, which according to the equation of Idso[29] is 0.839.[22] Now this clear-sky value of e_a suggests that, starting from a condition of no clouds, i.e., the solid diagonal line of Fig. 4.3, Earth's mean surface temperature could warm by no more than about 9 °C due to a complete "blackening" of the atmosphere, i.e., as a result of moving vertically to the dashed horizontal line of Fig. 4.3. In reality, however, clouds cover about 50% of the planet at any given time;[31] and Kimball et al.[32] have demonstrated by direct measurement that the extra downward-directed thermal radiation due to the presence of clouds is, in the mean, a linear function of percent cloud cover. Hence, since the upper boundary of the data distribution of Fig. 4.3 represents the maximum greenhouse warming possible under full cloud cover, the maximum warming possible under current mean global conditions of cloudiness and surface albedo is only half of that calculated for clear-sky conditions, or approximately 4.5

^{O}C, which is identical to the maximum greenhouse warming predicted from th
completely independent considerations related to Fig. 4.1(2), as described i
Sec. 4.5.

4.7 Climatological Evidence for Upper Limits to Surface Air Temperature

Are there any other types of evidence which can be marshalled to suppor
the existence of such a modest limit to the greenhouse warming o
contemporary Earth? Indeed there are. Over two decades ago, for example
C.H.B. Priestley[33] published a paper wherein he suggested that there was a
fairly well-defined upper limit to which air temperature could rise over
moist soil or well-watered vegetation; and from an extensive body of climatic
data, he determined that upper limit to be approximately 33 ^{O}C. He also
outlined the basic energy balance considerations responsible for this
apparent fact and, with a colleague, went on to refine them a few years
later,[34] showing that sensible heating of the air over tropical oceans
essentially vanishes at a temperature of about 32 ^{O}C and that it actually
reverses its sign above that temperature.

The physics behind this phenomenon is the nonlinear increase in
evaporation which occurs with rising temperatures, whereby more and more of
the net radiation incident upon a moist surface is dissipated as latent heat
as the surface temperature rises, resulting in a reduction in the amount of
energy available for sensible heat transport and heating of the surface air.
In addition, the latent heat removed from the surface by this means does not
reappear until it has risen above a good deal of the mass of the Earth's
atmosphere,[35] where it can be more effectively radiated to space.[20]

Newell et al.[36] and Newell and Dopplick[37,38] presented similar evidence
for an evaporation-induced limit to ocean surface temperature (and, hence,
that of the overlying air), at low latitudes, finding a value for this limit
of 30°C. A review of Bowen ratio data by Brutsaert[39] also identified this
temperature as being the most likely value for the sensible heating sign-
reversal described above. Newer evidence for the rapid intensification of
deep tropical convection above 27°C is probably linked to this phenomenon as
well.[40,41] It should also be noted, however, that this means of constraining
surface temperature effectively limits surface warming only in the tropics,
and that such equatorial buffering may actually lead to an amplification of
polar warming.[42]

4.8 Micrometeorological Evidence for Minimal Warming of Vegetated Surfaces

In a detailed experimental investigation of the effects of various nvironmental factors on plant foliage temperature, Idso et al.[43] determined he foliage temperature sensitivity of an assemblage of water hyacinth plants loating with their roots continuously immersed in water. Over the range of tmospheric vapor pressure deficits likely to be experienced in nature (0-10 Pa), this sensitivity ranged from approximately 0.014 $^{o}C/Wm^{-2}$ to 0.004 C/Wm^{-2}. Multiplying the smallest of these numbers by 1 Wm^{-2} and the largest y 10 Wm^{-2} (to cover the probable range of extra thermal radiation received t the surface of the Earth as a consequence of a 300 to 600 ppm doubling of he atmospheric CO_2 concentration, remembering that the U.S. National esearch Council[5] has suggested a value of 4 Wm^{-2} for this number), thus ives a plausible range for the foliage temperature increase likely to be xperienced by potentially-transpiring plants as a result of a 300 ppm ncrease in the CO_2 content of the air. The answer? Only 0.004 to 0.14 ^{o}C.

The same study also produced foliage temperature sensitivities for dead lants. In this case, absolutely non-transpiring water hyacinth leaves armed at a rate of 0.020 $^{o}C/Wm^{-2}$, while similarly desiccated alfalfa plants armed at a rate of 0.007 $^{o}C/Wm^{-2}$. And again, multiplying the smaller and arger of these two numbers by 1 and 10 Wm^{-2}, respectively, yields a doubled-O_2-induced temperature increase on the order of only 0.007 to 0.20 ^{o}C.

.9 Other Evidence for Upper Limits to Surface Air Temperature

Moving even further from the moist-surface extreme of tropical oceans, wet oil, and well-watered vegetation, it is also possible to demonstrate the xistence of an upper limit to the temperature of air overlying hot dry urfaces.

The basis for this demonstration is the equation of Idso[29] for the ownward-directed emittance of the cloudless atmosphere (e_a), defined as e_a = D$/c_{SB}T_a^{4}$, where R_D is the downwelling thermal radiation received at the urface of the Earth, c_{SB} is the Stefan-Boltzmann constant, and T_a is the creen-level (near-surface) air temperature in ^{o}K (^{o}C + 273):

$$_a = 0.70 + 5.95 \times 10^{-5} \, e_o \, \exp(1500/T_a) \tag{4.2}$$

here e_o is the screen-level vapor pressure in mb (0.1 kPa).

That this equation is a valid basis for what follows, it should be noted hat it has performed well in: (1) describing the extensive data set from hich it was derived in Phoenix, AZ,[29] (2) describing an extensive data set rom Sidney, MT,[44] (3) describing an extensive data set from 15 different ites scattered across the United States over a latitude range from 26° 12' N o 47° 46' N and over an elevation range from -30 m to +3342 m relative to sea level,[45] (4) giving meaning to a variety of apparently inconsistent data set from several places around the world,[46] which had previously posed a unresolved dilemma,[47] and (5) resolving yet another set of inconsistencies[4] from several points around the globe.[49]

Consider, then, the case of a completely saturated atmosphere. Setting e = 1 in Eq. 4.2, and solving for T_a under this constraint, produces a maximu T_o just slightly less than 37 °C. That is, analogously to the situatio described in Sec. 4.6, for a cloudless atmosphere which is completel saturated with water vapor, e_a will increase with T_a until it reaches a valu of unity at an upper limiting air temperature of just under 37 °C, at whic point, since e_a can rise no further, neither will T_a. In the case of the rea world, however, the atmosphere has a mean global relative humidity o approximately 75%;[50] and the same procedure produces an upper limiting ai temperature of about 47 °C for this situation. Of course, it is possible t attain air temperatures in excess of even this value;[51] but it can only happe when the relative humidity of the air is less than 75%.

For the globe as a whole, it has previously been calculated[22,52] that suc limitations as these allow for a maximum increase in screen-level ai temperature of just a little over 4 °C. In addition, it has also bee demonstrated[53] that somewhat analogous climate modeling considerations[54,5] imply a maximum increase in screen-level air temperature of just a littl under 5 °C, due to the almost perfect linear dependence of the model climat sensitivity upon the area-averaged surface air temperature of the globe. An once again, the mean of these numbers is the same as the maximum possibl greenhouse warming calculated in both Sec. 4.5 and 4.6.

4.10 GCM Evidence for an Upper Limit to Greenhouse Warming.

Subsequent to the demonstration[53] of the significance of the above mentioned climate modeling work, Cess and Potter[56] published a paper wher they investigated the subject in even more detail, utilizing results obtaine from five different GCMs: UKMO (United Kingdom Meteorological Office),[57] GIS (Goddard Institute for Space Studies),[58] GFDL (Geophysical Fluid Dynamic Laboratory),[59] NCAR (National Center for Atmospheric Research),[56] and OS (Oregon State University).[56] Plotting the ultimate change in global mea surface air temperature predicted by each of these models for a 300 to 60 ppm doubling of the atmospheric CO_2 content as a function of each of th models' global-mean control climates (i.e., their initial area-average surface air temperatures), they obtained the five data points reproduced i Fig. 4.4, to which they fit the solid diagonal line which has bee extrapolated in this figure to approximately 18.2°C.

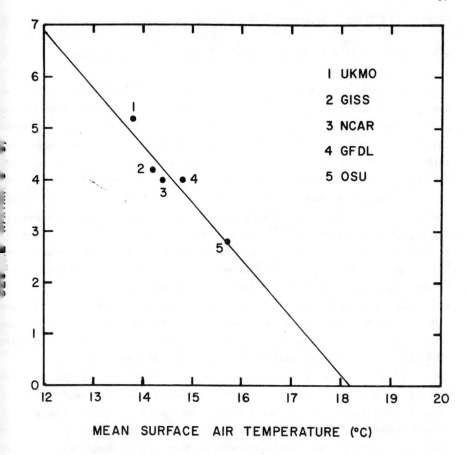

ig. 4.4 The increase in mean global air temperature predicted to result from 300 to 600 ppm doubling of Earth's atmospheric CO_2 concentration vs. initial ean global air temperature, as reported by Cess and Potter[56] for five ifferent GCMs.

This relationship clearly indicates that, as the mean surface air emperature of the Earth rises, the climate sensitivity of the planet rapidly ecreases, which Cess and Potter[56] have also shown to be a characteristic of everal other climate models as well.[55,60-67] In fact, the solid diagonal elationship of Fig. 4.4 suggests that -- if the linear extrapolation of that igure is valid -- by the time the Earth has achieved a mean surface air emperature of just a little over 18°C, its climate sensitivity will have gone o zero, implying that no matter how much CO_2 is pumped into the atmosphere, he Earth will warm no further, much like the limitation to warming produced y the sign reversal in sensible heating of the air over tropical oceans escribed in Sec. 4.7. And once again, the magnitude of the maximum reenhouse warming implied by this analysis is essentially the same as that suggested by all of the previous analyses of this section.

4.11 Climatological Evidence for the Stability of Earth's Climate

To demonstrate the reality of this implied climate stability on a time scale of several years, we turn to the study of volcano-climate relationships which generally suggest a surface cooling of the planet following major volcanic explosions,[63,64] especially those which release large amounts of sulfur into the stratosphere,[65] where sulfuric acid particles may form and reflect away a significant amount of incoming solar radiation.[66]

The veracity of this concept had never been well supported by actual data[67-69] until Sear et al.[70] published a paper based upon the technique of superposed epoch analysis,[71,72] which provided strong evidence for immediate Northern Hemisphere surface cooling following Northern Hemisphere eruptions and delayed (by 6 to 12 months) Southern Hemisphere surface cooling following Southern Hemisphere eruptions. From the two relationships of Fig. 4.5,[73] which are based upon these findings, it can be seen that Earth's surface temperature behavior following a major volcanic eruption is of a damped oscillatory nature. That is, instead of a gradual return to pre-eruption conditions following the initial cooling in each hemisphere, there are a number of rather abrupt heating and cooling reversals. In the Northern Hemisphere there are six such temperature trend changes before equilibrium is finally achieved; while in the Southern Hemisphere -- due to the greater thermal inertia of that ocean-dominated half of the globe -- there are but three such heating and cooling reversals over the 48-month span of data availability, with the ultimate return to equilibrium still some distance in the future.

The point to be made from this oscillatory behavior, which is also evident in the data of Bradley,[74] is that there is a strong tendency for Earth's climate system to not only resist but to correct for anomalously-induced temperature changes in any direction. Thus, the initial rapid cooling of the Northern Hemisphere does not gradually level off but stops abruptly between months 2 and 3, after which almost equally rapid warming begins. In like manner, this warming does not gradually taper off either; it stops abruptly between months 9 and 10, after which another rapid cooling begins. And this pattern of one temperature trend being terminated and replaced by another of almost equal intensity but of opposite sign is repeated again and again until, finally, the original equilibrium state is attained.

As the Southern Hemisphere data appear to tell much the same story, it is difficult to avoid the conclusion that Earth's climatic machinery is programmed, so to speak, to maintain the status quo. That is, it does indeed

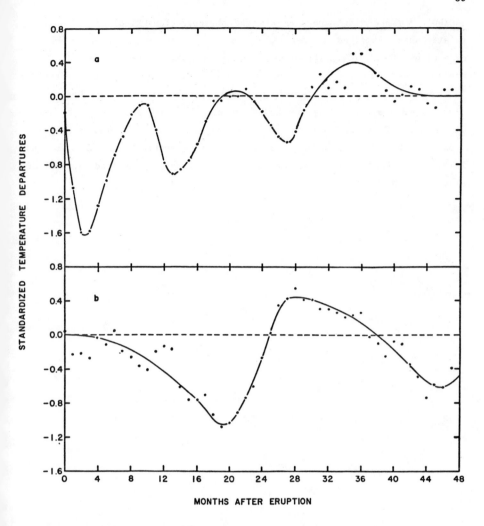

Fig. 4.5 Superposed epoch analysis of standardized monthly surface air temperatures for: a, Northern Hemisphere response to Northern Hemisphere volcanic eruptions; and b, Southern Hemisphere response to Southern Hemisphere volcanic eruptions. Adapted from Idso.[73]

appear that the planetary climate system actively counteracts both heating <u>and</u> cooling perturbations; and although these negative feedbacks produce readily detectable short-term deviations from Earths's mean climatic state, they clearly have a tendency to perpetuate the long-term climatic stability of the planet.

What could possibly be the cause of this phenomenon? At the present time, any answer to this question must remain highly speculative. However, it is possible that the global thermostat which provides for the stabilization of

Earths's surface air temperature is the phytoplankton-DMS-cloud reflectivity relationship described in Sec. 3.3, for it could readily operate in both directions, i.e., to counteract both heating and cooling trends, and it has an appropriately variable time constant. Then again, perhaps the answer resides in processes yet unknown, which will only be revealed through Broecker's[1] "long, hard job of working out how Earth's climate operates." In any event, it is clear from the physical evidence presented in this section (and even from certain model inferences) that the climate sensitivity of our planet has in all likelihood been overestimated by a full order of magnitude in most GCM studies of the CO_2 greenhouse effect.

5 LOOKING FOR THE PREDICTED WARMING

5.1 1988: Year of the Greenhouse?

Midway through 1988, the world was claimed by those who track its temperature to be warmer than at any time in recorded history,[1,2] and the remainder of the year appeared to prove them correct.[3] There was a searing drought across much of the United States;[4-6] and record high temperatures were broken almost daily, and broken with a vengeance.[7] On 17 July, for example, San Francisco, CA exceeded its previous high temperature for that date by almost 20°F (11°C); while on the following day, Death Valley, CA recorded a maximum temperature of 127°F (53°C).

As a result of such dramatic weather extremes, television and radio programs regularly hosted climatologists (climate modelers, in most instances), who explained the basics of the CO_2 greenhouse effect to increasingly concerned audiences.[8] Major newspapers and magazines covered the story for weeks on end;[9-12] and in testimony given in the U.S. Senate, James Hansen of NASA's Goddard Institute for Space Studies declared that he was "99% certain" that what we were experiencing was a direct result of the increasing CO_2 content of Earth's atmosphere.[1,2,7]

As a result, a tremendous "bandwagon" phenomenon was effectively set in motion,[13-15] and people ever since have been clamoring to climb aboard.[16-25] But, we must ask ourselves, is it headed in the right direction?[26,27] Clearly, one year does not a climate make, nor even ten a trend.[28-30] Yet with each passing day, Earth's steadily rising atmospheric CO_2 concentration has been repeatedly hailed as the direct cause of our unusual spate of extreme weather events.[31-33]

As calls to slow the buildup of CO_2 in Earth's atmosphere have increased in frequency and intensity,[34-44] however, a clear voice of reason has emerged above the din of speculation. There has been "too much orthodoxy in the CO_2 and climate problem," according to Soviet climatologist Kirill Kondratyev,[45] as well as "irresponsible predictions of climate change." Indeed, as clearly mandated by the implications of the many physical evidences presented in the preceding section, it is imperative that we maintain a healthy skepticism about the widely publicized claims that the CO_2 greenhouse effect is upon us. As Kellogg[45] has wisely said of Kondratyev's cautionary comments, "let us heed his words!"

5.2 Taking Earth's Temperature

But how can one argue with the facts, i.e., the data? Of course, one

can't, _if_ they're correct.[27]   And this is one of the major problems
associated with the current concern over potential greenhouse warming. Do we
really have the true facts in the matter of Earth's surface air temperature
trend over the last hundred or so years?

As indicated in Sec. 2.5, there have been several reports of an apparent
global warming of about $0.5^{\circ}C$ over the past century. But a number of recent
studies[46-50] have raised some serious questions about the magnitude of this
warming on the basis of the well-known urban heat island phenomenon,[51-55]
which contaminates many of the temperature records used to determine Earth's
climatic history,[56] and which produces a signal similar to that predicted to
result from the planet's increasing burden of atmospheric CO_2.[49]
Consequently, since the urban heat island effect has been detected in cities
with as few as 1,000 inhabitants,[57,58] and since it has been found to produce
localized urban warming of as much as 0.1 to $0.3^{\circ}C$ _per_ _decade_,[59-62] it is
clearly a major impediment to extracting the true surface air temperature
trend of the past century from historical records, as most of the available
data come from urban sites.

Although many attempts at climate reconstruction acknowledge this fact,
but few of them properly correct for it. In the global temperature trend
studies of Jones _et al._,[63-65] for example, only 1.5% of the stations utilized
were explicitly identified as having been contaminated with an urban warming
bias (although they had all passed a number of screening procedures which
could have eliminated urban-heat-island-contaminated data on other
grounds[66,67]). In the global temperature trend study of Hansen and
Lebedeff,[68] on the other hand, where urban heat island effects were evaluated
by determining the difference between the global warming computed over the
past century from (a) all of their data and (b) only those stations located
within urban centers of less than 100,000 inhabitants, the total station
number was reduced by about one-third and produced a 100-year warming bias of
approximately $0.1^{\circ}C$. Hansen and Lebedeff then "subjectively estimated" that
the urban warming bias in the remainder of their data was also about $0.1^{\circ}C$.
Consequently, they concluded that urban heat island effects contributed a
total of only $0.2^{\circ}C$ to the $0.7^{\circ}C$ global warming derived from their unadjusted
data set for the last 100 years, suggestive of a true global warming over the
past century of $0.5^{\circ}C$.

Clearly, this procedure is not satisfactory. For one thing, it should be
self-evident that there is no validity to a "subjective evaluation" of urban
heat island effects for cities with less than 100,000 inhabitants, especially
in light of the documented fact that such effects can be found in cities with
as few as 1,000 inhabitants.[57,58] Secondly, it is population _growth_, not some
level of absolute population size, which causes spurious warming in urban

temperature time-series, so that eliminating stations with populations greater than a certain fixed number does not solve the problem. Indeed, some such stations could even be _declining_ in population and thus producing a spurious _cooling_ trend, which could significantly reduce the net warming bias produced by the technique of Hansen and Lebedeff. Clearly, the only way to overcome this problem is to (a) work with relatively rural data sets and (b) analyze the effects of station population changes over the same period of time for which changes in temperature are being sought.

A most important contribution to the first of these approaches to the problem has been the development of the U.S. Historical Climatology Network (HCN).[69] This network is comprised of 1219 stations located within the conterminous United States which have at least 80 years of continuous monthly temperature and precipitation data. To be included in the network, the stations also had to have experienced few instrument and site changes over the period of data collection; and they were chosen so as to not include large cities (their median population in 1980 was 5,832). In addition, all stations were subjected to an exacting set of quality control measures which were used to appropriately adjust their data for time of observation biases, what few station and instrument changes they did experience, and relative inhomogeneities.[70,71]

In a detailed study of this important data base, Karl _et al_.[49] demonstrated that even the relatively rural HCN temperature record contains an urban warming bias of approximately 0.06°C over the 83-year period extending from 1901 to 1984. Furthermore, they noted that results from subsets of the network and from other studies indicate that "much larger biases are present in rapidly growing urban areas."

In an attempt to quantify these larger biases and assess their significance for the global temperature trend of Hansen and Lebedeff, Balling and Idso[50] used the HCN data base to evaluate the 100-year temperature trend of a large area for which a comparison could be made. This region was defined by box 16 of Hansen and Lebedeff's Fig. 2 to be the portion of the United States located east of 90° W longitude. The comparison indicated that this portion of Hansen and Lebedeff's global data set contains a 100-year urban warming bias in excess of 0.5°C. And if this result is typical of the rest of their data, it is clear that it accounts for the major part of their apparent 100-year global warming trend. In an independent study of these two data sets over the _entire_ United States, for example, Karl and Jones[72] found much the same thing: an urban warming bias in the data of Hansen and Lebedeff of 0.3 to 0.4°C over the period 1901-1984. Noting that this bias "is larger than the overall trend in the United States during this period," and that urban heat islands in other parts of the world can be just as large, or larger, than those observed in the United States,[73] they too were forced to question the reality of recent reports of global greenhouse warming.

Clearly, then, and especially in light of the world population explosion depicted in Fig. 1.1, as well as the fact that some urban heat islands in excess of even 5°C have been documented,[74,75] the global warming trends typically reported in the literature[56] may well be considerably larger than what has actually taken place in the real-world countryside. And what little real warming may have occurred appears to be centered on the middle latitude belts of both the Northern and Southern Hemispheres, with practically no change in either polar region over the past several decades,[76] in clear contrast with what is generally suggested by state-of-the-art GCMs, where polar regions are predicted to warm the most.[77] In fact, almost all of the highly touted north Alaskan warming of the past century, which has been inferred from temperatures measured in boreholes in permafrost,[78] took place prior to 1948, well before major releases of thermally active trace gases, prompting Michaels et al.[79] to conclude that "popular accounts that cite the borehole profiles as evidence for anthropogenerated warming are in error." Furthermore, even if there had (or has) been a real warming in high northern latitudes over the past century or so, it would be equally reasonable to attribute such warming to the intensification of Arctic haze over this period, as the latter phenomenon has been predicted to increase air temperatures in an analogous fashion.[80,81]

On top of the problem of temperature change over land, there is also the need to properly evaluate the surface temperature status of the world's oceans, which cover approximately 70% of the Earth's surface. Here, again, there are some potentially serious biases in the data sets of the past century.[82] And as Barnett[83] has recently noted, "the biases are largely uncorrectable ... at least with a high level of confidence." Hence, it is difficult, indeed, to speak of truly global climate change over the past several decades with much surety. In the words of Oort et al.,[84] "the problem has become almost intractable because of changes in instruments, methods of observation, exposure and height of the instruments, time of observation and the size of ships, as well as because of errors in ship location, relocation of land stations, and urbanization effects."

5.3 Formula for a Fever

But why was the summer of '88 so abnormally warm across so much of the United States and Canada, breaking records which would have been broken even without the help of the urban heat island effect? The answer is to be found, not so much in the absolute temperature of the globe, but in the pattern of shifting air currents over the planet.

It has been known for some time, for example, that there are large-scale interactions between the ocean and the atmosphere which operate over extended distances of both space and time.[85-89] One of the first of these "teleconnections" to be discovered dealt with the influence of warm summer waters in the tropical Pacific on subsequent North American winters.[90-95] Over the past several years it has also been recognized that there is a high correlation between wintertime cooling in the North Pacific and the North American temperature field the following summer.[96-99] Kawamura[100,101] and Iwasaka et al.[102] have demonstrated that the center of the sea surface temperature (SST) anomaly which drives this relationship is located in the central North Pacific; and Park and Kung[103] have recently shown that negative anomalous SST departures in this region during the winter and spring presage high summer temperatures across much of the North American continent.

In their quantitative analysis of this phenomenon, Park and Kung determined the mean negative SST departure of the central North Pacific for the seven years most representative of this condition between 1964 and 1980, finding an average value of $-0.4^{\circ}C$ in the winter and $-0.2^{\circ}C$ in the spring. As early as November of 1987, however, the negative SST departure of this region was already $-1^{\circ}C$; and by April of 1988 it had dropped still further to the unusually low value of $-2^{\circ}C$. Hence, these extreme SST departures would be expected to foreshadow an anomalously warm summer across much of North America, such as was indeed experienced. Consequently, it is likely that the blistering North American summer of 1988 was not at all due to the effects of an ever-intensifying CO_2 greenhouse effect, but that it was somehow linked to an unusually cold antecedent winter and spring over the central North Pacific. And as Reifsnyder[104] has thus noted, "the fact that this past summer has been the warmest in the history of systematic weather records is really irrelevant" to the question at hand.

But what about the drought?

5.4 Drought: A Recurring Phenomenon

There is no longer any question that drought is a periodically recurring phenomenon the world over;[105-107] and the United States is no exception to this rule.[108-110] Hence, it should not surprise us that after the droughts of the 1930s, 1950s and 1970s we should so suffer again as the 1990s are approached. But is it possible that the rapidly rising CO_2 content of Earth's atmosphere is exacerbating the problem, as many claim?[111] To answer this question, we turn to the study of tree rings.

Dendroclimatology, or the reconstruction of past climatic regimes from tree ring data, is a well developed science;[112-114] and its usefulness to the study of past precipitation patterns,[115-117] streamflow histories,[118-120] and droughts in general[121-123] is clearly established. Consequently, the recent tree ring study of Blasing et al.,[124] which touches upon the potential interaction of rising atmospheric CO_2 concentrations and recurrent droughts in the south-central United States from 1750 to 1980, is of extreme importance.

Over this 230-year period, the analyses of Blasing et al. clearly demonstrate that "the probability of severe and prolonged drought similar to 20th century dry spells does not appear to have increased ... as a result of the Northern Hemisphere warming which began about 1900." In fact, they note, "droughts before 1870 appear to have lasted longer than droughts after that date;" while the data of Stockton and Meko[122] are suggestive of a lengthening out of the time between droughts, from approximately 15 years before 1890 to about 22 years after that date. Hence, if it is possible to say anything at all about CO_2 and drought, one would have to conclude that the rise in Earth's atmospheric CO_2 content would appear to be ameliorating the situation rather than exacerbating it, at least in the United States. And, surely, one year's weather conditions, or even those of an entire decade, are much too meager to offer as proof of an impending climatic change, as many are attempting to do.[125] Indeed, Earth's climatic variability is predominantly irregular,[126-128] being governed by strong nonlinearities[129,130] which can induce significant regional droughts without any changes in external forcing factors.[131,132] And one such event may well have played a role in the North American drought of 1988.[133]

The details of this scenario have been widely publicized.[134] They begin with a dramatic drop in Pacific SSTs off equatorial South America,[135] which pushes a large region of warmer water northward and displaces the intertropical convergence zone in like manner.[136] The response of the jet stream under such circumstances is to establish a pattern of highs and lows which reduces the normal flow of oceanic moisture into the heartland of North America.[137] And hence the drought.[138]

As it represents the antithesis of an El Nino in the El Nino-Southern Oscillation (ENSO) cycle,[139-141] which is known to influence rainfall variability,[142] this phenomenon has recently been dubbed La Nina.[143] And as it has not made a major appearance over the past decade, while some of the strongest globe-warming El Ninos of the past century have occurred during this period, it is now being theorized that what many have claimed to be evidence of CO_2 greenhouse warming may be nothing more than the thermal ramifications of the random behavior of this great system of global climate modulation,[144] which may itself be perturbed by such independent phenomena as id-ocean magma production and submarine lava flows.[145,146]

In summary, then, although it was especially severe in the southeastern United States,[147] when the North American drought of 1988 is put into istorical perspective, in the words of R.R. Heim, Jr.,[148] "it is simply the latest in a long series of similar fluctuations that characterize the climatic history of our country," or as Karl et al.[149] have put it, "the drought of 1988 cannot be distinguished from the normal climate variability of the past century." And most assuredly, "greenhouse gas effects," to quote renberth et al.,[150] "almost certainly were not a fundamental cause." In act, in one of the most rigorous studies ever made of U.S. climate records, anson et al.[151] found "that there is no statistically significant evidence of an overall increase in annual temperature or change in annual precipitation for the contiguous U.S., 1895-1987 [nor] is there evidence of change in winter or summer precipitation on the northern plains during that period." Hence, as summed up by Jerome Namias,[152] the U.S. drought of 1988 "was a consequence of normal atmospheric variability, and had no connection whatever with the greenhouse effect."

5.5 Hurricane Gilbert

Next to the heat and drought experienced in 1988, the single most dramatic weather phenomenon of the summer was probably Hurricane Gilbert, one of the most intense Atlantic hurricanes of this century.[153,154] At the peak of its fury, the news media were touting it as typical of what we would commonly experience in a high-CO_2 world of the future. Such was also the implication of a simple modeling study conducted by Emanuel,[155] where a doubling of the atmospheric CO_2 concentration caused a 40 to 50% increase in hurricane intensity; and it could probably also be inferred from the modeling study of Hobgood and Cerveny,[156] who concluded that ice-age hurricanes were substantially weaker than contemporary storms.

It is perhaps not surprising, then, that the governing council of the American Meteorological Society and the board of trustees of the University Corporation for Atmospheric Research subsequently issued a policy statement[157] suggesting that CO_2/trace gas-induced warming over the next 50 years would likely lead to both "a higher frequency and greater intensity of hurricanes." In reviewing the available evidence, however, such a conclusion does not seem readily justified.

For one thing, state-of-the-art GCMs predict that tropical ocean surface temperatures will increase very little with a doubling of the atmospheric CO_2 content, but that the tropical tropopause will warm more than the tropopause anywhere else, and much more than the tropical ocean surface.[77,158,159]

Consequently, the tropical troposphere should be more stable in a higher CO_2 world than it is now, if the models are correct; and this observation would lead one to believe that hurricanes may be <u>less</u> intense under such circumstances than they are presently. In addition, with polar surface warming more than tropical surfaces in the model climate,[77,158,159] there would also appear to be less of a need for hurricanes to transport heat from tropical regions to polar regions in a higher CO_2 world,[160] if the models are correct. Hence, greater hurricane intensities are not necessarily compatible with the predictions of contemporary GCMs.

5.6 A Millennial Perspective

No matter how one interprets the climatic extremes of the current year or decade, however, it is extremely dangerous to view them too narrowly.[161] Indeed, even the course of a century or more can in certain instances be misleading; and the past 130 years of apparent global warming is a case in point.

It is generally assumed, for example, that since the apparent mean global air temperature rise over this period has coincided with a similar rise in atmospheric CO_2 content (Sec. 1.4), that the latter phenomenon must be the cause of the former, particularly since there is a very simple theory which suggests that that is the way it should be. This view, however, is rather myopic, as it does not encompass the broader environmental context within which both the CO_2 and temperature trends of the past century are embedded, and as it closes its eyes to the existence of equally simple hypotheses which can just as readily cast CO_2 in the role of an <u>inverse</u> greenhouse gas.[162-165]

Consider, then, the thousand-year sweep of history portrayed in Fig. 5.1. The first three plots there depicted clearly delineate a period of relatively cool temperatures between about 1530 and 1900, which has come to be known as the Little Ice Age;[172] while the fourth plot reveals the course of atmospheric CO_2 concentration over this period. As can be readily gleaned from this material, the termination of the Little Ice Age corresponds very closely in time and relative magnitude to the modern buildup of CO_2 in Earth's atmosphere; and proponents of GCM simulations of greenhouse warming would have us believe, as mentioned above, that the latter phenomenon is the cause of the former.[173] Such need not be the case, however, for there are two other possible explanations: (1) the former phenomenon could be the cause of the latter, or (2) the great similarity of the two phenomena may be nothing more than a "global geophysical coincidence."[171]

How can one determine which of these three alternatives is correct? Very simply. If either of the first two hypotheses is right, Earth's atmospheric

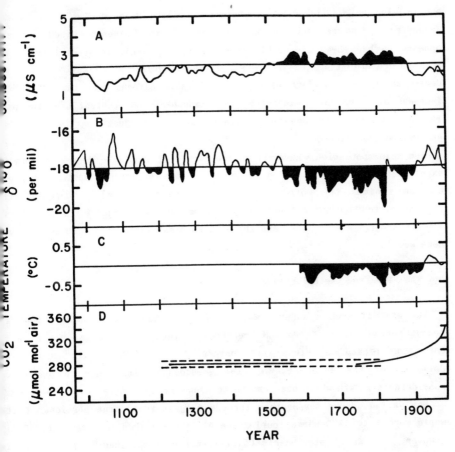

Fig. 5.1 Decadal variations in conductivity (A) and oxygen isotopic ratios (B) determined by Thompson et al.[166] for an ice core taken from the summit of the Quelccaya ice cap in southern Peru; decadal variations in temperature (C) determined by Groveman and Landsberg[167] for the Northern Hemisphere; and similar time scale variations in Antarctic atmospheric CO_2 content (D) determined by Friedli et al.[168] from a South Pole ice core and by Friedli et al.[169] from a Siple Station core. The solid horizontal lines of parts A and B represent averages over the entire 1,000-year records of those parameters; while the solid horizontal line of part C represents the 1880-1980 mean of the data there presented, and the solid horizontal line of part D represents the mean CO_2 content of air extracted from a South Pole ice core representative of the time period 1220-1560. The two dashed horizontal lines of part D are based upon the analyses of Friedli et al.,[169] which lead them to conclude that there is "no reason to assume any secular fluctuation exceeding 10 ppm between the years 1200-1800," a conclusion which is echoed by Oeschger and Stauffer.[170] Adapted from Idso.[171]

CO_2 concentration should decline in phase with the planet's temperature record at the beginning of the Little Ice Age, producing a mirror image of the way in which the two rise together at its termination; for if the two phenomena bear a cause-and-effect relationship to each other at the Little Ice Age's end, they

should also be so linked at its start. As the data of Fig. 5.1 clearl
demonstrate, however, such is not the case: Earth's atmospheric CO_2 conten
appears to have been incredibly constant for a period of several centuri
both before and after the initiation of this significant climatic excursion
Consequently, we are left with but one viable alternative: the coinciden
increase of global air temperature and atmospheric CO_2 content over the pas
130 or so years is probably nothing more than just that -- a globa
geophysical coincidence.[171]

And this conclusion makes all of the dickering about the degree of warmin
experienced over this period a rather moot exercise; for no matter how grea
or how minimal the warming has been, none of it can be unequivocall
attributed to CO_2! None, that is, until we have warmed _above_ the pre-Littl
Ice Age temperature plateau; and there is no clear indication that we have ye
done so.[174]

5.7 The View from the Holocene

But what if such a warming were to occur? Would we then be forced t
accept the validity of the GCM greenhouse effect predictions? Not at all
For if we expand our view still further, to cover the entire 10,000-yea
duration of the Holocene, we find embedded at its center a period o
considerable warmth which has come to be known as the Altithermal.

Now various proxy temperature records suggest that large portions of th
world were 1 to 2°C warmer during the Altithermal than they are now.[174-17]
However, records of atmospheric CO_2 concentration obtained from ancient ai
trapped in polar ice,[180-182] as well as from carbon isotope data extracte
from marine sediments,[183] reveal that the CO_2 content of the air at that tim
was about equivalent to that of Earth's more recent pre-industrial era (se
Fig. 1.2), or about 80 ppm less than what it is today. Consequently, it is
clearly possible to have significantly warmer temperatures than those of the
present with considerably lower concentrations of atmospheric CO_2. And this
fact suggests that the Earth would have to warm above even Altithermal
conditions before the GCM greenhouse predictions could be given much
credence.

5.8 The Late Quaternary

Going back even further in time, Bryson[184] has developed a global
volcanicity index which he has combined with a simple model of Pleistocene
glaciation[185] to produce a glacial ice volume time-series, from which he has
derived an ice-albedo time-series which he has combined with a Milankovitch-

pe solar irradiance model to drive a hemispheric heat budget model which
elds a 40,000-year time-series of Northern Hemisphere surface temperature.
e climatic episodes thus depicted match the known field data quite well,
ggesting that the Little Ice Age and other similar climatic excursions are
imarily driven by variations in volcanicity and not CO_2. In fact, Bryson
tes in his discussion of the model that

> The carbon dioxide emissivity was taken to be dependent on the am-
> bient hemispheric temperature, in opposition to the current scien-
> tific consensus. However, preliminary investigation indicates that
> if the carbon dioxide level of the atmosphere were to be included as
> determined from ice core studies and commonly used sensitivities em-
> ployed, the resulting model would depart very far from the present
> results and fail to give as good a match with the field data.

addition, Kondratyev and Khvorostyanov[186] have demonstrated how documented
ermal contrasts between the Labrador current and the Gulf Stream[187,188] may
ve supported more Atlantic cloudiness during the time of the Little Ice
e, which would also have tended to cool the Earth during that period, again
ndependent of any considerations related to CO_2.

9 A Glacial Cycle Perspective

In spite of the obvious problems raised by these several observations,
roponents of GCM greenhouse simulations continue to ascribe the termination
f the Little Ice Age to the modern buildup of CO_2 in Earth's atmosphere. So
lso have they tried to attribute much of the waxing and waning of full
lacial cycles to ups and downs in the CO_2 content of the air;[183,189] and
owhere has this been more evident than in the aftermath of the successful
ecovery of the famed Vostok ice core.[190-192]

Extracted from a Russian drilling site in Antarctica between 1980 and
985, the Vostok ice core has provided a 150,000-year record[193] of both
urface air temperature[194] and atmospheric CO_2 concentration.[195]
haracteristically, one of the original reports[196] of the project's findings
uggested that the core data implied that "CO_2 changes have had an important
limatic role ... in amplifying the relatively weak orbital forcing;" while a
ubsequent report[197] went so far as to suggest that the observed CO_2-
emperature correlation provides "remarkable confirmation" of current CO_2
reenhouse effect predictions. A careful examination of the pertinent data,
owever, suggests otherwise.

To begin with, changes in atmospheric CO_2 content never precede changes in
ir temperature, when going from glacial to interglacial conditions; and when
oing from interglacial to glacial conditions, the change in CO_2
oncentration actually lags the change in air temperature.[196] Hence, changes
n CO_2 concentration cannot be claimed to be the cause of changes in air

temperature, for the appropriate sequence of events (temperature change following CO_2 change) is not only never present, it is actually violated in half of the record.[198]

Other newly discovered evidence also suggests that it is climate which determines CO_2 content, and not vice versa, in the Vostok ice core. Noting that there currently exists a significant excess of certain plant nutrients in the offshore waters of the Antarctic and Pacific Subarctic Oceans, for example, Martin and Fitzwater[199] demonstrated -- by direct experimental means -- that the addition of small amounts of dissolved iron to those waters greatly increases their phytoplanktonic growth rates.[200] This observation establishes the core principle of the biological pump phenomenon described by Broecker[201,202] and others,[203,204] whereby CO_2 is withdrawn from the atmosphere by enhanced phytoplanktonic productivity and transferred to the deep ocean.[205] When this pump operates at maximum efficiency, i.e., when all essential nutrients for maximum marine phytoplanktonic photosynthesis are present, it has been shown that it can lower atmospheric CO_2 levels by a factor of three relative to levels associated with extreme pump inefficiency,[206] which Martin and Fitzwater[199] and Martin et al.[207] have demonstrated to be a characteristic of present-day interglacial Earth. Hence, an important question relative to what drives what, with respect to CO_2 and climatic change, has to do with the relative supply of iron to the oceans during glacial and interglacial times.

Here, again, the Vostok ice core plays an invaluable role; for it provides convincing evidence for much greater iron concentrations in surface waters during glacial epochs. For both the present and the previous interglacial, for example, aluminum levels -- which are directly proportional to iron levels[208] -- are fully 50 times less that what they are during the intervening glacial period.[209,210] And this is exactly what one would expect as the Earth oscillates between glacial and interglacial conditions; for there is considerable evidence indicative of a several-fold increase in exposed arid land areas,[211,212] significantly enhanced wind speeds,[213-216] and atmospheric dust loads ten to twenty times greater,[209-211,217,218] during glacial as opposed to interglacial periods.[219] Hence, as a result of factors dependent upon climatic change, Martin and Fitzwater suggest "that the enhanced supply of [iron] from the atmosphere stimulated photosynthesis, which led to the drawdown in atmospheric CO_2 levels during glacial maxima."

Additional evidence for the validity of Martin and Fitzwater's hypothesis comes from the study of organic carbon burial rates in both the Atlantic and Pacific Oceans.[220-223] In particular, to quote Lyle,[224] "two to five times higher organic carbon accumulation rates have been observed for the 18 kyr BP glacial maximum than for the Holocene in several short sediment cores from th oceans." And it is generally acknowledged that the mass accumulation te of organic carbon in ocean sediments is due primarily to increases in ceanic productivity.[223,225] In addition, a recent independent assessment of eanic ice-age productivity based on a transfer function[226] relating modern lanktonic foraminifera found in deep-sea sediments to modern productivity has roduced nearly identical results.[227] Consequently, and with much ustification, Lyle too concludes that "changes in primary productivity in the quatorial oceans probably result from, rather than cause, climate changes," lthough enhanced oceanic productivity may well amplify such changes via the hytoplankton-DMS-cloud reflectivity phenomenon discussed in Sec. 3.3.

Clearly, then, these observations of real-world biological phenomena, ombined with the Vostok ice core analyses and the biological pump mechanism or exchanging CO_2 between the atmosphere and the oceans, suggest that limate changes over the last glacial cycle determined contemporary tmospheric CO_2 levels, rather than the other way around.

But if CO_2 variations do not amplify the relatively weak effects of rbital forcing, the significance of which has itself been questioned,[228] hat does? Several possibilities have recently become real probabilities in his regard. The first candidate mechanism emerges from the work of Saigne nd Legrand,[229] who analyzed portions of two Antarctic ice cores for ethanesulphonic acid (MSA), which is one of the primary oxidation products f DMS and an unequivocal indicator of marine biogenic activity.[230] Their indings -- that "during the last ice age, MSA contents were 2-5 times higher han today" -- strongly suggest that the phytoplankton-DMS-cloud reflectivity elationship was likely responsible for a good deal of the ice-age cooling. ndeed, without even allowing for an increase in cloud cover, Legrand et l.[231] have calculated that the likely DMS-induced albedo increase of a onstant areal coverage of marine clouds during the last glacial maximum hould have produced a global cooling of approximately $1^{\circ}C$. And, of course, xpansion of cloud cover may have been even more important.

In addition, as both the frequency and latitudinal extent of mid-latitude yclones are coupled to the latitudinal temperature gradient,[232] and as this radient is known to have been significantly enhanced during the last ice ge,[233] a sizeable equatorward shift of associated storm track clouds and tratus decks should have further amplified the initial impetus for cooling. or a shift from $45^{\circ}N$ to $35^{\circ}N$, for example, the observations of Ramanathan et l.[234] suggest that a global cooling on the order of $3^{\circ}C$ may have ensued. And inally, Harvey's[235] study of the climatic effects of the increased burden of erosols in the ice-age atmosphere suggests that this latter phenomenon robably produced most of the remainder of the cooling, which he calculates to e 2 to $3^{\circ}C$. Consequently, the Vostok ice core data actually weigh in against

the great magnitude of CO_2-induced greenhouse warming predicted by curren
GCMs, as other more readily documented processes appear to be capable o
totally explaining ice age cooling.

Just recently, however, the phytoplankton-DMS-cloud reflectivit
phenomenon has been challenged.[236]   Based upon the likelihood tha
anthropogenic emissions of sulfur dioxide (SO_2) are approximately twice a
great as equivalent emissions of DMS from marine phytoplankton, plus the fac
that Northern Hemisphere emissions are significantly greater than those o
the Southern Hemisphere, and that they have developed this preeminance ove
only the past century or so, S.E. Schwartz[237,238] has suggested that, if th
phytoplankton-DMS-cloud reflectivity phenomenon is true, there should hav
been a gradually increasing cooling of the Northern Hemisphere relative t
the Southern Hemisphere over the last hundred years, which, he claims, ha
not happened. This contention, however, is debatable.

Consider, for example, the land-based annual mean surface air temperature
trends for the Northern and Southern Hemispheres displayed in Fig. 5.2. About
a century ago, according to these data displays, both hemispheres began to
warm; and in the Southern Hemisphere, this warming has continued unabated
right up to the present day. In the Northern Hemisphere, however, it is clear
that there has been no net warming of any significance from about 1940 onward,
in spite of the great increase in Northern Hemispheric population experienced
over this period and the demonstrable impact of this phenomenon on the
temperature histories of the cities which are the source of much of the
Northern Hemispheric temperature record (recall Sec. 5.2). Yet, it is during
this latter half-century, according to the indurtrial carbon production data
of Keeling,[240] that industrial productivity has experienced a several-fold
increase over what it was at the end of the first half-century since the
initiation of global warming. Hence, it is clear that something has indeed
radically altered the temperature history of the Northern Hemisphere relative
to that of the Southern Hemisphere and that it has done so over the period of
greatest industrial sulfur dioxide emissions and in such a way as to provide a
positive result for the test of the phytoplankton-DMS-cloud reflectivity
phenomenon proposed by Schwartz.[237,238]

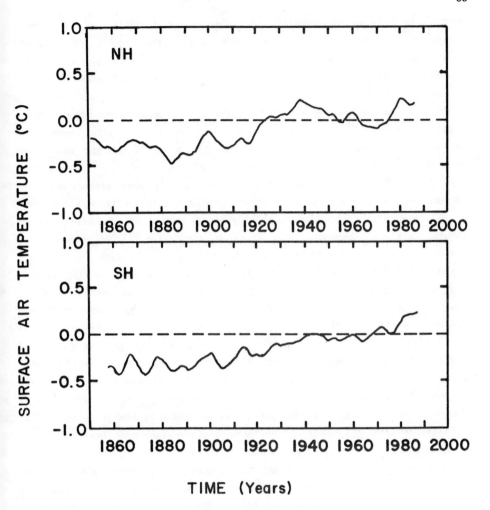

Fig. 5.2 Land-based mean annual surface air temperatures for the Northern (upper) and Southern (lower) Hemispheres, expressed as departures from the 1951-1970 means. Smooth curves show 10-year Gaussian filtered values. Adapted from Jones.[239]

5.10 The K/T Boundary Event

Expanding our horizons still further, consider the phenomenon of the Cretaceous/Tertiary (K/T) boundary event, which occurred some 65 million years ago and led to the ultimate extinction of perhaps three-fourths of the world's then-existing plant and animal species.[241,242] One of the most stimulating scientific hypotheses of the past decade suggests that the cause of this planetary trauma was the impact of a huge extraterrestrial bolide,[243-245] such as a comet or asteroid,[246] which would have injected so much particulate matter into the atmosphere, both directly and via induced volcanism,[247] that the amount of sunlight reaching the Earth's surface would have been reduced for several months below that required for photosynthesis, leading to the collapse of the marine food chain and massive oceanic extinctions.[248,249]

In addition to the original evidence for this catastrophic scenario, which included the worldwide occurrence of the now famous iridium anomaly in marine[243,250] and freshwater[251] sediments, as well as the concurrent enrichment of certain other noble elements and the depletion of several rare earths,[252,253] a number of subsequent discoveries have strengthened the hypothesis. Of particular note is the finding of stishovite, a dense phase of silica widely accepted as an unambiguous indicator of terrestrial impact events, in the K/T boundary layer at Raton, New Mexico.[254] At several other places around the world there is also a soot layer which coincides with the iridium layer that is 100- to 10,000-fold enriched in elemental carbon, suggestive of an intense global fire triggered by the impact,[255,256] a conclusion which is supported by organic geochemical studies which reflect a pyrolytic origin for the soot.[257] Another piece of supporting evidence is the discovery of a tsunami deposit directly beneath the iridium anomaly in Texas, which has been interpreted as being caused by a bolide-water impact.[258] And then there is the high K/T $^{87}Sr/^{86}Sr$ seawater ratio, which has been attributed to increased continental weathering in the aftermath of the event as a result of acidic precipitation caused by the large amounts of nitrogen oxides which would have been produced by shock heating of the atmosphere.[259]

Although other causes of the K/T boundary event have been postulated,[260-266] and both the impact and volcanism theories have been questioned with respect to their ability to account for the observed dinosaur extinctions,[267] no one disputes the fact that <u>something</u> of a very dramatic nature occurred at that point in time, and that it had vast and far-reaching climatic and biological consequences.[268-270] Specifically, there is considerable evidence to suggest that oceanic productivity was severely depressed for hundreds of thousands of years following the event,[271-274] during which time ocean surface waters may have been as much as 10°C warmer than normal[275-277] and ocean bottom waters as much as 5°C warmer.[278] And, again, this response is exactly what one would expect from the phytoplankton-DMS-cloud reflectivity relationship.

Consider, first of all, that the K/T extinctions included about 90% of the marine calcareous nanoplankton,[279] and that calcareous nanoplankton are one of the primary, if not the dominant, producers of DMS.[280] With 90% of them gone, global DMS production rates could have dropped by an equivalent amount; and

Rampino and Volk[280] have calculated that such a reduction in atmospheric DMS would have decreased atmospheric CCN concentrations and marine cloud albedos by amounts sufficient to produce a global surface warming of almost $10^{\circ}C$. Hence, the phytoplankton-DMS-cloud reflectivity relationship is consistant with documented climatic phenomena operating across a whole spectrum of time scales, from the very short, as described in Sec. 4.11, to the very long, as described here. Consequently, its climatic role in glacial-interglacial oscillations is thereby strengthened, while that of CO_2 is reduced.

5.11 Searching for Fingerprints

As noted in Sec. 2.5, no one has ever been able to clearly establish the validity of enhanced global warming due to the rise in atmospheric CO_2 content experienced over the past few centuries. Consequently, the newest approach to the problem has been to look for what have come to be referred to as CO_2 greenhouse "fingerprints."[281-283] That is, rather than concentrate on just one aspect of the problem -- which generally reduces to a search for an upward trend in the mean surface air temperature of the planet and is itself a difficult problem[284,285] -- a number of different data bases are being simultaneously analyzed in light of the full spectrum of CO_2 greenhouse predictions. As R.A. Kerr[283] has described it, "the approach now being pursued is more like developing a composite picture of a culprit rather than arresting the first suspect who has the same color eyes."

Included within this expanded investigation are searches for latitudinal differences in changes in both surface air temperature[286,287] and precipitation,[288,289] air temperature changes at different heights in the troposphere,[290,291] as well as in the stratosphere,[292,293] sea surface temperature (SST) changes,[294,295] changes in the water vapor content of the troposphere,[296] changes in the spectral composition of outgoing thermal radiation at the top of the atmosphere,[297,298] changes in the atmospheric pressure field at sea level,[290,299] changes in sea level itself,[300-303] changes in lake levels,[304-306] changes in permafrost characteristics,[78] changes in extent and depth of snow cover,[307-309] extent and concentration of sea ice,[310-313] extent, flow and mass balance of glaciers[314-318] and polar ice sheets,[319,320] changes in the dates of lake and river freeze-up and break-up,[315,321,322] and changes in the stable-isotope ratio of hydrogen in tree cellulose.[323-325] Commenting on the philosophy of this approach, Barnett and Schlesinger[290] note that the "use of two or more [of these] fields in a detection study minimizes the likelihood that a nearly significant result in one field (air temperature) will be trusted when the signal in the other fields (e.g., SST) is far from that expected."

Obviously, this new approach is still in its infancy and has yet to produce a clear and convincing demonstration of CO_2-induced warming.[290,326] Continued sustained efforts in all of the many aspects of this endeavor are clearly warranted, however, for even the 0.8°C equilibrium warming suggested by Fig. 4.1(2) to accompany a 300 to 600 ppm doubling of the air's CO_2 content (0.4°C due to CO_2 and 0.4°C due to other trace gases) is something which may yet prove to be of significance for the planet, if the atmospheric CO_2 concentration eventually more than doubles and if other radiatively-active greenhouse gases continue their similar upward concentration trends.

5.12 Sea Level Trends

As just one example of the suite of environmental parameters which comprise the total CO_2 greenhouse fingerprint, let us consider the status of global sea level research.

In 1980, Emery[327] proposed that eustatic sea level was rising at a rate of 3 mm/yr, fueling speculation that a CO_2-induced rise of 50 to 100 cm would be sustained over the next century. The 1985 Villach Conference in Austria reaffirmed this prediction but widened the range of potential rise to 20 to 140 cm.[328] When all of the available data are viewed in their entirety, however, such predictions appear to lack a substantive basis.

Consider, for example, the observational results of Fig. 5.3, which demonstrate that, over the period 1960-1979, sea level has actually fallen over large expanses of coastline, notably, western Europe, western North America and Eastern Asia. Some of the causative phenomena that have been linked to these various changes in sea level are long-term tectonic and isostatic behavior,[330,332] variations in ocean current intensity,[333] crustal cooling along rift margins and overriding of adjacent tectonic plates,[331] river discharge,[334] El Nino-Southern Oscillation events,[335-337] rainfall, storminess and sea surface temperature,[338,339] and salinity, on-shore wind stress and vertical mixing of surface and deep waters.[328] Indeed, worldwide sea level has such great temporal and spatial variability[340] that Bryant[328] has concluded that the reality of global sea level rise is neither confirmed nor easily confirmable. Like most of the other fingerprint components, it will still require a great deal of research to decipher the story which it has to tell.[83,341]

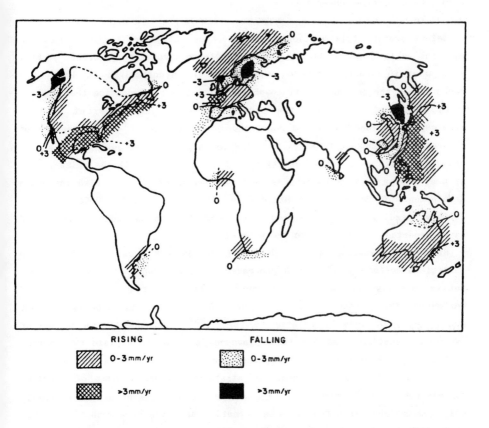

Fig. 5.3 Average annual rates of change in worldwide sea level between 1960 and 1979, as determined from data supplied by the Permanent Service for Mean Sea Level, Bidston Observatory, UK and a number of other sources.[327,329-332] Adapated from Bryant.[328]

Yet even if we did know what the sea level was doing on a mean global basis, it would not be immediately clear what that knowledge implied with respect to climate.[342]   For example, although sea level certainly drops during major glacial-epoch coolings,[343] there is evidence that it sometimes rises by as much as one to three meters during more minor cooling events.[344] Consequently, it is possible that a minor CO_2-induced warming may actually lead to a decrease in sea level. And, of course, this type of response can only occur if water is being sequestered as snow and ice, which in turn suggests that the albedo of the Earth's surface may possibly increase somewhat with a minor global warming. Hence, whereas the ice-albedo feedback is believed to be positive for large-amplitude climatic excursions, it may well be negative for less dramatic perturbations.

5.13 The CO_2-Ozone Connection

Before leaving this section, it is perhaps obligatory to touch upon the evidence related to the potential reality of the phenomenon described in Sec. 2.6, wherein atmospheric CO_2 enrichment is hypothesized to lead to a reduction in the protective stratospheric ozone layer surrounding the Earth.

Originally, of course, the chief culprit in the apparent loss of stratospheric ozone was man's manufacture and release to the atmosphere of chlorofluorocarbons or CFCs.[345-347] And to this day, that assessment remains unchanged, with near total cessation of CFC emissions still the primary goal of most groups seeking to reverse this ominous trend.[348-350] With the rapid development of the Antarctic Ozone Hole (AOH) in the early 1980s, however, Earth's steadily rising atmospheric CO_2 concentration has come to be viewed as an important contributor to the problem.

By way of explanation, the deepening of the AOH from the time of its initial detection up until 1988 proceeded at such a rapid rate that it could not be directly attributed to the much slower rate of increase in the CFC content of the atmosphere.[351] And, obviously, neither could the significant AOH recovery of 1988 be related to it.[352,353] Consequently, a number of dynamical considerations had to be invoked in order to explain the extreme and erratic behavior of the AOH phenomenon, among which were the potential for the isolation of the Antarctic stratosphere into a sort of chemical containment vessel[354] and the potential for the development of the extremely cold temperatures required for the formation of polar stratospheric clouds, as described in Sec. 2.6.

Now the first of these requirements is readily satisfied by the Antarctic vortex, which forms over the continent in winter and allows very little mixing with air from lower latitudes; while the latter requirement is believed to be met by the standard CO_2 greenhouse effect, whereby increases in atmospheric CO_2 are predicted to lower air temperatures above the tropopause. Hence, the standard CO_2 greenhouse effect was thereby identified as a major player in this complex scenario.

But do we really know enough to have much confidence in this conclusion? Consider the common observation that ozone and temperature are negatively correlated in the upper stratosphere,[355,356] plus the fact that the temperature dependencies of the various chemical reactions involving ozone are such that, all else being equal, ozone amounts could well increase if stratospheric temperatures decrease.[357] Working with a model of the entire stratosphere which incorporates these relationships, Brasseur and Hitchman[358] found that a doubling of the atmospheric CO_2 content may increase the ozone content of the upper stratosphere by as much as 20 to 30 percent, significantly compensating for predicted losses at lower levels. Hence, they concluded that this "self-healing" effect makes the total ozone column amount "somewhat resilient to vertically local changes," as recent ozone profile measurements seem to suggest.[359] Consequently, even though an AOH-type phenomenon may be occurring in the Arctic as well,[359-363] scientists studying these processes are still a long ways away from being able to totally explain them.[364] As D.F. Heath[359] has put it, "none of the current theories or model predictions for either natural or anthropogenic sources of stratospheric ozone depletion appear capable of explaining the [observations]."

A second example of this nature involves the discrepancy between observed and calculated values of ozone concentrations in the upper stratosphere and lower mesosphere. Invariably, measured O_3 concentrations in these regions have been about 50% greater than those predicted by state-of-the-art physical-chemical models.[365-369] As a consequence, Rusch and Clancy[370] have concluded that if the measurements are correct, "our understanding of ozone chemistry is disturbingly incomplete" and that "major processes are either seriously misunderstood or are absent from our chemistry entirely." Finding no reason to doubt the data, McElroy and Salawitch[371] have thus postulated the existence of an "as yet unidentified source of high-altitude O_3," which could well prove of considerable significance to the overall stratospheric ozone budget.

More recently, it has also been found that the Arctic does not require the extreme cold which, from experience in the Antarctic, had been thought to be necessary to liberate the active chlorine which attacks and destroys ozone in the presence of sunlight.[372] If true, this observation would tend to suggest that the ozone problem is _not_ linked to greenhouse temperature perturbations.

Then, too, there is the phenomenon of historical ozone concentration increases in the troposphere,[373-377] due primarily to increases in anthropogenically-produced nitrogen oxides and other pollutants.[378] Long-term data bases from France, for example, suggest that rural ozone levels there have essentially doubled over the past century.[379] And just as the anthropogenically-driven increase in CO_2 has made its presence felt throughout the stratosphere,[380] as has the anthropogenically-driven increase in CFCs,[381] so also is it not inconceivable that the anthropogenically-driven increase in tropospheric ozone may do likewise, perhaps through tropical upwelling,[380] convective thunderstorm activity,[382] or the effects of large fires,[383] in spite of the fact that ozone transport across the tropopause is generally downward-directed.[384-386] Hence, if man is indeed responsible for AOH-type phenomena which destroy stratospheric ozone -- which does indeed appear to be the case -- he may also be fueling counterforces which partially restore those losses at the same time. Indeed, between 1974 and 1985, UVB radiation for eight geographic regions in the United States actually decreased,[387] indicative of just such a phenomenon.

Of course, this whole discussion becomes rather moot if there is no significant greenhouse effect due to man's production of CO_2, as most of the evidence presented herein would appear to suggest. And last of all, it must always be remembered, as F. Sherwood Rowland[381] has emphasized, that "no [ozone] hole would exist without chlorofluorocarbons and other anthropogenic organochlorine compounds." Hence, there is a straightforward solution to the problem which is independent of any linkage to CO_2, i.e., stopping the emission of CFCs.

5.14 A Final Comment on Clouds

By now it should be clear that clouds are perhaps the single most important link in the CO_2-climate connection.[388] As Ramanathan et al.[234] have put it, "we must be able to predict changes in cloud forcing if we are to predict the total response of climate to various perturbations." And to this end they have recently made a major contribution in analyzing data obtained from the Earth Radiation Budget Experiment (ERBE) system of satellites.[389-391]

Based upon direct measurements of both reflected shortwave and emitted longwave radiation, for example, they have provided unambiguous evidence that the presence of clouds reduces Earth's mean surface air temperature. In fact, they found that "the size of the observed net cloud forcing is about four times as large as the expected value of radiative forcing from a doubling of CO_2" and, of course, that it acts in the opposite direction. As a result, it is clear that even a slight increase in planetary cloud cover could readily offset much of the warming impetus of a doubling or tripling of the air's CO_2 content; and as indicated in Sec. 3.2 and 3.3, this is the type of synergism which appears to exist in the real world, i.e., cloud coverage increases when air temperature rises.

Other evidence supports this latter conclusion as well. Ann Henderson-Sellers, for example, has extensively studied the relationship between mean surface air temperature and total cloud amount by means of historical analogue climate scenarios, where blocks of relatively warm and cold years are extracted from the historical climate record of the past century and scrutinized for potential cloud coverage response to surface air temperature change. For both the United States[392] and Europe[393] her findings demonstrate that cloud amount increases over both areas in all seasons when going from the block of cold years to the block of warm years; while similar studies of the Canadian Arctic and mid-latitudes yield much the same result.[394] Hence, it is difficult to avoid the conclusion that cloud feedback phenomena should argely compensate for any warming tendency produced by increases in CO_2 and ther greenhouse gases. In addition, there is also evidence to suggest that loudiness considerably lessens latitudinal contrasts in planetary albedo cross the polar ice boundary, and that less ice would lead to increased loudiness and a conteracting of the highly-touted ice-albedo feedback henomenon which amplifies polar warming in most climate models.[395]

6 INTERIM ASSESSMENTS

6.1 The First Five Sections

In the preceding five sections, we have reviewed the status of curren
scientific thinking with respect to the role played by atmospheric CO_2 in th
multi-faceted problem of climate change. On a cosmological time scale, an
based upon a comparative planetary climatology, the evidence presente
therein identifies CO_2 as one of the major determinants of surface ai
temperature regimes on the terrestrial planets of the solar system. It
relative atmospheric abundance, for example, is what makes Mars too cold
Venus too hot, and Earth just right, for the sustenance of life (th
"Goldilocks" phenomenon[1]). Likewise, its gradual removal from Earth'
atmosphere over the eons is probably what has allowed our planet to maintai
a continuously functioning biosphere in the face of the steadily increasin
luminosity of the sun.

Viewed from this broad planetary perspective, however, it is clear that
any further removal of CO_2 from the atmosphere would be counter-productive.
Indeed, it could straightway lead to the demise of practically all life or
Earth, as CO_2 is one of the primary raw materials required for plant
photosynthesis, and not too much more of a concentration drop would reduce
its atmospheric abundance below the level where but few plants could extract
enough of it from the air to survive, which would also doom the animal
population of the planet by eliminating the ultimate source of almost all of
its food.

At this critical point in Earth's history, we are witness to a unique
event: man has suddenly appeared on the scene, possessed of the appropriate
level of technology and employing it in activities of sufficient magnitude
that the consequences of his presence are beginning to overwhelm those of the
forces of nature in determining both the absolute abundance and the relative
composition of the trace chemical constituency of the atmosphere. And the
most impressive global consequence of this unparalleled biological prominence
would appear to be man's reversal of the long-term downward trend of
atmospheric CO_2 concentration, which for so long was the salvation of life on
our planet, but which has recently (geologically speaking) become a major
threat to its continued survival. In truly Gaian fashion, then, the
biosphere appears to have acted to forestall its eminent extinction by
producing a new and unique species (man) endowed with both the capacity and
the inclination to create, as an ubiquitous by-product of its supra-metabolic
activity, enough of the required substance (CO_2) needed to counteract the
physical forces of planetary evolution and return the Earth to a state of

ristine biological vigor,[2] thereby gaining sufficient time for man to cquire the knowledge and expertise needed to resolve the problem on a much ore permanent basis than that provided by the burning of our remaining lanetary reserves of fossil fuels, for Sundquist[3] has calculated that the eochemical perturbations resulting from anthropogenically-produced CO_2 may ersist for tens of thousands of years.

But herein lie the seeds of a complex societal dilemma; for that which ould preserve us biologically, some say, may doom us climatically. And it s this juxtaposition of beneficial and detrimental consequences which cuts o the very core of the current carbon dioxide conundrum.

The worry, of course, is that the increased CO_2 content of Earth's tmosphere, caused by the activities of man, will greatly intensify the urrent level of greenhouse warming. The climate model predictions reviewed n Sec. 2, for example, suggest that a simple 300 to 600 ppm doubling of the ir's CO_2 concentration will ultimately raise the mean surface air emperature of the planet by about $4^\circ C$, and that simultaneous increases in he concentrations of other radiatively-active trace gases will boost this igure still higher to fully $8^\circ C$, with all manner of attendant ill effects ollowing in the wake of the warming. And since considerably more than a ere doubling of the atmospheric CO_2 content is expected to eventually ccur,[4-6] many people are deeply concerned about the ability of modern ivilization to cope with the climatic consequences. In the words of loughton and Woodwell:[7]

> The world is warming. Climatic zones are shifting. Gla-
> ciers are melting. Sea level is rising ....
> The warming, rapid now, may become even more rapid ... and
> it will continue into the indefinite future unless we take de-
> liberate steps to slow or stop it ...
> There is little choice. A rapid and continuous warming
> will not only be destructive to agriculture but also lead to
> the widespread death of forest trees, uncertainty in water sup-
> plies and the flooding of coastal areas ... The consequences
> are threatening enough so that many scientists, citizens and
> even political leaders are urging immediate action to halt the
> warming.

As indicated in Sec. 3, however, the mathematical models responsible for these predictions are still in their infancy: there are many real-world climatic processes which they represent but crudely, others which they completely ignore, and undoubtedly still others which have yet to even be discovered, and which also, therefore, are not a part of the present prediction scheme. Indeed, the models fall so far short of reality in so many important respects that their use may actually pose a serious danger; for in the hands of those unacquainted with their limitations, or unwilling to acknowledge them, their predictions could readily be used to justify the institution of policies which may actually be detrimental to our future well-being.

There are two approaches to the solution of this problem, both of whic
should be pursued concurrently. First, and most obvious, we must continu
our quest to develop ever more sophisticated representations of the globa
climate system; for as we improve our mathematical models of the phenomena o
nature, we should certainly improve the predictions which flow therefrom-
at least if we adhere to the philosophy of J.L. Reid, i.e., "his outrageou
insistance that ocean circulation models should bear some resemblance t
reality."[8] Simultaneously, however, we must continually look to the rea
world for quantitative data which will better illuminate the many differen
aspects of the total problem and provide the observational bounds withi
which the climate models must be constrained to operate if they are t
produce correct views of the future.

In pursuing the latter of these two primary courses of action in Sec.
and 5, there are numerous indications that the current predictions of today'
most advanced climate models are inadequate representations of what is reall
to come. For example, the $8^{\circ}C$ warming computed to result from a 300 pp
increase in the CO_2 content of the air and from concomitant increases i
other radiatively-active trace gases is almost a _quarter_ of the tota
greenhouse warming currently provided by the entire atmosphere! And i
tripling of the atmospheric CO_2 concentration (along with appropriate furthe
increases in other trace greenhouse gases) is predicted to produce a mea
global warming well in excess of a _third_ of that provided by the rest of th
current atmosphere!! Clearly, there is no way that this could possibl
happen, as indicated by the several empirical analyses of the proble
presented in Sec. 4, which suggest that today's climate models are predictin
a greenhouse warming which is fully ten times greater than what is implied b
real-world observations. Fortunately, we thus have sufficient time at ou
disposal to develop the base of knowledge which will someday be required t
deal with the much reduced global warming which our burning of fossil fuel
may ultimately bring to pass at some future date.

6.2 Coming Attractions

With this bit of a respite, then, we shall leave the topic of CO_2 and
climatic change and shift to a consideration of the direct effects of
atmospheric CO_2 enrichment upon the biology of the planet, saving for the
concluding section a final synthesis which will combine both aspects of the
subject in a broader analysis of carbon dioxide and global _environmental_
change.

7 PLANT RESPONSES TO ATMOSPHERIC CO_2 ENRICHMENT

1 The Basic Growth Response

For about a century now,[1-15] and as a result of well over a thousand laboratory and field experiments,[16] scientists have known that increasing the CO_2 content of the air around a plant's leaves nearly always leads to a significant increase in vegetative growth and development.[17,18] Indeed, the plant science literature is replete with reviews and analyses of the subject;[19-29] and fully fifty years ago, thousands of commercial nurseries were already enriching greenhouse air with extra CO_2 for the purpose of enhancing crop production,[30] as they continue to do today.[31-34]

The basis for this profound biological response to atmospheric CO_2 enrichment is to be found in the mechanics of the primary plant process of photosynthesis,[35] whereby light energy is converted into chemical energy in the presence of certain plant pigments (chlorophyll, carotenoids, etc.) to produce carbohydrates from a substrate pool of carbon dioxide.[36,37] As the CO_2 for this process comes primarily from the atmosphere and must traverse a number of serial resistances on its way to the site of its biochemical transformation within the plant tissue,[38-40] increasing its concentration in the atmosphere increases its concentration gradient across these resistances and, hence, its rate and magnitude of transport to the site of photosynthesis.[41] And all of the experimental evidence of the past hundred or so years suggests that with more CO_2 available for participation in the photosynthetic process, more carbohydrates will consequently be produced, which has, of course, been verified by direct experimental observation.[42-47] In addition, it has also been demonstrated that less carbon will be lost to the process of photorespiration under such conditions.[48-51] Hence, as noted in Sec. 1.5, these observations would seem to suggest that the atmospheric CO_2 concentration on the Earth of today may well be far below what should be considered normal for optimal functioning of the planet's vast array of vegetation.

There are many consequences of this affinity of plant life for high levels of atmospheric CO_2.[52] First, and most obvious, is the fact that plants grown in air enriched with CO_2 are generally larger than similar plants grown in ambient air. They are usually taller,[53-56] have more branches or tillers,[57-60] more and thicker leaves[61-67] containing greater amounts of chlorophyll,[68] more and larger flowers,[69-73] more and larger fruit,[74-77] and more extensive root systems.[78-81] The bottom line, however, at least in terms of economic productivity, is the harvestable yield produced; and Kimball's comprehensive reviews of this subject[82,83] indicate that for all of the many plant species which have been studied in this regard, the mean increase in harvestable yie] produced by a 330 to 660 ppm doubling of the air's CO_2 content approximately 33%.

Further increases in the CO_2 content of the air augment plant productivit even more, although generally at a somewhat diminishing rate. Allen et al., for example, have analyzed a number of CO_2-enrichment studies performed wit soybeans, finding that the growth response of this important crop specie takes the form of a rectangular hyperbola, which is suggestive of but a slc productivity rise above a CO_2 concentration of 1,000 ppm. In his survey o the entire field of CO_2 enrichment work, however, Kimball[83] found severa plants which continued their strong positive responses to atmospheric CC enhancement beyond a concentration of fully 3,000 ppm; while others[85-87] hav presented evidence for hyperbolic growth responses with non-orthogona asymptotes at similar high values of atmospheric CO_2. Indeed, some plant have shown continued positive growth responses to CO_2 concentrations as hig as 10,000 ppm.[69] And in photoautotrophic experiments, where higher plan tissue cultures are grown with CO_2 and light as the sole carbon and energ sources, respectively,[88] CO_2 concentrations of one to five percent (10,000 t 50,000 ppm) are generally a common requirement.[89,90] In fact, there is t date only one known species which can be grown in this manner at ambient CO levels.[91] Consequently, the response characteristics of Earth's plant life t atmospheric CO_2 enrichment would appear to allow the planet's blanket o vegetation to express its fullest growth potential under even the mos "adverse" (highest) CO_2 concentrations which have been predicted to resul from mankind's unrestricted burning of fossil fuels.[92-94] Indeed, full use o all of the world's recoverable fuel reserves would likely return Earth' atmospheric CO_2 content to levels characteristic of the "optimal environments"[95] of the early and mid-Cretaceous[96-98] and the Permian anc Carboniferous periods,[99] when vegetation flourished from pole to pole[100-10: and great coal beds were laid down on almost every continent,[103,104 producing a remarkable "greening of the Earth."[105]

7.2 CO_2 as an Antitranspirant

In addition to enhancing vegetative productivity, atmospheric CO_2 enrichment tends to reduce the amount of water transpired by plants and thereby lost to the atmosphere.[106,107] The mechanism responsible for this phenomenon involves the progressive partial closure of the stomatal pores of plant leaves as the CO_2 content of the air around them is increased.[108-111] Also well known for almost a century,[112,113] this process too has been studied intently for the past several decades;[62,114-123] and in a review of

experiments involving 18 different species, Kimball and Idso[124] found that
a approximate 330 to 660 ppm doubling of the atmospheric CO_2 concentration
roduced a mean transpiration reduction of 34%. And like the stimulation of
lant growth by atmospheric CO_2 enrichment, this phenomenon has also been
ound to continue far beyond a CO_2 concentration of 1,000 ppm,[55,108,125,126]
dditionally indicating that plant water relations, like plant growth itself,
hould continue to improve as the CO_2 content of the Earth's atmosphere
ontinues to rise, under even the "worst case" scenario of future fossil fuel
tilization.

.3 The Upward Trend of Plant Water Use Efficiency

So just how good can things get? An illuminating answer to this question
s provided by an analysis of the effects of atmospheric CO_2 enrichment on
lant water use efficiency, defined as the ratio of plant photosynthesis or
ry matter production to the amount of water transpired per unit area of leaf
urface. Based upon the mean results of the preceding two subsections -- a
ultiplicative productivity enhancing factor (P) of 1.33 and a multiplicative
ranspiration reducing factor (T) of 0.66, for an approximate 330 to 660 ppm
oubling of the air's CO_2 content -- it is readily appreciated that such a
oubling of the atmospheric CO_2 concentration fully doubles the water use
fficiency of nearly all plants (P/T = 1.33/0.66 = 2.02), as has been
emonstrated in numerous experiments.[55,56,127-130]

What is especially intriguing about this result is its constancy. Whereas
3 plants achieve a doubled water use efficiency by means of nearly
quivalent percentage changes in P and T, C_4 plants achieve the same result
ith somewhat smaller increases in P and somewhat larger decreases in
.[131,132] Hence, per-unit-leaf-area plant water use efficiency responses to
tmospheric CO_2 enrichment are extremely consistent across most types of
egetation. And again, they continue that way to CO_2 concentrations of 1,000
pm or more. Indeed, Rogers et al.[133] have shown that the positive water use
fficiency responses of both C_3 and C_4 species to atmospheric CO_2 enrichment
re nearly as strong between 600 and 900 ppm as they are between 300 and 600
pm. And this result suggests that a 330 to 990 ppm tripling of the
tmospheric CO_2 concentration may almost triple the per-unit-leaf-area water
se efficiencies of most plants.

.4 A Shifting of Species

What will be the result of this significant improvement in the ability of
lants to utilize water? For one thing, it is almost self-evident that in

the high CO_2 world of the future, many plants will be able to live in area
which are presently too dry for them.[134-136] Consequently, typically mes
vegetation will in all likelihood migrate towards more xeri
conditions.[137,138]

As an illustrative example of what this phenomenon may mean to a specif
area, Idso and Quinn[139] made a number of detailed assessments of change
likely to occur in twelve biotic communities in Arizona and New Mexico i
response to a 330 to 660 ppm doubling of the air's CO_2 content. The basis fo
their projections was the assumption that if the water use efficiency of
plant is doubled, it can survive with only half the amount of effectiv
precipitation that it presently receives at the water-limiting edge of it
range, where the effective precipitation was calculated as per the procedur
of Willmott,[140] based upon Thornthwaite's water budget approach[141] t
obtaining the difference between potential evapotranspiration an
precipitation.

Graphical representations of Idso and Quinn's predicted changes in plan
altitudinal distributions for northern and southern Arizona are presented i
Fig. 7.1 and 7.2; while their predicted changes in the areal distributions o
these same plant communities are presented in Table 7.1 for both Arizona an
New Mexico.

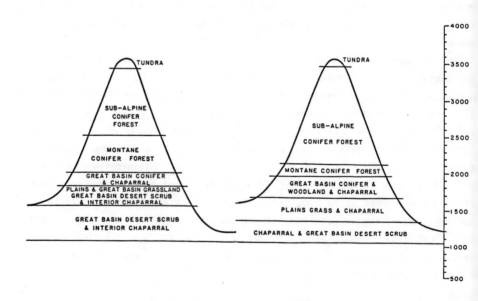

Fig. 7.1 Present elevation ranges of biotic communities in northern Arizona
(left) and predicted elevation ranges of the same plants for a 330 to 660 ppm
doubling of the atmospheric CO_2 concentration (right). Adapted from Idso and
Quinn.[139]

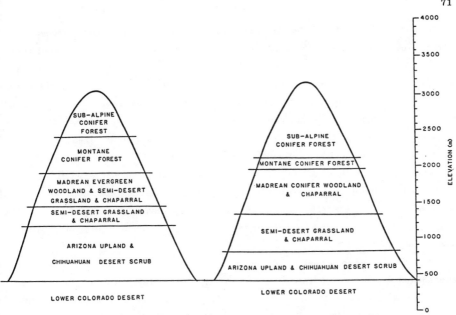

Fig. 7.2 Present elevation ranges of biotic communities in southern Arizona (left) and predicted elevation ranges of the same plants for a 330 to 660 ppm doubling of the atmospheric CO_2 concentration (right). Adapted from Idso and Quinn.[139]

Table 7.1 Areas (km^2) of several different types of vegetation under current conditions and as predicted by Idso and Quinn[139] for a doubling of plant water use efficiency.

Plant Community	Arizona				New Mexico			
	Actual	%	Predicted	%	Actual	%	Predicted	%
1	19	.01	19	.01	397	.13	397	.13
2	878	.30	12,006	4.07	8,192	2.60	31,859	10.11
3	19,602	6.64	32,132	10.89	25,286	8.57	80,135	25.43
4	52,460	17.77	69,584	23.58	70,573	22.40	66,787	21.19
5	13,265	4.49	24,578	8.33	460	.15	0	0.00
6	7,536	2.55	25,164	8.53	1,811	.57	14,946	4.74
7	23,734	8.04	4,550	1.54	6,099	1.94	0	0.00
8	5,300	1.80	1,800	.61	29,024	9.21	0	0.00
9	31,454	10.66	13,350	4.52	58,337	18.51	47,106	14.95
10	39,154	13.27	20,090	6.81	114,934	36.47	73,883	23.45
11	41,884	14.19	52,165	17.67	0	0.00	0	0.00
12	46,584	15.78	30,862	10.46	0	0.00	0	0.00
13	13,276	4.50	8,846	3.00	0	0.00	0	0.00

Plant Community Types: (1) Tundra, (2) Subalpine Conifer Forest, (3) Montane Conifer Forest, (4) Great Basin Conifer Woodland, (5) Interior Chaparral, (6) Madrean Evergreen Woodland, (7) Great Basin Desertscrub, (8) Chihuahuan Desertscrub, (9) Semidesert Grassland, (10) Plains Grassland, (11) Arizona Upland, (12) Lower Colorado, (13) Mohave Desertscrub.

As can be seen from the figures, the vegetation types that expanded in elevation range under doubled CO_2 were primarily the wet forests, particularly the Subalpine Conifer Forest, but also to some extent the Madrean Evergreen Woodland; while communities that contracted vertically were mostly the Desertscrubs, particularly Great Basin Desertscrub, Arizona Upland, and Chihuahuan Desertscrub. Much the same was also true of changes in area, as indicated in Table 7.1. However, changes in areal extent of vegetation are dependent, not only on new elevational limits to growth, but also on the presence of expanses of land at the appropriate elevations. Thus, even though the elevational range of the Montane Conifer Forest decreased significantly, its areal extent increased to almost twice its original size in Arizona and to more than three times its original size in New Mexico.

Reviewing these changes in more detail, the Subalpine Conifer Forest expanded the most in Arizona, to about thirteen times its original extent; while in New Mexico it increased to almost four times its original area. Similarly, the Madrean Evergreen Woodland expanded into the Semidesert Grassland, increasing its original area in Arizona by more than a factor of three and its original area in New Mexico by a factor of eight. In addition, Interior Chaparral (which can be quite wet in winter) nearly doubled in areal extent in Arizona; but it completely disappeared in New Mexico.

Vegetation types that changed but little in areal extent, for both Arizona and New Mexico, were Tundra and Great Basin Conifer Woodland. Since the location of Tundra depends chiefly on severe winter conditions limiting tree growth below it, the lower elevation limit of Tundra would not be expected to change materially with atmospheric CO_2 enrichment; and, hence, in this example it was held constant. On the other hand, although Great Basin Conifer Woodland did increase its elevational range, its total areal extent changed but little, due to a compensatory reduction in available area at its new vertical location.

At the low end of the elevational spectrum, Desertscrub and grassland vegetation were significantly reduced in areal extent under high CO_2. In Arizona, for example, Semidesert Grasslands decreased to only two-fifths of their original size. Similarly, Plains Grasslands in New Mexico declined to a little over half of their original extent, due to displacement by Great Basin Conifer Woodland. Also, both Great Basin Desertscrub and Chihuahuan Desertscrub were dramatically reduced in Arizona and totally eliminated in New Mexico, due to displacement by Plains Grasslands and Semidesert Grasslands; while the extreme deserts of Arizona were both reduced to three-fifths of their original area, the Lower Colorado subdivision of the Sonoran Desert being replaced at its higher elevations by Arizona Upland and the

Mohave Desertscrub being replaced by Chaparral.

In general, then, with increasing CO_2 there is a strong tendency for forests to increase at the expense of shrubs,[142] shrubs at the expense of grasslands,[143] and grasslands at the expense of deserts.[136]

7.5 Competition or Cooperation?

Although the preceding example is presented in some detail, it is still but a gross simplification of the many complex ecological changes which will actually occur as the CO_2 content of the atmosphere continues to rise.[144-149] For one thing, although water use efficiency should increase rather uniformly among the many diverse plants which presently populate the planet, there are likely to be subtle differences which may ultimately produce profound competitive advantages and disadvantages for certain closely associated species.[134,150-158] Strain[159] has pushed this point to the extreme by claiming that "you cannot benefit one species in a closed system without detriment to another;" but as Goldsmith[160] has effectively countered, this is an assumption that is not substantiated by real-world data.

Nevertheless, special concern arises where C_3 and C_4 plants typically coexist.[161-165] In the case of agricultural systems, for example, C_3 crops generally fare better than C_4 weeds when the atmospheric CO_2 content is increased,[166,167] as do C_3 weeds in competition with C_4 crops.[168] Fortunately, however, the former situation is the more common of the two.[169,170] Also, as indicated in the preceding subsection, there would appear to be a tendency for natural plant communities at the dry end of the available water spectrum to be replaced by more mesic vegetation as the CO_2 content of the air climbs ever higher. Indeed, the comments of Strain[159] raise the specter of extinction for many species of desert flora. But that point of view, as logical as it may first appear, is actually the antithesis of the great bulk of biological transformations likely to be experienced as a result of atmospheric CO_2 enrichment.

Consider, for example, the fact that enhanced levels of atmospheric CO_2 invariably enable plants to grow better, or to better cope with the extremes of their environment; and then consider where many plants grow under such conditions today, without the benefit of high CO_2. The answer, of course, is the tropics, which contain a far greater diversity of species than any other part of the planet.[171,172] Now it is very possible, as R.M. May has suggested,[173] "that tropical ecosystems are so much richer largely because their environment is more predictable than that of temperate and boreal places," an external advantage which is much akin to the internal advantages conferrred by atmospheric CO_2 enrichment. Consequently, this observation suggests that higher atmospheric CO_2 levels should not lead to significant species extinctions, but rather to a global extension of the type of mutualism characteristic of tropical ecosystems.[174]

Considerable evidence may be mustered to support this optimistic view of the effects of atmospheric CO_2 enrichment on species diversity. Looking to the past, for example, several studies of Tertiary floras have demonstrated that many montane taxa of that period regularly grew among mixed conifers and broadleaf schlerophylls,[175-179] whereas today these forest zones are separated from each other by fully 1,000 m in elevation and 10-20 km or more in distance.[180] Indeed, during this many-million-year period, when the CO_2 content of the atmosphere was generally much greater than it is today,[181] all three forest zones merged to form a "super" ecosystem, which, in the words of Axelrod,[180] "was much richer than any that exists today." Even under current conditions, in fact, modern forestry experiments have demonstrated that trees planted in mixtures sometimes grow better than they do in single-species plantings.[182-185]

One mechanism by which this type of mutualism may be fostered has to do with the stimulation of vesicular-arbuscular mycorrhizal (VAM) fungi, which are ubiquitous in most terrestrial ecosystems[186,187] and the most prevalent of all soil fungi.[188] These unseen inhabitants of the soil provide a number of benefits to the plants they "service." They increase the absorption of water and nutrients by the plant, protect the plant from soil-borne diseases, and reduce the incidence of nematode infection of roots.[189] And as Johnson and McGraw[190] have noted, "their vigor may be expected to reflect the vigor of their hosts," which with CO_2 enrichment would be expected to increase. Consequently, whereas community ecology paradigms of the past, based largely on data pertaining to above-ground interactions, have tended to stress relatively short food chains with competition and antagonism as major organizing forces in community development, Edwards and Stinner[191] note that, today, "ecologists studying biotic interactions in soil systems generally have observed complex food webs, a great diversity of organisms, and a wide range of symbiotic interactions." Indeed, many endomycorrhizae,[192] as well as certain ectomycorrhizae,[193,194] have even been demonstrated to actively transfer nutrients between individual plants of both the same and different species. In fact, seedlings of some plants will not grow at all unless they interact with the mycorrhizae of an adjacent host plant,[195] while in other situations, both seed germination and initial plant growth rates are greatly stimulated by the presence of such fungi.[196,197] As a result, Moore[198] contends that mutualism is common below ground and that it "can have profound effects on the structure and activity of soil microbial communities, the decomposition of organic matter, and ultimately plant growth."

As one specific example of this phenomenon, Arachevaleta et al.[199] have recently demonstrated that the "infection" of tall fescue grass by a compatible fungal endophyte results in "growth stimulation, improved survival, and drought tolerance to the host plant." This relationship is clearly one of mutualistic symbiosis,[200,201] as are most all endophytic fungus-grass associations,[202] in that the fungus also benefits from the association by receiving nutrients, protection, and assistance in reproduction and dissemination.[199]

But the plant response is what we most notice; and in a study of endophyte-infected (EI) and endophyte-free (EF) perennial ryegrass, Latch et al.[203] found that EI plants produced considerably more herbage, leaf area, tillers and roots than did EF plants, perhaps due to endophyte production of plant growth hormones,[203,204] which must in turn stimulate endophyte growth even more. In addition, many of the benefits derived from endophytic associations are passed on to future generations through the seeds produced by EI plants.[205]

Finally, it should be noted that as some plants become crowded, they alter their rooting pattern, so that more roots grow downward rather than horizontally,[206] again reducing competition and allowing all members of the community to be maximally stimulated by atmospheric CO_2 enrichment. And as atmospheric CO_2 enrichment tends to enhance all of these many mutualistic interactions, it may stimulate whole ecosystems to pull themselves up by their own bootstraps, so to speak, and significantly rise above the state of "disclimax" in which so many of them find themselves today.[207] (See also Sec. 7.6, 7.8.6 and 8.10.)

7.6 Soil Improvements

Up to this point, we have dealt primarily with the direct effects of CO_2 on plants. Consider next what consequences some of these effects may have for the soil, as well as what synergistic feedbacks these phenomena may produce relative to subsequent vegetative productivity.

First, and most obvious, is the stabilizing effect of enhanced plant cover on the world's valuable topsoil.[208-210] Each and every year, billions of tons[211,212] of this most important resource are lost to the ravages of wind[213-215] and water.[216,217] However, as plants grow ever more vigorously with increasing concentrations of atmospheric CO_2 (as described in Sec. 7.1), and as they subsequently expand their ranges to cover previously desolate and barren ground (as described in Sec. 7.4), both of these types of erosion should be significantly reduced, as has been demonstrated in numerous experiments with both natural and managed ecosystems.[218,219]

Consider next the effect of enhanced productivity on soil organic matter.[220] As more plant material is produced, both above- and below-ground (and in most natural ecosystems most of it is produced below-ground[221]), soil organic carbon levels should likewise rise.[222] And as Lowrance and Williams[223] have noted, soil organic carbon provides a large portion of the soil's cation exchange capacity, as well as serving as a source of mineral nitrogen and as a sink for fertilizer nitrogen.[224,225] It also tends to reduce the damage done by soil-borne plant pathogens;[226-228] and its ultimate decay releases more CO_2 to the soil, where more of it may be dissolved in the soil's reservoir of water. In addition, since CO_2 is removed from the atmosphere by precipitation processes (as described in Sec. 1.2), higher atmospheric CO_2 levels tend to bring more dissolved CO_2 to the soil's store of moisture by this means too.

So what are the advantages of CO_2-enriched soil moisture? Although several investigators have claimed that plants should receive little direct benefit from dissolved CO_2,[229-232] a number of experiments have produced significant increases in root growth,[233-236] as well as yield itself,[237-241] with CO_2-enriched irrigation water. Early on, Misra[242] suggested that this beneficent effect may be related to CO_2-induced changes in soil nutrient availability; and this hypothesis may well be correct. Arteca et al.,[243] for example, have observed K, Ca and Mg to be better absorbed by potato roots when the concentration of CO_2 in the soil solution is increased; while Mauney and Hendrix[244] found Zn and Mn to be better absorbed by cotton under such conditions, and Yurgalevitch and Janes[236] found an enhancement of the absorption of Rb by tomato roots. In all cases, large increases in either total plant growth or ultimate yield accompanied the enhanced uptake of nutrients. Consequently, as it has been suggested that CO_2 concentration plays a major role in determining the porosity, plasticity and charge of cell membranes,[245,246] which could thereby alter ion uptake and organic acid production,[236] it is possible that some such suite of mechanisms may well be responsible for the plant productivity increases often observed to result from enhanced concentrations of CO_2 in the soil solution.

A very similar role may be played by CO_2-induced increases in VAM fungal activity. In soils severely deficient in important trace metals, for example, experimentally-stimulated VAM colonization of plant roots has been documented to lead to an enhanced absorption of those elements.[247-251] And where heavy metals are present in the soil in concentrations which are toxic to plants, VAM fungi tend to suppress their absorption.[252-254] Consequently, as has been noted by Heggo et al.,[255] enhanced VAM activity generally ameliorates both problems: "heavy metal uptake is increased in soils with low or moderate metal concentrations [while] foliar uptake is decreased in soils with inherently high metal concentrations." In addition, enhanced colonization of legume roots by VAM fungi has been found to lead to greater rates of nitrogen fixation.[256]

Consider next the lowly earthworm, whose soil-forming activities were studied by Darwin[257] well over a century ago, and which was honored with a special symposium at the 80th annual meeting of the American Society of Agronomy convened at Anaheim, CA over the period 27 Nov. - 2 Dec. 1988. As C.A. Edwards[258] reminded everyone in attendance, "earthworms play a major role in improving and maintaining the fertility, structure, aeration and drainage of agricultural soils."

Elucidating the first of these points, Sharpley et al.[259] noted that "by ingestion and digestion of plant residue and subsequent egestion of cast material, earthworms can redistribute nutrients in a soil and enhance enzyme activity, thereby increasing plant availability of both soil and plant residue nutrients," as was also reported by several other groups of researchers at the meeting.[260-262]

The soil-forming and structural-enhancement activities of earthworms were likewise lauded by several scientists,[263,264] as were their drainage-enhancing characteristics. In particular, W.D. Kemper[265] described how "burrows opened to the surface by surface-feeding worms provide drainage for water accumulating on the surface during intense rainfall." As he explained it, "the highly compacted soil surrounding the expanded burrows has low permeability to water which often allows water to flow through these holes for a meter or so before it is sorbed into the surrounding soil."

Hall and Dudas[266] also reported that the presence of earthworms appears to mitigate the deleterious effects of certain soil toxins; while Logsdon and Lindon[267] described a number of other beneficial effects of earthworms, including 1) enhancement of soil aeration, since under wet conditions earthworm channels do not swell shut as many cracks do, 2) enhancement of soil water uptake, since roots can explore deeper soil layers by following earthworm channels, and 3) enhancement of nutrient uptake, since earthworm casts and channel walls have a more neutral pH and higher available nutrient level than bulk soil.

So what do all of these beneficial effects of earthworms have to do with atmospheric CO_2 enrichment? Simply this: higher levels of atmospheric CO_2 promote greater levels of plant productivity, and greater levels of plant productivity promote greater levels of earthworm activity.[268,269] As Edwards[258] has described it, "the most important factor in maintaining good earthworm populations in agricultural soils is that there be adequate availability of organic matter." Consequently, with the greater plant productivity which accompanies atmospheric CO_2 enrichment, earthworm populations should be significantly augmented. This phenomenon should then lead to the many beneficial changes in soil properties described in the preceding paragraphs, with subsequent amplification of the initial positive effects of atmospheric CO_2 enrichment which are directly experienced by the plants. And where earthworms perform these valuable functions in humid and sub-humid regions, termites and ants perform analogous services in the arid and semi-arid tropics.[270] In addition, ubiquitous microarthropods[271] tend to comminute,[272] or physically restructure,[273] organic matter in many soil ecosystems,[274-276] increasing the surface area of detritus and facilitating microbial activity,[277-279] again in direct proportion to the amount of organic input to the soil system.

7.7 Water Resource Improvements

One of the major environmental problems currently confronting much of the world is the ominous threat of widespread contamination of the planet's dwindling supply of pristine groundwater,[280-282] due primarily to the improper disposal of industrial waste products[283,284] and the excessive agricultural use of fertilizers,[285,286] herbicides,[287,288] and pesticides.[289,290] So serious has the problem become, in fact, that young children in many U.S. cities cannot drink the local water for fear of their developing methemoglobinemia (blue baby syndrome) as a result of the water's high nitrate content;[291] while across the country literally thousands of wells have tested positive for harmful levels of noxious chemicals.[292] And the worst is probably still to come, due to the slow rate of travel of the various pollutants down to the water table.[293]

There are several ways in which higher atmospheric CO_2 concentrations may help to partially alleviate this regrettable situation. First, as noted in the preceding subsection, as plants respond to atmospheric CO_2 enrichment and grow more vigorously everywhere and more profusely in stressful environments, they should gradually increase the amount of organic matter in the soil; and as soil organic matter effectively adsorbs and immobilizes most water-born chemicals,[294,295] its greater presence in the soil should slow the rate at which contaminants move downward to the groundwater. Second, as plant roots penetrate deeper into the soil profile with atmospheric CO_2 enrichment, the amount of time available for detainment and modification[296] of potential groundwater contaminants should likewise be lengthened. And finally, associated with the augmented concentration and vertical extent of the enhanced soil organic matter should be correspondingly augmented populations of soil microorganisms which attack and detoxify would-be water pollutants.[297-299] Considered in their totality, these several interrelated henomena should play a major role in helping to reduce the amount of roundwater contamination likely to otherwise be experienced in the coming ears.

In much the same manner, these phenomena should similarly be of importance n situations where rivers[300] and lakes[301] are fed by groundwater discharge or ven surface water drainage.[302,303] For that matter, atmospheric CO_2 nrichment should also enhance the purifying properties of flow-through aquatic systems, both natural[304,305] and man-made,[306,307] by analogous means, as well as by stimulating oxygen transport to the root zone,[308,309] as many aquatic plants respond to atmospheric CO_2 enrichment with increased growth rates comparable to those exhibited by terrestrial vegetation.[56,310-312]

7.8 Other Responses

Besides growing bigger and better and transpiring less water, plants often exhibit a number of other responses to atmospheric CO_2 enrichment, as described in the following subsections.

7.8.1 Water Use Efficiency Revisited

Atmospheric CO_2 enrichment generally enhances plant water use efficiency by increasing the plant's rate of photosynthesis and/or decreasing its rate of transpiration.[313] Recently, however, a third means by which CO_2 accomplishes this feat has also been discovered.

Normally, under conditions of stress which reduce photosynthetic rates, stomatal conductance decreases in phase with photosynthesis, so that water is not needlessly lost from the plant.[314,315] But in some species there is a marked lack of stomatal control of water loss.[316-318] Cotton is one example of this group which has a rather unresponsive (to photosynthetic rate) stomatal conductance and therefore often exhibits less-than-optimum water use efficiencies under ambient CO_2 conditions.[319] Just recently, however, Radin et al.[320] have demonstrated that when the CO_2 content of the atmosphere is approximately doubled, the stomatal behavior of cotton is much more tightly coupled to mesophyll photosynthetic capacity, thereby significantly alleviating this problem.

7.8.2 Tissue Dry Matter Content

Do the enhanced growth characteristics of plants exposed to CO_2-enriched air produce more substantial plant tissues,[321] or -- as has sometimes been claimed[322,323] -- do they merely cause plants to load up with water and become

more succulent? In a recent survey of the literature relative to thi
question, Idso et al.[324] found that plant percentage dry matter conten
appears to change but little with atmospheric CO_2 enrichment when plant dr
matter contents are normally low. Under conditions conducive to th
development of high plant dry matter contents (high carbohydrate source-to-
sink ratios, high light intensities, and high levels of water stress)
however, atmospheric CO_2 enrichment tends to _increase_ the percentage dr:
matter contents of plant tissues. Hence, most plants benefit from CO_2
enrichment of the atmosphere in this way also.

7.8.3 Root:Shoot Ratios

Generally speaking, all plant organs benefit from the aeria]
"fertilization effect" of atmospheric CO_2 enrichment.[325] What is not as wel]
established, however, is whether some parts of plants benefit more than
others from this phenomenon; and probably the most-asked question in this
regard has to do with the above- and below-ground portions of a plant's
anatomy. That is, do plant shoots or plant roots benefit significantly more
than the other from atmospheric CO_2 enrichment?

In a recent review of this topic, Idso et al.[326] found a number of
conflicting reports for plants whose primary yield component is produced
above-ground. In the case of root crops, however, the literature was fairly
consistent in suggesting a positive effect of atmospheric CO_2 enrichment on
root:shoot ratios. And this situation is only to be expected; for as noted
by Knecht,[327] when considering source/sink aspects of photosynthesis and
photosynthate storage, it is logical to presume "that increased
photosynthesis due to added CO_2 will be reflected by a positive response at
the 'sinks' of the plants," especially since such sinks may be limited by
source capacity under present CO_2 concentrations.[328]

The results of the literature review of Idso et al.[326] thus suggest that a
300 ppm increase in the CO_2 content of the air generally increases the
root:shoot ratios of most root crops by about a third, while leaving non-root
crops basically unaffected in this respect. New work has also substantiated
this general observation: a 300 ppm increase in the CO_2 content of the air had
no effect upon the root:shoot ratios of either cotton or soybeans, but it
increased the root:shoot ratios of carrots and radishes by approximately 36%
throughout their entire life cycles, as shown in Fig. 7.3. In addition, in a
long-term CO_2 enrichment study of an alfalfa variety which produces a large
tap root, it was found that a 300 ppm increase in the air's CO_2 content had
little effect on above-ground biomass production, but that it increased the
dry weight of apical root segments by more than 90%.[329]

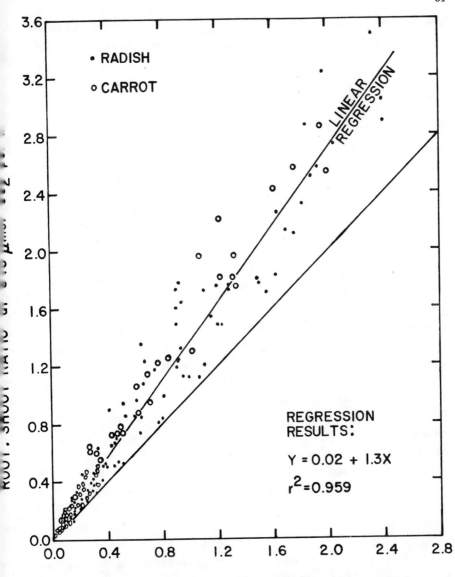

Fig. 7.3 Effects of increasing atmospheric CO_2 concentration from 340 to 640 μmol CO_2 per mol air on root:shoot ratios calculated from weekly destructive samples of successive crops of carrots and radishes grown over a period of 3 years.   The lower diagonal line is the one-to-one correspondence line. Adapted from Idso et al.[326]

In the case of natural vegetation, atmospheric CO_2 enrichment may have an even more pervasive effect on root:shoot ratios; for natural ecosystems are often low in available nutrients, and plants in infertile habitats generally respond to nutrient demand (as would be created by the enhanced grow potential conferred by atmospheric CO_2 enrichment) by increasing their ro mass,[330] as has recently been specifically demonstrated for this scenario.[33

7.8.4 Germination and Seedling Establishment

Results of studies of the effects of atmospheric CO_2 enrichment on se germination have been largely positive,[332-337] although there have been a f experiments where additions of CO_2 to the seed environment have produced noticeable response;[338,339] and under some circumstances CO_2 can even have inhibitory effect on the germination process.[340] In spite of this variety responses, however, it is generally accepted that CO_2 enhancement promote seed germination by a number of different means: first, by direct increasing the initial growth of seed tissues,[341-343] second, by enhancir the production of ethylene (C_2H_4), which often stimulates germination,[344-34 and third, by a variety of complex interactions with ethylene and a number c plant hormones,[347-354] which have yet to be fully understood.

Once seedlings appear, however, the effects of atmospheric CO_2 enrichmen become more noticeable and consistant.[336,337,355-357] Indeed, severa investigators have shown that the growth-enhancing properties of added CO_2 ar expressed most strongly when plants are in the seedling stage.[358-36 Similarly, promotion of rooting in cuttings of ornamental and forest tre species in response to atmospheric CO_2 enrichment has also bee documented.[328,362-365] And again, all of these responses, like the man others described previously, continue strong out to CO_2 concentrations o several thousand ppm. Indeed, the promotion of oat coleoptile elongatio continues to occur at CO_2 concentrations of several percent.[366,367]

7.8.5 Plant Life Cycles

In addition to hastening and enhancing germination and seedling establishment, atmospheric CO_2 enrichment has often been observed to modestly accelerate entire plant life cycles.[368,369] Anthesis and flowering, for example, are usually[370-375] -- but not always[376] -- slightly earlier in plants continuously exposed to CO_2-enriched air than in plants not exposed to elevated levels of CO_2. Plants also generally mature faster in CO_2-enriched as opposed to ambient air,[328,377-379] and they often senesce earlier[369,380 and at a more rapid rate.[328,381] One notable exception, however, is the mesohaline salt marsh ecosystem on Chesapeake Bay,[382] where atmospheric CO_2 enrichment over an entire growing season delayed senescence and led to an increased carbon uptake by all plant communities in late summer.[383]

8.6 Nitrogen Fixation

Another biological process which is stimulated by atmospheric CO_2 richment is that of nitrogen fixation,[384-388] whereby gaseous nitrogen (N_2) removed from the atmosphere and made available to plants, i.e., chemically duced for utilization in the production of proteins and other important ant constituents.[389] This significant feat is accomplished by several fferent groups of bacteria[390]: _Azobacter_ in aerated soils, _Clostridium_ in aerated soils, and _Rhizobium_ in legume roots,[391] as well as by tinomycetes[392] and certain of the blue-green algae in a number of other rrestrial habitats[389] and by filamentous and unicellular cyanobacteria in e world's oceans.[393]

In the case of terrestrial plants, perhaps the most important means by ich this activity is accomplished is through the symbiotic association of e responsible bacteria and their plant hosts.[394] The bacteria infect the ant and grow in clumps or nodules on its roots, where they receive rbohydrates from the plant and in return fix nitrogen from the soil air and nvert it into various nitrogenous compounds, which are then absorbed by the jacent vegetative cells and transported to all parts of the plant.[395,396]

Atmospheric CO_2 enrichment has been shown to directly stimulate the tivity of certain of these nitrogen-fixing bacteria[397,398] In addition, e capacity of the bacteria to fix nitrogen seems also to be limited by the ant's rate of carbohydrate production.[399-401] Hence, practically anything at stimulates plant growth,[402] including atmospheric CO_2 enrichment,[403] imulates bacterial nodule growth and activity; and several-fold increases the CO_2 content of the air have thereby lead to several-fold increases in fixation,[404-407] again with indications that the positive response rsists over a CO_2 concentration range in excess of 1,000 ppm.[408-410]

The beneficial effects of this phenomenon for agriculture are obvious; but t is equally important for natural ecosystems, and especially for arid egions, where desert legumes are about the only source of nitrogen[411-414] nd the nitrogen-fixing capacity of their nodulating rhizobia is strongly ied to host plant activity.[415] Enhanced nitrogen fixation due to tmospheric CO_2 enrichment may also play a major role in the rejuvenation of ragile tropical ecosystems, where vast reaches of land lie barren due to oor soil conditions and lack of rain.[416] Likewise, atmospheric CO_2 nrichment may be especially important to nitrogen fixation in Arctic egions, as the N_2 fixation rates of Arctic legumes are generally lower than hose of more temperate lands.[417]

84

7.8.7 Interaction with Global Warming

Due to the fact that there may possibly be a greenhouse warming of the planet in the years to come (albeit a much less severe warming than the generally predicted), it is important to consider what effect such a warming may have on the basic growth response of plants to atmospheric CO_2 enrichment.

Over the past several years, a number of experiments have shed considerable light on this topic; and in reviews of the pertinent literature published in 1985[106] and 1986,[107,418] there were indications that the stimulatory effects of atmospheric CO_2 enrichment may be significantly augmented as air temperature rises. Subsequent studies have clarified this point even more.[419-424]

The story starts with the inhibitory effect of normal oxygen concentrations on CO_2 uptake by plants.[425-429] Since this phenomenon is known to increase with increasing air temperature,[430-432] and since it has been established by a number of experiments that the O_2-inhibition of photosynthesis can be almost totally eliminated by raising the CO_2 content of the atmosphere to 900 ppm,[48,433-436] it logically follows that there is greater potential for benefits due to atmospheric CO_2 enrichment at higher air temperatures.[437] In the words of Mortensen,[438] "the [positive] effect of CO_2 enrichment will increase with increasing temperatures," and "the optimum temperature for photosynthesis will, as a consequence, increase in CO_2 enriched air," as has indeed been experimentally observed.[49,129]

As an example of this phenomenon, Fig. 7.4 shows how the CO_2 assimilation rates of the leaves of two agricultural crops and a tree respond to atmospheric CO_2 enrichment at three different temperatures. At the relatively cool temperature of 10°C, there is little change in CO_2 uptake as the intercellular CO_2 content is raised above about 300 ppm (or 300 μmol CO_2 per mol air). At 20°C, however, there is a continued increase in net photosynthesis far beyond this point, which at 30°C is even more pronounced.

To clarify these relationships, the data of Fig. 7.4 have been recast in the format of Fig. 7.5, where the ratio of net photosynthesis at 600 ppm CO_2 to net photosynthesis at 300 ppm CO_2 has been plotted as a function of leaf temperature for each plant. And as this figure clearly indicates, the stimulatory effects of atmospheric CO_2 enrichment rise dramatically with increasing temperature.

This interactive phenomenon has also been observed in experiments with a number of different plants in open-top, clear-plastic-wall, CO_2-enrichment chambers at Phoenix, AZ.[312] In the naturally-varying environmental conditions of the out-of-doors, however, it is sometimes difficult to

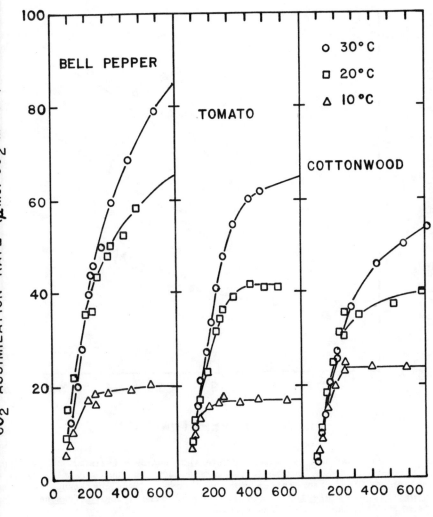

ig. 7.4 Net photosynthesis (CO_2 assimilation rate) as a function of ntercellular CO_2 concentration at three different leaf temperatures for bell epper (<u>Capsicum annuum</u>), tomato (<u>Lycopersicon esculentum</u>), and cottonwood ree (<u>Populus fremontii</u>). Adapted from Sage and Sharkey.[439]

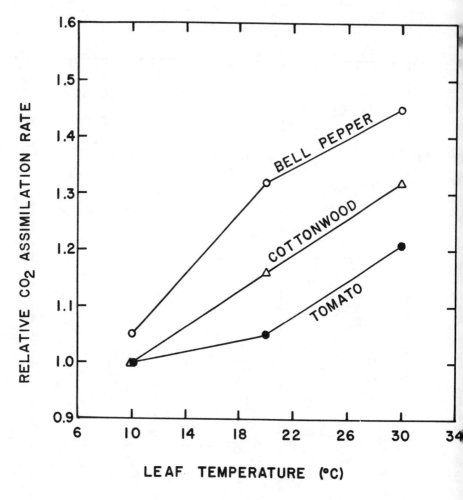

Fig. 7.5   Relative CO_2 assimilation rate (net photosynthesis at 600 μmol CO_2 per mol air divided by net photosynthesis at 300 μmol CO_2 per mol air) as a function of leaf temperature for bell pepper (<u>Capsicum annuum</u>), tomato (<u>Lycopersicon esculentum</u>), and cottonwood tree (<u>Populus fremontii</u>), as determined from the data of Fig. 7.4.

separate the effects of air temperature variability from those of solar radiation variability.   Thus, the mean net photosynthetic rate of the different CO_2 treatments has sometimes been used as a surrogate for the combined influence of <u>all</u> environmental parameters which may impact plant growth.

Two examples of this type of analysis are depicted in Fig. 7.6, where individual treatment net photosynthesis rates are plotted as functions of their combined mean rate.   As can be seen there, the short-term effects

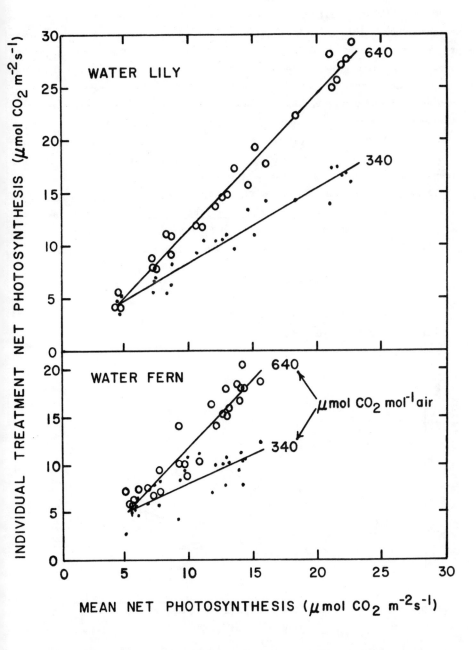

Fig. 7.6 Regressions of individual treatment net photosynthesis rates upon the mean net photosynthetic rate of both treatments (340 and 640 μmol CO_2 per mol air) for water lily (<u>Nymphea</u> <u>marliac</u> <u>carnea</u>) leaves sampled over a several-day period (top) and water fern (<u>Azolla</u> <u>pinnata</u> <u>pinnata</u>) plants sampled over a several-month period (bottom). Adapted from Allen <u>et</u> <u>al</u>.[310,311]

observed with water lily leaves over several diurnal cycles are of the same
basic nature as the long-term effects observed with water fern plants over
the greater part of a year. That is, when environmental conditions are not
conducive to high growth rates, atmospheric CO_2 enrichment has little effect
on net photosynthesis, as was also evident in the data of Fig. 7.4 and 7.5;
but as conditions improve, as manifest by the increased photosynthetic rates
themselves, the stimulatory effects of added CO_2 become ever greater.

A recent year-long study of this nature dealt with carrots and
radishes.[440] And as it had previously been demonstrated that the air
temperature variability over the year was the primary factor responsible for
the type of response displayed in Fig. 7.6,[312] its results were presented as
shown in Fig. 7.7, where the growth modification factor is the weekly dry
matter production at 640 ppm CO_2 divided by the weekly dry matter production
at 340 ppm CO_2, averaged over a large number of individual experiments.

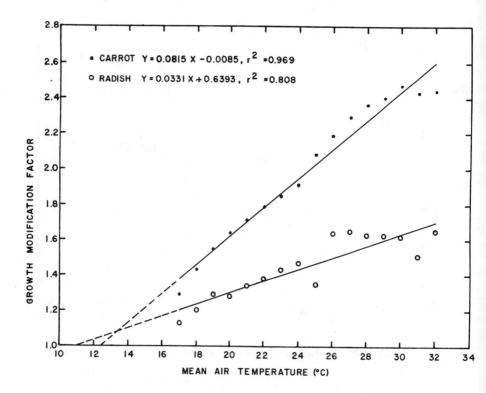

Fig. 7.7 Growth modification factors (weekly dry weight production at 640
μmol CO_2 per mol air divided by weekly dry weight production at 340 μmol CO_2
per mol air) for carrot (Daucus carota) and radish (Raphanus sativa) plants
as functions of the mean air temperature for a number of experiments conducted
over the year at Phoenix, AZ. Adapted from Idso and Kimball.[440]

As Fig. 7.7 demonstrates, the CO_2-temperature interaction in carrots and radishes is just like that of the preceding studies. The effects of added CO_2 are minimal at relatively cool temperatures, but they become greater and greater as temperature rises. In addition, it is clear from this figure that there may be significant differences among species with respect to the rate at which the stimulatory effect of atmospheric CO_2 enrichment increases with increasing air temperature.

So what is the significance of this effect? Consider the plant growth data of Fig. 7.7, which although pertaining to only two plants are representative of the range of results displayed by a number of others.[423] Locating the temperature at which the mean trend of the two relationships there depicted corresponds to a growth modification factor of 1.3 -- the mean response for a 300 ppm increase in the CO_2 content of the atmosphere -- and moving to the right another $0.8^{\circ}C$ -- the mean global warming likely to accompany such a CO_2 increase ($0.4^{\circ}C$ due to CO_2 and $0.4^{\circ}C$ due to other trace gases) -- it is noted that the 30% increase in plant productivity expected to result from such a CO_2 doubling under unchanging climatic conditions will likely be raised to a 35% increase in the face of a $0.8^{\circ}C$ warming. And as the interactive effects of increases in air temperature continue to be experienced as atmospheric CO_2 concentrations climb above the 1,000-ppm mark,[441-444] very large productivity enhancements due to this synergistic relationship may ultimately be realized. Indeed, the response of cassava, a tropical root crop, is more than twice as great as even the carrot response of Fig. 7.7.[445] In addition, there is evidence to suggest that the tendency for atmospheric CO_2 enrichment to reduce weed/crop ratios is also augmented by warmer air temperatures.[446]

7.8.8 Miscellaneous Effects

7.8.8.1 Photosynthetic Quantum Yield. In addition to providing more substrate for the basic photosynthetic reaction, atmospheric CO_2 enrichment also tends to increase the efficiency of the process or its "quantum yield."[48,447]

7.8.8.2 Dark Respiration. Very high CO_2 concentrations (5,000 to 100,000 ppm) have long been known to decrease dark respiration.[448-450] More recently, it has additionally been established that atmospheric CO_2 enrichment in the range of 300 to 1,000 ppm is also effective in this regard.[451-454] This phenomenon thus results in an increase in net primary production.

7.8.8.3 Root Respiration. Data on effects of atmospheric CO_2 enrichment on root respiration are rather scarce. The few studies which have been conducted, however, suggest that this process may also be somewhat reduced under high CO_2 concentrations.[455-457] But even if it were unaffected, such a non-effect would still tend to enhance net primary production in the face of augmented photosynthetic rates by not utilizing as large a percentage of the primary production as it does under current levels of atmospheric CO_2.

7.8.8.4 <u>Induction</u> <u>of</u> <u>CAM</u>. A number of succulent plants shift their mode of carbon metabolism from C_3 (Calvin-Benson cycle) to CAM (Crassulacean acid metabolism) in response to shortages of water[458-460] and increases in salinity.[461] This phenomenon, also, does not appear to be greatly affected by atmospheric CO_2 enrichment. Hence, this other non-effect should also be beneficial, as higher levels of atmospheric CO_2 should thus not interfere with this important stress avoidance mechanism.[462,463]

7.8.8.5 <u>Low-O_2-Induced</u> <u>Sterility</u>. C_3 plants generally grow much better under low oxygen concentrations (atmospheric O_2 contents of a few percent) than they do under the ambient concentration of approximately 21%[464-467] In spite of this growth enhancement, however, plants grown at low O_2 concentrations are often infertile and produce no seed or fruit.[466,467] But when CO_2 concentrations are raised by a factor of two or three, normal fruiting functions are restored.[468]

7.8.8.6 <u>Sex</u> <u>Ratio</u> <u>Changes</u>. Enoch <u>et</u> <u>al</u>.[69] have reported that atmospheric CO_2 enrichment causes cucumbers to produce a higher percentage of female flowers; while unpublished experiments of Anderson and Idso have shown atmospheric CO_2 enrichment to lead to a greater proportion of female sporocarps in water ferns.

7.8.8.7 <u>Sporulation</u>. As a general rule, fungus sporulation is induced by starvation or nutrient depletion;[469-472] and there is considerable evidence to suggest that this effect is due to reduced CO_2 concentrations due to reduced respiration under such conditions.[473-477] Consequently, higher atmospheric CO_2 concentrations may well mean less above-ground fungal activity, which is beneficial in terms of plant-disease relationships.

7.8.8.8 <u>Self-Incompatibility</u>. Many plants will not produce hybrid or selfed seeds,[478-481] i.e., they are "self-incompatible." Atmospheric CO_2 enrichment, however, can effectively eliminate this problem in most such plants,[482-487] although concentrations about two orders of magnitude greater than ambient are generally employed.

7.8.8.9 <u>Tissue</u> <u>Culture</u> <u>Growth</u>. Atmospheric CO_2 enrichment is a great boon to the growth of plant cells in aqueous suspension.[488,489] When the CO_2 content of the air is very low, for example, growth is inhibited;[490] while if it is increased, growth is enhanced,[491,492] particularly for cultures with low cell densities.[490,493]

7.8.8.10 <u>Stomatal</u> <u>Stimulation</u>. As explained in Sec. 7.2, under normal circumstances atmospheric CO_2 enrichment generally causes a reduction in stomatal apertures. When atmospheric oxygen concentrations are very low, however, just the opposite occurs.[494-496]

7.8.8.11   In Vitro to Ex Vitro Growth Stimulation. In vitro-propagated plantlets often experience excessive mortality when transferred to nonsterile media;[497-500] and those which survive are generally slow-growing[501,502] and extremely sensitive to environmental stresses.[503]   Under atmospheric CO_2 enrichment, however, most of these problems are eliminated.[504]

7.8.8.12   Stock Plant Cultivation. Cutting production and quality are both enhanced by atmospheric CO_2 enrichment in the cultivation of stock plants.[505,506]   The effects are most pronounced under conditions of low light,[507] and they continue to increase in magnitude as the CO_2 concentration of the air rises above 1,000 ppm.[508]

7.8.8.13   Unidentified Mechanisms. Certain cyanobacteria and green algae,[509-511] as well as some higher plants,[512-515] will only grow in media in equilibrium with atmospheres which are highly enriched in CO_2 (to 10,000 to 20,000 ppm).[512-514]   There are a number of different reasons for this phenomenon.[515,516]   In some instances, however, explanations are lacking, suggestive of an unidentified role for CO_2 in one or more of the biochemical processes which give these plants their unique properties.[517]   One thing which is clear, however, is that such plants have a great affinity for CO_2; and when its concentration in the air is below their liking, they actually concentrate it within their tissues by a powerful CO_2 "pumping" mechanism.[518-521]

7.9 CO_2 and Plant Temperatures

As indicated in Sec. 7.2, plant stomates do not open as wide in full sunlight under conditions of atmospheric CO_2 enrichment as they do under current ambient CO_2 concentrations. As a result, plant evaporative water loss can be much reduced on a unit leaf area basis as atmospheric CO_2 concentrations rise; and as this phenomenon represents a reduction in plant evaporative cooling power, leaf temperatures must consequently rise as well. Of course, higher leaf temperatures produce greater saturated vapor pressures in the substomatal cavities of the leaves, which effect increases the vapor pressure gradient between the leaf and the air and promotes an increase in transpiration which acts as a negative feedback to the primary impetus for foliage warming,[522] as does the generally larger leaf size of plants grown in CO_2-enriched air.[523]   In experiments designed to assess the net effect of these several interrelated phenomena, however, there has generally always been a net warming of the foliage as the CO_2 content of the air about the plant leaves is increased, ranging from 1 to $3^{\circ}C$ for a 330 to 660 ppm doubling of the atmospheric CO_2 concentration.[524-526]

This impetus for warming is basically beneficial, as indicated by the material presented in Sec. 7.8.7, which suggests that the positive effects of atmospheric CO_2 enrichment on plant productivity are significantly magnified when foliage temperatures increase concurrently. At first glance, however, it might also be thought that higher leaf temperatures would provide an effective force for global warming, which is generally perceived as undesirable; and if plant growth were not stimulated by atmospheric CO_2 enrichment, this reasoning would have some merit. But because of the fact that plants generally produce more leaf area when growing in air enriched with CO_2 (Sec. 7.1), the total amount of water transpired per unit area of land is not nearly so much reduced as is the amount of water transpired per unit area of leaf surface. Hence, we essentially have a situation where we can "have our cake and eat it too:" CO_2-induced increases in foliage temperatures magnify the biological benefits of atmospheric CO_2 enrichment without producing a significant impetus for global warming.

8 EFFECTS OF ENVIRONMENTAL STRESSES

8.1 A Tax on the Benefits of CO_2?

Up to this point, most of the information we have reviewed has pertained to plants growing under rather innocuous environmental conditions devoid of significant stress. So, one might ask, what happens when there are different impediments to full and unrestricted plant growth? Are the beneficial effects of atmospheric CO_2 enrichment reduced, as some have claimed?[1-5] Or are they unaffected or augmented even more, as others have claimed?[6,7] These are questions which will be broached in the following subsections, as we review what we know about a number of phenomena which have the potential to severely limit plant growth.

8.2 Water Stress

Lack of water is perhaps the single most important impediment to plant growth, as almost any soil will support a modest stand of some type of vegetation if it has a sufficient supply of moisture. But without adequate water, no soil, however fertile or otherwise beneficially endowed, can long sustain a significant level of biological productivity. Consequently, the interaction between the lack of soil moisture and the growth-enhancing effect of atmospheric CO_2 enrichment must rank uppermost in the hierarchy of potential limits to biospheric rejuvenation due to the increasing partial pressure of atmospheric CO_2.

The past decade has seen a number of studies designed to address this topic; and several of them have demonstrated that atmospheric CO_2 enrichment may largely compensate for significant deficiencies of soil moisture.[8-13] What is even more impressive, however, is what happens when more severe water shortages occur. Then, even though the growth of CO_2-enriched plants may be somewhat impaired, growth rates of ambient treatment plants are generally so strongly suppressed that the relative plant growth enhancement due to atmospheric CO_2 enrichment is even greater than what it is under optimum growth conditions of plentiful soil moisture.[6,14-27] In the case of C_4 and CAM plants, in fact, there is sometimes very little photosynthetic enhancement due to atmospheric CO_2 enrichment until the plants are stressed for water![28,29]

The primary reason for this augmented relative response, of course, is the tendency for atmospheric CO_2 enrichment to constrict stomatal apertures and reduce the rate at which water is lost from the plant via the transpiration process, as described in Sec. 7.2. This phenomenon allows the plant to get significantly more mileage, as it were, out of the limited supply of water available to it. And in situations where economic yield is the parameter used to evaluate the effectiveness of atmospheric CO_2 enrichment, the relative growth enhancement can actually become infinite, if the water stress is so severe that ambient-treatment plants die before producing any marketable product.

An evaluation of the magnitude of this phenomenon for intermediate values of plant water stress may be obtained from a consideration of the plant dry matter content data (recall Sec. 7.8.2) presented in Fig. 8.1, which were

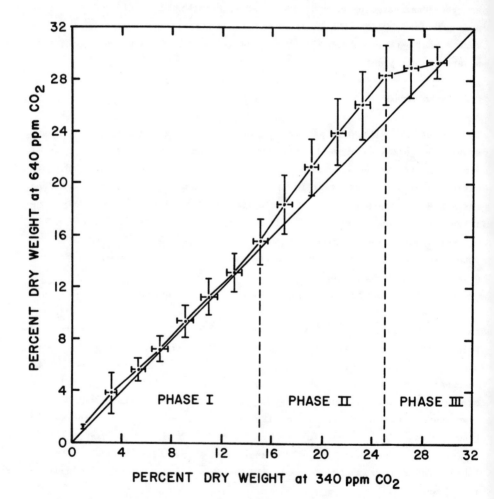

Fig. 8.1 Percent dry matter contents of plants grown at an atmospheric CO_2 concentration of 640 ppm vs. percent dry matter contents of corresponding control plants grown at an atmospheric CO_2 concentration of 340 ppm. Each point on the graph represents an average of 57 original data points, each one of which was itself the mean of from 20 to 800 individual plant measurements. Adapted from Idso.[30]

derived from numerous experiments performed on five terrestrial plants (carrot, cotton, radish, soybean, and the leaf succulent Agave vilmoriniana) and two aquatic species (water fern and water hyacinth).[31-40]

Fig. 8.1 consists of three distinctly different regions of plant response to water stress. The first region, denoted Phase I, represents the situation which prevails when plants are well supplied with water and transpiring at the potential rate. Here, the percent dry matter contents of plants grown in air of approximately 640 ppm CO_2 are essentially identical to those of control plants grown in ambient air of about 340 ppm CO_2; and such CO_2-enriched plants generally experience a mean growth enhancement on the order of 30%, recalling the results of Sec. 7.1.

As water becomes progressively less available, however, and as plant dry matter contents rise above a value of approximately 15%, the dry matter contents of CO_2-enriched plants begin to rise above those of their ambient-grown counterparts. This second region of plant response to water stress, which extends to an ambient-grown plant dry matter content of approximately 25% and which is denoted Phase II, has previously been documented to be a region of significant correlation between plant dry matter content and the severity of plant water stress;[41,42] and a linear regression analysis (r^2 = 0.997) of the six mean data points which delineate this phase of plant response suggests that as the dry matter contents of ambient-grown plants increase within this region, the dry matter contents of matching CO_2-enriched plants rise about 30% faster. Hence, by the time a water stress sufficient to drive ambient-grown plants to a dry matter content of 25% has been reached, CO_2-enriched plants have an approximate 14% advantage in dry matter content -- computed as (28.4% - 25.0%)/25.0% x 100% = 13.6% -- over their ambient-grown counterparts. Furthermore, multiplying this dry weight enhancement factor (1.14) by the basic growth enhancement factor (1.30) experienced by non-water-stressed plants as a result of a 300 ppm increase in the CO_2 content of the air (see Sec. 7.1) yields a total relative productivity enhancement factor for such severely stressed plants of 1.48. Consequently, the relative productivity advantage experienced by severely stressed plants growing in an atmosphere enriched with an extra 300 ppm of CO_2 may be as much as 60% greater than that experienced by similar well watered plants, computed as (48%-30%)/30% x 100% = 60%, although a greater absolute enhancement of plant growth may still be experienced by well watered plants.

The third region of Fig. 8.1 is indicative of even more extreme conditions of plant water stress. Denoted Phase III, it is here where further shortages of water lead to rapid dehydration of plants grown under ambient CO_2 conditions. As such plants senesce and ultimately die, the plot of the figure shifts sharply to the right, as the dying ambient-grown plants lose water and therefore increase in dry matter content at an abnormally rapid rate which has nothing to do with plant growth. The CO_2-enriched plants, on the other hand, are initially able to maintain their turgor and continue their basic life processes for some distance into this region; and it is under such conditions as these that relative productivity enhancement factors rise rapidly to infinity, as ambient-grown plants succumb to the rigors of the environment and die. Ultimately, however, there comes a point where CO_2-enriched plants die as well. But before such a condition of water shortage is reached, they perform much better than ambient-grown plants, again on a relative as opposed to an absolute basis.

8.3 High Temperatures

Closely associated with plant water stress is the stress created by unduly high temperatures, which can also be lethal. But, like water stress, the stress of high temperatures can be significantly mitigated by atmospheric CO_2 enrichment. In fact, the life-saving attributes of high atmospheric CO_2 concentrations are so exceptional that atmospheric CO_2 enrichment is often employed in standard heat therapy treatments designed to rid many types of useful vegetation of viral diseases.[43-45] Idso et al.[46] have also found it to be helpful in preventing the thermal death of water fern plants. Hence, and once again, we have a situation where atmospheric CO_2 enrichment is relatively more significant in the face of an environmental stress than it is under optimal growth conditions, as it can sometimes mean the difference between life and death for certain plants.

8.4 Low Temperatures

At the other end of the temperature spectrum is the specter of death-dealing stress induced by low temperatures. But again, atmospheric CO_2 enrichment often comes to the rescue. It has been shown to ameliorate adverse chilling effects in several plants;[47-50] it promotes the overwinter survival of rooted cuttings of ornamental nursery stock;[51] and it has been demonstrated to enable okra plants to grow to maturity at temperatures so low as to lead to their demise under current atmospheric CO_2 concentrations.[52] Thus, in the face of this stress too, atmospheric CO_2 enrichment often promotes the continuance of life, where death looms as a very real alternative without it.

5 Salinity

Soil salinity is another problem with which many plants must contend;[53-55] ad because it requires the expenditure of energy to remove or exclude salts rom their tissues, plants growing on saline soils generally have greater aintenance respiration rates than do similar plants growing on non-saline oils.[56] Hence, it is reasonable to expect that atmospheric CO_2 enrichment-- hich enhances the production of carbohydrates needed to provide such energy - may better enable plants to cope with this problem.[6,57] And, in fact, this esponse is just what experiments have demonstrated to be the case. tmospheric CO_2 enrichment has generally increased the salt tolerance of most lants studied in this regard;[58] and it again produces a greater _relative_ CO_2- nduced growth enhancement effect under conditions of salt stress than it does nder more optimal conditions for plant growth.[56,59-61]

.6 Air Pollution

A largely man-made stress with which many plants must contend is air ollution,[62-65] which can wreck havoc with vegetation in a number of ifferent ways.[66,67] Ozone, for example, can disrupt the permeability of ell membranes,[68-70] produce chlorotic or necrotic lesions in leaf tissue,[71] educe photosynthesis rates,[72] alter respiration rates,[73] and lower crop yields.[74,75] Sulfur dioxide additionally affects cell metabolism;[70] while nitrogen dioxide not only inhibits photosynthesis by itself,[76] but also produces ozone and other noxious oxidants via photochemical reactions with numerous hydrocarbons.[77] In addition, there is also some concern about the possibility of genetic damage to trees and crops as a result of foliar exposure to various air pollutants.[78]

Now the leaf cuticles of most plants are generally impermeable to most types of gaseous air pollution.[79] Hence, as a number of investigators have clearly demonstrated,[6,80-83] stomates provide the main route for entry of air pollutants into plant leaves, which is most fortunate; for as Acock and Allen[57] have noted, "because high CO_2 concentrations reduce stomatal conductance, high CO_2 concentrations probably will decrease damage done by pollutants," which has, in fact, been demonstrated in a number of CO_2- enrichment studies[84-92] and water stress experiments.[93-97] In fact, one would be hard pressed to come up with a better cure for the problem; for as atmospheric CO_2 enrichment closes the door to air pollutant entry into plants, it compensates for the increased difficultly which this phenomenon also provides for its own needed entry by virtue of its increased concentration gradient from the air to the leaf.

8.7 Acid Rain

One particular type of air pollution which attacks terrestrial plants in more insidious manner, both directly and via its effects on the chemistry the soil solution,[98,99] and which perturbs aquatic ecosystems perhaps ev more,[100-102] is acid rain.[103-105] However, as rain of any pH strikes a le surface, it wears away some of the leaf's epicuticular waxes,[106] leachin certain nutrients and metabolites from within the leaf;[107] and thes substances subsequently tend to neutralize the acidity of the rainwater,[108 [111] as do deposits of plant exudates,[112] cation exchange between the leaf an raindrop,[113] and cation diffusion from the plant transpiration stream.[11 Hence, CO_2-induced stimulation of plant growth should provide for a greate neutralization of acid rain by producing a greater amount of leaf surfac over already vegetated ground (Sec. 7.1) and by promoting the growth o plants over currently barren ground (Sec. 7.4). The exploration of greate soil depths by CO_2-induced stimulation of root growth will concurrentl provide for more acid-neutralizing cations (Sec. 7.8.3), as will the severa soil environment improvements described in Sec. 7.6 and the enhance propensity for soil weathering to be described in Sec. 8.10. These phenomen should also be of benefit in areas where heavily acidified cloud particles[11 come into direct contact with vegetation.[116] In addition, as the deleterious effects of acid rain appear to be more pronounced when plants are short of water,[117] the propensity for atmospheric CO_2 enrichment to somewhat alleviate drought stress (Sec. 8.2) should also tend to counteract the undesirable consequences of this insidious pollutant.

Closely related to this last point is the fact that plants under stress typically liberate large amounts of ethylene,[118] plus the observation that ethylene has been implicated as a major contributor to the toxicity of ozone,[119] very possibly as a result of secondary chemicals produced by reactions which are favored in acidic environments.[120] Hence, the capacity for enhanced acid neutralization provided by the CO_2-induced stimulation of vegetative productivity described above, as well as its ability to lessen the severity of the many other plant stresses described in this section, may well serve as an ubiquitous partial deterrent to the adverse effects of air pollution in general.

8.8 Disease

Just as most air pollutants gain entry into plant tissues through their stomates, so also do various plant pathogens infect plants in like manner, including all bacteria,[121-123] some fungi,[124-127] certain viruses,[128] and a

umber of nematodes.[129,130]   Indeed, Rich[131] has noted that "stomata are the
ost important infection courts for those foliage pathogens that cannot
enetrate unbroken epidermis," of which there are many, many types.   Hence,
s increasing CO_2 concentrations progressively constrict plant stomatal
pertures in the years ahead, the incidence of many plant diseases may drop
omewhat.   And in some instances they may even decrease independently of the
nfluence of CO_2 on stomata.[132]   In addition, decreased stomatal apertures
ould reduce plant guttation, which should curtail the amount of "leaf spot"
nd "tip burn" maladies typically suffered by a number of plants.[133-136]

.9 Insect Foraging

Next to disease, the major biological nemesis of Earth's vegetation is the
orld's vast array of foraging insects.[137-140]   How will elevated levels of
tmospheric CO_2 affect their appetites?

As to direct effects on the insects themselves, there is little evidence to
uggest major changes in their behavior, other than a more frequent opening of
heir spiracular valves,[141] which could, however, upset their water balance
nd possibly lead to a reduction in foraging efficiency.   Hence, indirect
ffects due to changes in the nutrient quality of the foliage upon which they
eed are likely to be more important.[142,143]   For example, the enhanced
arbohydrate production characteristic of plants grown in CO_2-enriched air
enerally leads to decreased nitrogen/carbon ratios in plant tissues,[144-146]
hich tends to decrease their nutritive value.   And this phenomenon could lead
o three different responses: (1) the insects may eat more of the foliage to
ake up for its nutrient deficiency, (2) they may suffer a number of
eleterious side effects due to this deficiency, which could lead to the
estruction of less foliage, or (3) they may maintain the status quo.[147]

One of the first studies designed to explore these several alternatives
eemed to suggest that the first possibility was the most likely: soybean
ooper larvae fed at significantly higher rates on plants grown at an
tmospheric CO_2 concentration of 650 ppm than they did on plants grown at 350
pm.[148]   Subsequent studies, however, have often favored the other
lternatives.   Butler et al.,[149] for example, could find no significant
ifferences in the population growth responses of four generations of
weetpotato whiteflies feeding on cotton growing in atmospheric CO_2
concentrations of 350, 500 and 650 ppm.   Neither were there any differences
in the populations of thrips living on the cotton.[150]   But for flea beetles,
leafhoppers, predaceous flies, and pink bollworms, populations were much
reduced in the higher CO_2 treatments compared to those in the ambient
treatment.[150]

In another study of beet armyworms feeding on cotton, Akey and Kimball[1] found that insects raised on foliage grown at an atmospheric CO_2 content of 640 ppm experienced significant increases in development times and mortality as well as significant decreases in growth, as compared to insects raised on foliage grown at a CO_2 concentration of 320 ppm. And this general response has tended to be the finding of many other such studies as well: plant-eating caterpillars have not seemed to do as well on a diet of CO_2-enriched foliage as they do on plants grown under ambient CO_2 conditions.[152-154]

Helping to harmonize these diverse observations is the recent study of Fajer et al.,[155] who found early instar larvae of the buckeye butterfly Junonia coenia, which fed on host plant foliage grown in an atmosphere of 700 ppm CO_2, to grow more slowly and suffer nearly three times the mortality of control larvae raised on ambient-CO_2-grown foliage, but who found late instar larvae to not be adversely affected by eating high-CO_2-grown leaves. Indeed, they noted that late instar larvae appear to compensate for the reduced nitrogen content of high-CO_2-grown foliage by consuming more of it. Fortunately, all late instar larvae must first have been early instar larvae, however, so that the number of late instar larvae in the natural environment may be expected to be somewhat reduced as a result of the enhanced mortality of early instar larvae in a CO_2-enriched world of the future. In addition, as Fajer et al. also note, "delayed growth of the smaller and less vagile, early instar larvae may reduce the fitness of insects by increasing their exposure to parasitoids and predators[156] and may diminish their likelihood of completing development on the same host in climatically limited environments."[157,158]

A closely related phenomenon is the effect of plant nutrient/carbon balance on plant herbivore defense systems.[159,160] The working hypothesis in this field of research is that anything which alters the nutrient/carbon balance so as to reduce soluble carbohydrate reserves -- such as the demands imposed by increased growth, for example (and, hence, the tie to atmospheric CO_2 enrichment) -- may reasonably be expected to lead to a reduction in the synthesis of carbon-based defensive chemicals.[161,162] Several studies have shown this hypothesis to have considerable merit.[163-165] However, the defensive mechanisms of some plants have been shown to be largely determined by genotype and not the environment[166,167] or resulting carbon supply.[168] Hence, there is as yet no consensus as to the effects of atmospheric CO_2 enrichment on plant defense mechanisms. But even if certain insects would feed more voraciously on the extra vegetation produced by atmospheric CO_2 enrichment, this phenomenon is itself one of the classic defense strategies of many natural plant communities, i.e., the synchronous production of an excess of vegetation which may be safely sacrificed to herbivores during ertain periods of the various plants' life cycles.[169]

Clearly, the final word on the matter is not yet in, as the subject is so omplex.[146] As a further example, some studies have shown that when aterpillars proliferate, nutrients from their faeces actually stimulate rees to replace foliage,[170,171] much like the nutrient recycling activities f larger browsing animals stimulate grassland production.[172] In addition, oth vertebrate and invertebrate herbivore damage generally stimulates the roduction of plant defensive chemicals;[173,174] and immature insects usually o not develop at as rapid a rate when the water content of the foliage upon hich they feed is reduced,[175] as is characteristic of plants grown in CO_2-nriched air (Sec. 7.8.2)

The preponderance of the available evidence thus seems to suggest a tipping of the scales in favor of plants over foraging insects, as atmospheric CO_2 concentrations continue to rise in the coming years. At the very least, it seems clear that the percentage destruction should be no more than it is at present, which would, of course, lead to a significant increase in net harvestable productivity. In fact, in a recent experiment where two species of western U.S. grasshoppers fed on sagebrush plants, the amount of defoliation at 350 ppm CO_2 was 33%, while at 650 ppm it was only 16%, prompting the principal investigator to conclude that "plants may be better able to sustain growth under herbivory at elevated CO_2 levels."[176]

8.10 Soil Infertility

Lack of needed nutrients has often been considered to be a major roadblock to the full realization of the positive biospheric benefits which may occur as a consequence of atmospheric CO_2 enrichment, as a large portion of the world's biota exists on nutrient deficient soils,[177] and as CO_2-induced increases in plant photosynthesis have long been thought to be incapable of being translated into increases in plant growth in the absence of sufficient quantities of certain nutrients.[178-183] In fact, such impediments to growth, especially when they retard the development of photosynthate "sink" organs, can sometimes lead to the accumulation of assimilates in "source" leaves,[184-186] characteristically resulting in end-product inhibition of photosynthesis.[187-190] And in extreme situations, such accumulations of starch can actually damage chloroplasts.[191-193]

Perhaps nowhere is this problem more important than in the case of the world's forests, which cover about a third of the land area of the globe and account for approximately two-thirds of the planet's terrestrial photosynthetic activity.[194] Many of these areas are deficient in essential nutrients,[195-198] which would appear to limit their potential for growth.[199,200] Hence, it is not unreasonable to question the ability of atmospheric CO_2 enrichment to have a significant impact on this sector of the biosphere.[182] And, indeed, there is some evidence to support this contention, as greater stimulation of plant growth due to atmospheric CO_2 enrichment is often observed under more favorable conditions of soil fertility.[6,201-205]

But this type of response is by no means universal; for a number of similar experiments have produced significant growth responses to atmospheric CO_2 enrichment across a wide range of soil nutrient conditions,[206-210] including low levels of both nitrogen[211-213] and phosphorus.[214] And sometimes the _relative_ growth response has even been greater when essential nutrients were limiting,[215-217] much like the relative growth response to water stress described in Sec. 8.2.

To understand how such situations may arise, it is necessary to consider several matters related to root growth and associated soil mycorrhizal activity.[218-220] First of all, when the soil is poor in nutrients, increased photosynthesis due to atmospheric CO_2 enrichment generally results in proportionally greater carbon allocation to roots.[221-224] And since most of the microorganisms in the rhizosphere depend upon plant roots for a supply of easily metabolized carbon compounds,[225] this enhanced flow of carbohydrates to the plant roots stimulates nearby soil microbial growth as well.[226,227] Indeed, any environmental factor that affects photosynthesis can potentially affect rhizosphere functioning,[228-230] as rhizosphere populations have been shown to respond to variations in light, temperature, ozone and foliar fertilization,[231-235] as well as to atmospheric CO_2 enrichment. As a result, Lamborg et al.[236] have concluded that "the activity of all rhizosphere microorganisms will be enhanced by the larger photosynthetic activity of plants at higher CO_2."

Now the significance of rhizosphere stimulation arises from a number of different phenomena. First of all, returning to the plant itself for just a moment, enhanced plant growth due to atmospheric CO_2 enrichment stimulates the root cap to produce more mucigel,[237] a chemically complex substance consisting of carbohydrates, amino acids, vitamins, organic acids, nucleotides, and enzymes,[238] to name but a few of its many important components. This viscous slimelike material functions as an effective lubricant and thereby reduces the resistance of the soil to root penetration, allowing a greater volume of soil to be explored and exploited.[239] In addition, certain of its chemical components increase the solubilities of important plant nutrients, such as iron and phosphorus,[240,241] making them more readily available for uptake by the roots.

Mucigel also functions as a protective niche for soil microbes.[242]

Consequently, as its presence in the soil is enhanced, so also will the size and activity of the rhizosphere be increased. And enhanced mycorrhizal activity generally allows more nutrients to be removed from the soil by the plant roots,[219,220,223,224,226,243,244] while at the same time it acts to exclude certain plant toxins and parasitic fungi.[245,246]

One way by which these feats are accomplished is through the promotion of soil fungal growth as a consequence of increased plant root exudation.[247] As the hyphae of symbiotic fungi extend outward from plant roots, for example, they greatly extend the life of absorptive roots beyond the few days that root hairs are normally active,[248] and they effectively increase the area of root surface available for nutrient uptake,[249] which is especially advantageous for the aquisition of slowly diffusible ions.[250,251] In addition, the fungi involved in mycorrhizal associations generally produce plant growth hormones which regulate root growth in ways best suited to the conditions of the soil environment,[252] the most obvious of which is to stimulate greater root growth.[253,254] Certain soil bacteria which metabolize carbon compounds from plant roots function in much the same manner, producing and excreting phytohormones[255,256] which stimulate the formation of root hairs and additional lateral roots.[257,258]

All of these phenomena, along with the CO_2-induced enhancement of nitrogen fixation described in Sec. 7.8.6, promote the growth of the whole plant, which then stimulates rhizosphere activity even more. And as larger and more active plant root systems allow nutrient mining from an enlarged volume of soil,[223,224,226,227] they likewise reduce the amount of nutrients lost from the soil by leaching.[259] Also, some of the more limited plant nutrients are utilized more efficiently by the plants.[223,224] And as the uptake of such needed ions as NO_3^- requires the expenditure of metabolic energy,[260-262] the enhanced availability of carbohydrates typically provided by atmospheric CO_2 enrichment should promote this process by yet another means.[263-265] Hence, it is not surprising that atmospheric CO_2 enrichment has been shown to exert a profound influence upon tree growth in nutrient-poor soils. Indeed, growth responses several times greater than those typically observed in well-fertilized agricultural crops have been obtained in several instances.[224,226,227] Likewise, the disappearance of mycorrhizal fungi has been identified as one of the primary causes of deciduous forest decline.[266]

On a longer time scale, some scientists have worried that if the C/N ratio of plant litter increases with increasing atmospheric CO_2 concentration, slower rates of microbial decomposition may result in the accumulation of insoluble nutrient stocks in the soil and lead to a decrease in soil fertility.[267-271] Of course, the beneficial effects of atmospheric CO_2 enrichment on nitrogen fixation and soil mycorrhizal activity will both tend to compensate for this possibility; and recent studies of both C_3 and C_4 plants have shown, in fact, that the C/N ratio of plant litter does not change with atmospheric CO_2 enrichment.[272] In addition, organic acids produced by plant rhizospheres generally hasten the chemical weathering of various minerals,[273-276] as does the higher soil CO_2 concentration produced by enhanced plant growth rates.[277]

As but one example of this phenomenon, phospho-microbes[278] secrete organic acids which play a vital role in the solubilization of tri-calcium phosphate and organic and rock phosphate;[279-281] and a number of studies have demonstrated that their activity has a significant effect on the mobilization of soil phosphates leading to large yield increases.[282-285] Hence, as indicated by Rosenberg et at.,[286] "CO_2 'fertilization' may initiate or stimulate an acceleration of soil formation through enhanced biological activity," so that a "more rapid release of many essential nutrients might follow." And Luxmoore et al.[233] have calculated that for the expected rate of increase in net primary production due to atmospheric CO_2 enrichment, only a 2% shift of nitrogen and a 1% shift of phosphorus from the soil to vegetation would be needed over a 40-year period to maintain current vegetative C/N/P ratios.

Although it must be emphasized that the situation is very complex and much more work must yet be done to clarify the picture, it would thus appear that soil infertility will not be an obstacle to significant enhancement of forest growth and productivity due to atmospheric CO_2 enrichment. Indeed, the luxuriant growth of contemporary rainforests is a testament to this fact, for the soils upon which they grow are typically very poor in nutrients; but because they recycle their litter so rapidly and have developed such effective means of extracting nutrients from it, they are essentially decoupled from the nutrient status of the soil itself.[287] And finally, as woody species encroach upon deserts and grasslands, as described in Sec. 7.4, the available nitrogen status of the soil should markedly improve, due to enhanced nitrogen fixation by such nodulating shrubs as mesquite,[288,289] and to enhanced soil mining by deeper-penetrating plant roots.[290,291]

The world's croplands should likewise benefit. It is a well established fact, for example, that differences in soil microbial populations due to different cropping and soil management practices result primarily from altered inputs of carbon related to changes in plant productivity and subsequent soil organic matter levels.[292-294] Some of the immediate benefits which may thus result from CO_2-induced productivity-stimulated increases in soil microbial populations are a greater magnitude of nutrient cycling,[295] a greater reserve of biological nitrogen,[295] and a more rapid decomposition of pesticide residues.[294] Longer term benefits may also include better soil tilt and improved infiltration of water.[296,297]

Greater microbial populations, of course, also mean greater populations of microbe-munching nematodes, which are found the world over in both aquatic and terrestrial ecosystems.[298] But even this phenomenon has its positive aspects, for bacteriophagous nematodes and mites indirectly stimulate the rate of organic matter decomposition by grazing,[299,300] enhance rates of mineralization and nutrient turnover,[301-303] and stimulate both terrestrial[304] and aquatic[305,306] decomposer microflora to maintain maximum growth rates. And their own excretion of ammonia and other inorganic nutrients directly stimulates plant growth.[307] In addition, nematodes may also play a role in controlling populations of insect pests which spend a portion of their life cycle in the soil.[308]

8.11 Soil Toxicity

For a variety of reasons,[309,310] certain chemicals sometimes accumulate in soils in concentrations which are toxic to both plants and animals.[311-313] One way by which this problem may be alleviated is through microbial uptake, metabolism and volatilization of the offending elements,[314-316] which can effectively detoxify the soil.[317,318] And as a number of experiments have demonstrated that the addition of organic carbon to the soil has a tendency to stimulate all of these processes,[316,319,320] the increase in soil organic matter resulting from enhanced plant productivity due to atmospheric CO_2 enrichment should thus lead to a significant reduction in soil toxicity via these mechanisms. In addition, there is evidence to suggest that certain soil fungi function in much the same manner.[321] Consequently, as atmospheric CO_2 continues to accumulate in the years to come, the soil-polluting aspects of man's activities should be somewhat balanced by this beneficial effect of the chief by-product of his industrial endeavors.

8.12 UV-B Radiation

One final stress deserving of mention, although it is currently but a potential stress, is that of enhanced UV-B radiation, which has been postulated to occur as a result of stratospheric ozone depletion (Sec. 2.6 and 5.12). Several laboratory and controlled-environment studies have indicated that as the intensity of this supposedly debilitating radiation is increased, plant photosynthesis is somewhat reduced.[322,323] In the field, however, adverse effects of this nature are rarely encountered;[324,325] and Lovelock[326] has gone so far as to suggest that the concept of stratospheric ozone operating as "Earth's fragile shield" is a myth. In fact, he claims that "it is an insult to the versatility of biological systems to assume that a weakly penetrating radiation like solar ultraviolet could be an insurmountable obstacle to surface life."

In support of this contention, Lovelock cites the case of "naked washed bacteria," which when suspended in air are easily destroyed by UV-B irradiation, but which when clothed in mucus secretions or the organic and mineral constituents of their natural environment are almost totally immune to such affronts. To this evidence can also be added the experimental observation that plants in the field appear capable of increasing their production of UV-absorbing pigments in the face of increased UV-B exposure.[324,325] And in a recent study of enhanced UV-B irradiation of wheat and oat plants growing in the field,[327] it was found that photosynthesis-- which is an integrative measure of such diverse processes as membrane function, enzyme activity and light harvesting reactions -- was not adversely impacted by UV-B exposures corresponding to 30 to 45% depletions in stratospheric ozone. It is but a simple step of logic, then, to suppose that the growth-promoting properties of enhanced concentrations of atmospheric CO_2 may intensify such self-preserving tendencies of the planet's plant life even more.

8.13 The Bottom Line

So what is the answer to the question posed at the start of this section? Is environmental stress the bane of atmospheric CO_2 enrichment? Or is it a further stimulant to even greater _relative_ growth response?

Clearly, if a particular stress is extreme enough, no amount of CO_2 enrichment can relieve it. And as conditions improve, it is also clear that even when extra CO_2 begins to have a beneficial effect, the consequent growth of the plant may still be considerably less than that of a comparable plant growing under ambient CO_2 concentrations but without the impediment of the stress. In addition, there have been a few experiments where the effects of atmospheric CO_2 enrichment have been rather minor in the face of significant limitations to growth imposed by the conditions of the environment.[209,328-332] Nevertheless, the vast majority of the studies which have been designed to probe this question have produced results which suggest that most environmental stresses are not significant impediments to plants receiving benefits from atmospheric CO_2 enrichment. Indeed, extra CO_2 often softens the severity of the stress; and in situations where this mitigation of deleterious effects allows the plant to live where it would otherwise die, the relative benefits are incalculable. In terms of a mathematical growth enhancement factor, for example, they may actually rise to infinity.[333] Consequently, under most conditions of stress, atmospheric CO_2 enrichment generally produces a greater <u>relative</u> growth enhancement than it does under optimal growth conditions. And by increasing a plant's overall vitality, it strengthens the plant's defenses, which should enable it to better withstand various diseases and insect pests.[334-338]

9 SIGNS OF VEGETATIVE REJUVENATION

9.1 Plant Productivity

Just as physical scientists have looked for signs of a CO_2-induced warming of the Earth, so also have students of the life sciences looked for signs of a CO_2-induced "greening" of the Earth. And since overall growth is the plant response of most interest, several researchers have noted that significant increases in primary plant productivity may have already occurred as a result of the increase in atmospheric CO_2 content Earth has experienced over the past few centuries.

In a study of Australian wheat yields over the period 1958-1977, for example, Gifford[1] suggested that a significant part of the upward trend "may be due to atmospheric [CO_2] change rather than to managerial or genetic causes." Similarly, Wittwer[2] has noted that "the 'green revolution' has coincided with the period of recorded rapid increase in concentration of atmospheric carbon dioxide, and it seems likely that some credit for the improved yields should be laid at the door of the CO_2 buildup." And in a quantitative analysis of this phenomenon, Allen et al.[3] conclude that soybean yields "may have increased by 13% from about 1800 A.D. to the present due to global carbon dioxide increases."

Of course, there is presently no independent way to confirm the reality of these estimates; but in the future there may be, as airborne and satellite remote sensing techniques continue to grow in sophistication and long-term data bases are acquired.[4,5] Over the past few years, for example, several studies have demonstrated the capacity of satellites to estimate the green-leaf biomass of plant canopies and, presumably, terrestrial photosynthesis and primary production on a global scale.[6,7] As these techniques are refined and new approaches developed, we should soon be able to track the yearly productivity of the terrestrial biosphere with considerable confidence.[8-11] In fact, Wessman et al.[12] have even demonstrated the potential utility of remote sensing for evaluating lignin content of whole forest canopies and nitrogen-cycling rates across forested landscapes; while Card et al.[13] have added sugar, starch, protein, cellulose and chlorophyll to the list of likely assessable parameters. In addition, it appears that we will shortly have the ability to monitor ocean productivity from space as well.[14-16] For the present, however, we must look to other sources of information for signs of the predicted planetary revitalization.

9.2 Tree Ring Data

In an important paper published in the fall of 1984, LaMarche et al.[17] presented what they called "the first direct evidence of CO_2-related growth enhancement in natural vegetation: increased widths of annual rings of trees in subalpine habitats in the western United States during recent decades." Indeed, widths of annual rings of bristlecone and limber pine trees near the timberline in central Nevada were 106% greater in the decade ending in 1983 than in the years 1850 to 1859 at one site and 73% greater at another; while similar dramatic increases were observed in bristlecone pines at high-altitude sites in California, Colorado, and New Mexico.

Predictably, such dramatic results, along with their bold interpretation, precipitated a number of probing commentaries.[18-21] And shortly thereafter, a similar trend in the tree-ring chronology of spruce trees growing at the tree-line in northern Quebec was claimed to be explainable largely in terms of Earth's climatic amelioration as it recovered from the chill of the Little Ice Age.[22] In the case of the data of LaMarche et al., however, there were no obvious climatic trends that could explain the positive trends in tree growth which they observed.

The aspect of LaMarche et al.'s data which most stretched the credulity of others, however, was the great magnitude of the apparent growth response, as it was much larger than what most people were willing to accept on the basis of laboratory experiments,[18] although some model studies have suggested that a 300 to 600 ppm doubling of the atmospheric CO_2 content could well result in a 1.8-fold increase in sustainable forest growth.[23] It is very possible, however, that several of the stress-amplification factors discussed in Sec. 8, as well as the synergistic feedback phenomena involving the soil biota discussed in Sec. 7, could be greatly amplifying the basic growth response of these timberline trees. In addition, the few tree species which have been studied generally exhibit larger basic growth responses than other types of vegetation,[24-27] sometimes approaching an actual doubling of growth for a doubling of the atmospheric CO_2 concentration.[28-30] But as in the case of plant productivity discussed in the preceding subsection, it is difficult to unequivocally demonstrate that the observed growth trends of the trees studied by LaMarche et al. are primarily due to the historical increase in atmospheric CO_2 content -- at least at the present time. Nevertheless, work goes on in an attempt to eventually provide just such a means of verification.[31]

One particular study which may help to clarify the situation is being conducted by D.C. West of the Oak Ridge National Laboratory on a virgin stand of Pinus palustris in Thomas County, Georgia. Its objective is "to produce tree-ring chronologies capable of quantifying tree response to increasing CO_2 during the past century." In a recent progress report,[32] West noted the following:

> Normal or expected tree growth shows a slight but monotonic decrease in annual ring increment with age. The longleaf pine stand in this study deviates significantly from this trend. Beginning about 1920 and continuing to the present, the annual increments of <u>all</u> age classes are increasing. Presently, annual growth is approximately 30% greater than in 1940 .... The increased growth cannot be explained by trends in precipitation, temperature, or Palmer Drought Severity Index.

By default, then, and much like the study of LaMarche et al.[17] and the contemporary work of Parker,[33] West's results would appear to be explainable only in terms of the rising CO_2 content of Earth's atmosphere. And again, the growth enhancement is much larger, by about 40%, than what most people have expected on the basis of simple CO_2 enrichment experiments. Thus, it seems very likely that the several synergistic phenomena discussed in the preceeding two chapters may well be operating to magnify the response of natural ecosystems to atmospheric CO_2 enrichment.

Additional support for this contention comes from the study of Hari and Arovaara,[34] which was designed to detect possible fertilization effects of atmospheric CO_2 enrichment on the growth of mature Scots pine trees growing in their natural range in northern Finland. Normally, the magnitude of the annual radial growth increment of these trees increases exponentially during the first stage of growth, after which it peaks and subsequently enters a slow decline phase.[35] However, in what they called a very unexpected development, even the oldest trees in their study area were still in the phase of increasing volume growth. Furthermore, the increase in growth could not be explained by anything except the increasing CO_2 content of Earth's atmosphere. In their own words, "CO_2 seems to be the only environmental factor that has been changing systematically during this century in the remote area under study." And again, the increase in growth attributable to CO_2 -- a basal area increment ranging from 15 to 43% over the period 1950 to 1983 -- was much larger than what would be expected on the basis of simple laboratory experiments, a finding which is becoming more and more common with each new study[36] and which is truly remarkable in terms of the widespread potential for forest dieback due to acidic precipitation[37-39] and direct vegetative contact with acidified clouds.[40,41]

9.3 Leaf Stomatal Densities

Paralleling the effort to discern CO_2 effects in high-altitude tree rings-- indeed, even stimulating it -- has been the study of changes in plant physiological properties with altitude,[42-45] since the partial pressure of CO_2 steadily declines with height above sea level.[46,47] One of the major findings of this work is the discovery that leaf stomatal density generally increases with altitude,[48-50] as plants make adjustments to compensate for this deficiency of their environment. Hence, it is logical to expect that plants of a few centuries ago likewise possessed more dense concentrations of stomatal pores on their leaves, as they had to cope with the similarly reduced atmospheric CO_2 concentrations of the pre-industrial era.

So how do you check the leaves of plants which grew 200 years ago? F.I. Woodward did it by studying preserved herbarium specimens in central England, finding that the number of stomates per unit area of leaf surface for eight arboreal species declined by about 40% over the past two centuries.[51-53] And he confirmed this observation by growing some of the same plants under various partial pressures of CO_2, obtaining a 67% decrease in stomatal density by increasing the atmospheric CO_2 concentration from 280 to 340 ppm.

Although these findings would appear to be incontrovertible evidence for the reality of significant real-world vegetative change induced by the increase in atmospheric CO_2 content over the past two centuries, it should be noted that some laboratory experiments suggest that stomatal densities do not respond significantly to _further_ increases in atmospheric CO_2 concentration;[51] and in one plant (rice), they actually appear to increase somewhat with increasing CO_2.[54] In addition, Korner[55] has been unable to demonstrate any significant change in the stomatal densities of a number of alpine species over the past several decades. Thus, this particular phenomenon may have already run its course.

On the other hand, O'Leary and Knecht[56] have observed the stomatal density of _Phaseolus vulgaris_ leaves to continue to decrease with increasing CO_2 up to a concentration of 1200 ppm; while Bristow and Looi[57] observed stomatal density reductions in _Marsilea_ at CO_2 concentrations as high as 6500 ppm. As a result, the future course of this phenomenon is not clear. Nevertheless, its past reality supports the future reality of the basic growth and transpiration responses described in Sec. 7.1 and 7.2, which are also manifest in differential altitude studies of plant physiological responses.[50]

9.4 Biogeographical Disturbances

In Sec. 7.4 it was hypothesized that a CO_2-induced enhancement of plant water use efficiency should lead to vast changes in the distribution of the planet's natural plant communities, with the ultimate conversion of many desert areas into grasslands and the replacement of contemporary grasslands by a cover of shrubs and trees. Clearly, long-term monitoring of the land

surfaces of the globe by satellite should allow the discrimination of suc
trends. In addition, detailed repetitive vegetation surveys of a mor
conventional nature may also be effective in detecting potential plan
migrations. In fact, a certain amount of such data is already available; an
it tells the story of what Mayeux and Johnson[58] call "the most remarkabl
phenomenon in the recent vegetational history of the American Southwest an
the Southern Great Plains," namely, "the large-scale increase in th
abundance and density of woody species and their progressive dominance o
former grasslands."

The story has a long prologue, beginning with the creation of th
savannas, grasslands, and deserts of the Southwest and Southern Great Plain
about 25 million years ago, when the climate began to dry during the middl
of the Tertiary and continued to become ever more arid throughout th
Pleistocene, and especially over the past 15,000 years following the retrea
of the Wisconsin glacial.[59] Even the appearance of European man had n
noticeable impact on the character of the vegetation,[60] nor did a period o
intense disturbance by humans about A.D. 1200,[61] but then, over the brie
period of the past 150 years, a change occurred which transformed the
savannas into what Malin[60] has called a "tangled jungle."

One of the first people to comment on this phenomenon was Josiah Gregg.[62]
In a book published in 1844, he noted that parts of the southwest "now thickly
set with trees" were once, according to the oldest inhabitants of the land,
"as naked as the prairie plains;" and by 1906, range scientists were already
deeply involved in studying the change.[63,64] Since that time, a number of
studies have updated the transformations occurring subsequently.[65-68] In one
of them, York and Dick-Peddie[69] determined that 75% of 31 townships in New
Mexico consisted of open rangeland in 1858, but only 5% of them remained
dominated by grasses a hundred years later. Likewise, Buffington and Herbel[70]
found that mesquite occupied only 26% of the area of the Jornada Experimental
Range in southern New Mexico in 1858, but that fully 70% of the area was
covered by it in 1963. A very similar change was also documented on the Santa
Rita Experimental Range in southern Arizona between 1904 and 1954.[71]

More recently, brush survey reports for the entire state of Texas indicate
a dramatic increase in the area infested with woody plants, ranging from 88
million acres in 1963[72] to 92 million acres in 1973[73] to 106 million acres in
1983.[74] In addition, this dramatic transformation of the land has occurred in
the face of substantial and sustained management efforts to diminish unwanted
brush through the use of both mechanical and chemical treatments.[58]

Much the same story is beginning to unfold in other parts of the world.
In South America, for example, where the floristic composition of the
striking vegetation gradient from the arid Patagonian steppe westward to the

temperate rain forests of the southern Andes was established between 5,000 and 3,000 years ago,[75-77] there has been "a dramatic invasion of the steppe by tree species ... as well as by many species of tall shrubs" over the past century.[78] Again, this transformation of the landscape has occurred in the face of heavy forest clearing by European colonists[79,80] and in the absence of significant climatic change.[81-83] And a very similar tale can be told of the brush encroachment phenomenon in South Africa.[84]

Although these shifts in vegetation patterns cannot be absolutely proven to be due to the concomitant increase in atmospheric CO_2 content, and other causes have been proposed,[85] they are very suggestive of such a relationship, as they are precisely what has been predicted to occur in the face of the rising CO_2 content of Earth's atmosphere, as explained in Sec. 7.4, and as all of the other postulated causes have one or more weaknesses which render them highly questionable. Hence, in the words of Mayeux and Johnson,[58] it is clear that "an experiment assessing the consequences of increased CO_2 levels has already been conducted on a global scale, and we should consider the possibility that the recent increases in abundance of woody plants on rangelands is its most visible result."

Perhaps we should also consider the possibility that it is increasing CO_2 which has curtailed the spread of the world's deserts. For close to three decades, now, there has been a widespread belief that the Earth's deserts were on the move, expanding into new territories. Within just the past few years, however, it has become clear that this perception is incorrect; and the "myth of the marching desert" has had to be radically revamped.[86]

9.5 Seasonal CO_2 Cycle

Probably the strongest evidence for the reality of global vegetative stimulation due to the steadily rising CO_2 content of Earth's atmosphere is to be found in the ever-recurring seasonal perturbation which is superimposed upon the mean upward trend of atmospheric CO_2 concentration.[87,88] This cyclical variation is caused by the photosynthetic and respiratory characteristics of seasonal plant growth and decay and is driven primarily by Northern Hemispheric terrestrial vegetation, dominated by the combined boreal forests of North America and Eurasia.[89] In brief, the CO_2 content of the northern half of the globe begins to drop in spring and early summer, as CO_2 is removed from the atmosphere by the year's new flush of vegetative activity; while in late summer and autumn it is returned to the atmosphere, as plants die and release their sequestered carbon when they decompose.

Suppose for the moment, then, that the productivity of Earth's vegetation is increasing slightly each year in response to the ever-increasing CO_2

content of the atmosphere. A direct consequence of this phenomenon would be an ever-increasing extraction of CO_2 from the air in spring and early summer as well as an ever-increasing release of CO_2 to the atmosphere in late summer and autumn, due to the decomposition of the ever-increasing seasonal biomass of the planet. Consequently, the net result of both processes should be an ever-increasing seasonal CO_2 cycle amplitude.

The results of the first search for this potential indicator of Earth's increasing vegetational vitality were published in 1975.[90] Based upon 1 years of meticulous CO_2 measurements at Mauna Loa, Hawaii, the study came up negative. Six years later, however, the long-sought signal was finally detected by G.I. Pearman and P. Hyson.[91] What is more, these investigators identified similar signals in atmospheric CO_2 data from Weather Station P in the north Pacific and from Barrow, Alaska. And they demonstrated that the could reject the null hypothesis that there had been no increases in the amplitudes of the seasonal CO_2 cycles at those locations at the 98, 81, and 99.5 percent confidence levels, respectively. Consequently, they concluded that "it is most probable that there has been an increase in the summer net ecosystem production of the Northern Hemisphere of 8.6 percent over the period 1958-1978."

Confirmation of this initial positive result followed quickly. A 1983 Bell Laboratories study not only reaffirmed the reality of the biological signal at Mauna Loa, but documented its existence at the South Pole as well.[92] In the words of the report's authors, the CO_2 data from both locations exhibited "exactly the type of behavior that one would expect if the activity of seasonally varying biomass were increasing." To be doubly sure of their results they also considered several other alternative explanations; but in all instances their analyses led them to decide that other causes were unlikely. Hence, they too concluded that "the CO_2 seasonal behavior reflects an increase in global photosynthetic activity."

Two other studies published in 1985 came to much the same conclusion.[93,94] These newer analyses again dealt with the ever-lengthening CO_2 records of Mauna Loa, Hawaii and Weather Station P. And again, after very detailed analyses, the first concluded, with respect to Weather Station P, that the data "leave little doubt that the principal cause of the annual cycle is the activity of terrestrial vegetation" and that "it seems likely that an increase in amplitude of the annual cycle reflects an increase in activity of terrestrial plants." Similarly, with respect to Mauna Loa, the second study noted that "it seems likely that the increase mainly reflects enhanced metabolic activity of the land biota."

Another study of interest, published in 1987, looked at certain fine-scale asymmetries in the Mauna Loa data set, considering many different enomena as possible causes of the unique interannual variations in the CO_2 cord.[95] Eliminating candidate mechanisms one-by-one, its author, I.G. ting, finally concluded that "of all the processes that significantly fluence atmospheric carbon dioxide, only biotic processes seem capable of oviding the requisite seasonal asymmetry." Hence, as he too had monstrated the trend toward larger CO_2 cycle amplitudes in recent years, ting joined with previous investigators of the phenomenon in ascribing the crease in the amplitude of the CO_2 cycle at Mauna Loa, and other sites as ll, "to biotic influences."

There is consequently little reason to doubt that the rapidly rising CO_2 ntent of Earth's atmosphere is indeed enhancing the productivity of the rrestrial biosphere. In fact, so obvious is the signal that Houghton[96] has cently suggested that it is too <u>large</u> to be due to aerial fertilization by $)_2$. It is very possible, however, that the amplifying effect of concurrent ir temperature rise described in Sec. 7.8.7, the greater <u>relative</u> hancement of plant productivity produced by atmospheric CO_2 enrichment in e face of various environmental stresses (Sec. 8), and the synergistic ffects described in Sec. 7.6 and 7.8.6 may well be responsible for the urprisingly strong response of the land plants of the globe to recent ncreases in atmospheric CO_2 concentration, which is manifest in the ever-ncreasing amplitude of the seasonal CO_2 cycle[97] and in the recent remarkable lleviation of forest dieback.[98]

But what about the oceans?

.6 Oceanic Productivity

Historically, the central regions of the world's oceans have been onsidered to be vast biological deserts.[99] Over the past few years, owever, several studies of primary photosynthetic production have yielded esults which suggest that marine productivity may well be twice or more as reat as what was believed only two decades ago,[100-107] causing some to peculate[108] that "the ocean's deserts are blooming."

Discussion related to this phenomenon has generally centered on potential nadequacies of early measurement techniques.[109-111] Very careful studies ave indicated, however, that it is not possible to clearly demonstrate that roblems associated with traditional methods were responsible for their early ow productivity estimates;[112,113] but the paradox thereby created appears to ave finally been resolved with the publication of the landmark paper of enrick et al.[114] in October of 1987.

As a result of a long time-series of measurements carried out in the entral North Pacific, Venrick and his associates demonstrated that the total chlorophyll a content of the water column there has nearly doubled since 1968, causing them to suggest that the recent tendency towards ever greater primary production rates reported in recent years "appears to reflect, at least partially, true interannual differences rather than previous methodological bias." Hence, the oceans too would appear to be experiencing a biological rejuvenation which is almost "too large" to be explained by atmospheric CO_2 fertilization alone.

Again it must be remembered, however, that strong synergistic feedbacks may be at work in the marine environment, just as they are in the terrestrial environment. As but one example of such a phenomenon, it is well known that episodic increases in oceanic nitrate concentrations can lead to tremendous phytoplanktonic "blooms" which are capable of supporting a significant fraction of annual gross productivity.[115,116] Thus, it is very possible that, as with terrestrial ecosystems, an ever-increasing input of CO_2 to the surface waters of the world's oceans via enhanced air-sea exchange,[117] rainfall,[118] and riverine organic matter discharge,[119] may well stimulate marine cyanobacteria to greater rates of nitrogen fixation (Sec. 7.8.6), which would subsequently enhance surface phytoplanktonic productivity, as would the enhanced receipt of nitrates directly resulting from the CO_2-induced increase in rainfall itself.[120] Indeed, the consistancy of the chlorophyll a data trend of Venrick et al.[114] argues strongly for some pervasive phenomenon of this type, as opposed to more episodic climatic and circulation events,[114,121] as the cause of the apparent increase in oceanic productivity over the past two decades.

10 CO_2 AND ANIMALS

0.1   Are There Direct Effects?

Up to this point, we have reviewed what is known about the direct effects of fluctuations in Earth's atmospheric CO_2 content on its inanimate environment (climate system) and its surface covering of plant life, as well as some of its indirect effects on the soil micro flora and fauna, finding that all of these interrelated components of the biosphere are indeed sensitive to such perturbations.   It is only natural, therefore, to think next of the higher orders of the animal kingdom.   Do its members also respond directly to atmospheric CO_2 enrichment?

Most studies of direct CO_2/animal interactions have focused on potentially deleterious effects of extremely high CO_2 concentrations, which can lead to a state of hypercapnia,[1-3] or an excessive amount of CO_2 in the blood, which generally produces acidosis,[4,5] a serious and sometimes fatal condition characterized in humans by headache, nausea and visual disturbances.   In this regard, however, it appears we have little to worry about within the context of expected increases in Earth's atmospheric CO_2 content; for several investigations of these phenomena have indicated that they do not seriously impact human welfare until the concentration of CO_2 in the air reaches a value of about 1.5%,[6,7] which is approximately 50 times greater than Earth's recent nominal base value of 300 ppm.   In fact, it is very possible that much smaller concentration increases may well be helpful.

Returning to plants for just a moment, as a springboard for this hypothesis, there is considerable evidence suggestive of the likelihood that CO_2-induced perturbations of intracellular pH may significantly influence transport processes at the plasma membrane, as well as enzyme activities and intracellular concentrations of hormones.[8]   Hence, it is reasonable to believe that similar phenomena in animal cells may be likewise affected.   In addition, since animals have also had to contend with the gradual lowering of Earth's atmospheric CO_2 content over geologic time, it is not unreasonable to suppose that they too may well be better adapted to a somewhat higher atmospheric CO_2 concentration than that of the present era.

But not too high, of course, for the anesthetic properties of extreme concentrations of atmospheric CO_2 are well documented.[9-11]   Nevertheless, there is sufficient evidence to suggest the existence of enough positive effects of relatively minor increases in atmospheric CO_2 concentration that W.O. Fenn[12] once went so far as to publically state that

> in emerging from the sea the ancestors of man had to adapt to a P_{CO2} concentration some 10 times higher than they were accustomed to ... And perhaps it is due to higher P_{CO2} that man's brain is

superior to that of the fish. It might even be that the necessity of adapting to a still slightly higher P_{CO2} due to fossil fuels might result in still further improvement.

In much the same vein, Albers[13] has stated that "the utilization of oxygen and the oxygen consumption of the whole organism could well be altered by carbon dioxide." In fact, certain experiments have suggested that coronary blood flow tends to increase somewhat with increasing carbon dioxide,[14] as does the supply of blood to the brain.[15] In a somewhat unrelated context, it has even been demonstrated that increasing CO_2 concentrations will cause Hydra littoralis[16,17] and the marine hydroid Podocoryne carnea[18] to reproduce sexually as opposed to their common method of budding.

Some of the early evidence for a possible beneficial response of animals to atmospheric CO_2 enrichment came from the work of Valley and Rettger,[19] with bacterial cultures, and from the work of Geyer and Chang[20] and Runyan and Geyer,[21] with cultures of human cells. All of these investigators found that the growth rates of their cell suspensions were greatly reduced, if not totally stopped, when they were incubated in a CO_2 deficient atmosphere. And it has subsequently been demonstrated that the growth responses of bacterial and animal cells to atmospheric CO_2 enrichment are generally stronger than those of most plant cells.[22]

There is also a wealth of information suggestive of a number of direct effects of dissolved unhydrated CO_2 gas as distinct from effects mediated by cell pH changes.[23-29] Indeed, the direct action of CO_2 on proteins may cause changes in their rates of solution, absolute solubility, reactivity, stability, charge and configuration.[30] And as CO_2 is 25 to 35 times more soluble than O_2 and 50 to 60 times more soluble than N_2 in water, with more than 99% of the dissolved "free" or molecular CO_2 actually in the physical form of CO_2,[30] it may well be one of the keys to understanding cellular control at the molecular level.[31-33]

In a provocative elucidation of this general concept, M.A. Mitz[30] wrote the following:

> Based upon the knowledge that carbon dioxide can reversibly change the physical, chemical and biochemical properties of the cellular constituents, it is proposed that transient localized concentration changes of carbon dioxide within part of the cell can markedly influence cell metabolism through dynamic changes in the constituents. It is postulated that the carbon dioxide released by oxidation or decarboxylation in the cell can rapidly build up highly localized concentrations of free CO_2 with water. The latter reaction of CO_2 and H_2O goes on in parallel and, with time, may supersede the initial rapid direct action of CO_2 and the cell constituents. Meanwhile the direct reaction can temporarily cause reversible changes in cell properties. In certain cases the increased production of CO_2 may initiate or control other reactions. High local CO_2 concentration gradients may be large enough to exert a local high pressure (physical effect) which can also play a role in cellular control. While the ad-

sorption of O_2 by a cell will be continuous and often uniform, the CO_2 released from a given cell is expected in steps or bursts ("puffs"). Equilibrium of metabolically produced CO_2 may be the exception rather than the general state in the living cell.

Comparing reported direct CO_2 effects, one notes certain characteristics in common. For example, direct CO_2 effects are highly selective.[34] There may be activation or inhibition of one enzyme and not of another closely related or even the same enzyme in another tissue.[35] One part of a membrane may respond to CO_2 but not another.[36] Another characteristic is that different specific partial pressures of CO_2 are required to initiate specific activities.[27,37] This represents a minimum concentration or what one may call a "threshold" effect. Another characteristic is a rapid almost instantaneous response which would almost certainly be a direct effect.[36,38,39] Furthermore in some of the experiments there is the suggestion of a control range.[40] As the CO_2 is slowly increased above a given value, the response is in one direction and at a still higher CO_2 concentration the response is in the opposite direction. Collectively high sensitivity, sharp threshold effects, rapid response and a finite control range are characteristics indicative of a sensing and control mechanism.

What are some specific examples of this general and wide-ranging phenomenon? First of all, analogously to the direct interaction of ribulose bisphosphate carboxylase-oxygenase (Rubisco) with the external gaseous environment of plants, the hemoglobin of vertebrates likewise interacts directly with the gases of Earth's atmosphere and may be thereby significantly influenced by atmospheric CO_2 perturbations via the common mechanism of carbamate formation, which occurs in both groups of organisms,[29,41,42] as does hemoglobin itself.[43-45] Indeed, experiments indicate that there is a significant linear relationship between blood CO_2 content and the CO_2 concentration of the ambient atmosphere,[46] which is further modulated by temperature.[47]

Now carbamate formation often results in the solubilization of heavy metal (calcium and barium) salts of amino acids.[48] Such a mechanism would facilitate the movement of metal ions across membranes and from one part of a cell to another, possibly enhancing the processes associated with bone building.[30] Similarly, certain proteins like gamma globulin, liver cathespin and insulin (which are insoluble in distilled water) may be solubilized by merely raising the partial pressure of CO_2.[37] And it is likely that enzymes having their active sites masked by interactions with other proteins or polymeric structural elements of the cell may be temporarily dissociated and activated by such a CO_2 tension increase.[49]

Biologically active amines associated with specific nerve action, hormone action or biocommunications in general are also known to have their biological activity modified by carbamate formation.[50] And in the words of Mitz,[30] "what simpler mechanism could anyone devise to stimulate fluid transport, as in the case of the bile from the gall bladder, than the action of CO_2 to lower the melting point for a solid to become a liquid at body temperature?" In fact, he suggests that CO_2-saturated lipoprotein could well constitute a special solvent for cholesterol and other lipids in the circulatory system. If so, the rising CO_2 content of Earth's atmosphere may possibly be implicated as a contributing factor to the significant worldwide downturn in circulatory heart disease experienced over the past two decades.

CO_2 also appears to play a major role in animal reproduction. For example, a certain (and variable-among-species) low-level threshold CO_2 concentration is essential for sperm survival and activity.[40,51] And in mammals, it is also important in the process of embryo implantation within the uterus. In experiments with mice, for instance, "puffs" of CO_2 produced by the blastocyst at the epithelium have been demonstrated to be the stimulus which initiates the sequence of events which leads to successful implantation;[26] while other experiments have shown that this response can be initiated by CO_2 injections into the uterus, with increasing effectiveness as the CO_2 concentration is increased from 300 ppm to 1000 ppm.[25]

Perhaps the most important aspect of this finding, however, is the shape of the total response curve over the full range of CO_2 concentrations investigated. As may be seen in Fig. 10.1, the decidual response of the pseudopregnant mouse to CO_2 enrichment steadily rises with increasing CO_2 to a concentration of 0.1% (1,000 ppm), whereupon it begins to monotonically decline as the CO_2 content of the air rises above that pivotal value.

The significance of this observation derives from the fact that many of the adverse effects of very high CO_2 concentrations have been observed to intensify as the CO_2 content of the air rises from approximately 0.3% to 3% or more.[7] And in the words of K.E. Schaefer,[7] "the question has been raised as to whether one can use a backward projection with data on the effects caused by higher CO_2 concentrations to predict effects of smaller CO_2 concentrations, assuming a linear hypothesis." In answering this question, Schaefer noted that his data "demonstrate the existence of concentration-dependent shifts in CO_2 effects on target organs, which rules out any backward projection." And Fig. 10.1 clearly shows that such a shift may occur at a CO_2 concentration which is just below the lowest ambient CO_2 values which have generally been used in chronic animal CO_2 studies: 2,000 to 3,000 ppm.[7]

Much like the productivity increases of plants described in Sec. 9.1, the physical capacities of a number of animals have also been steadily increasing over the past century or so, as manifest by such diverse measures as various livestock traits,[52] performance characteristics of thoroughbred race horses,[53] and Olympic track records.[54] Again, however, there is as yet no way to know

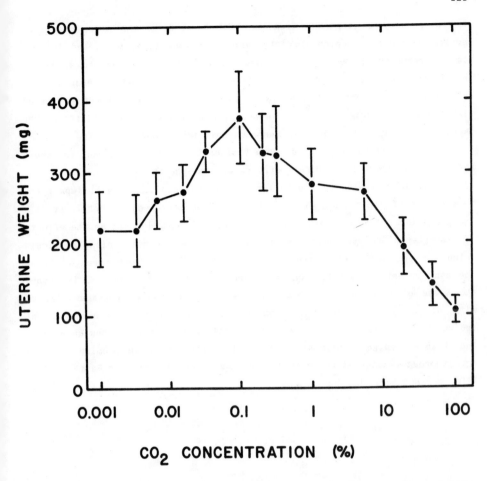

Fig. 10.1 The decidual response of the pseudopregnant mouse uterus to the intraluminal injection of various concentrations of carbon dioxide in air (above 0.2%) and nitrogen (below 0.2%, except for 0.03% which was ambient air). Adapted from Heatherington.[25]

what fractions of these gains, if any, may be attributable to the recent historical buildup of CO_2 in Earth's atmosphere. Clearly, we yet have much to learn in these exciting areas. Nevertheless, enough is currently known to indicate that atmospheric CO_2 enrichment may indeed have a number of significant direct impacts on the Earth's animal kingdom, as well as on its plant life, and that the net effect of these many impacts may well be of a stimulatory or positive nature.

10.2 Indirect Effects

Several indirect effects of atmospheric CO_2 enrichment on animal life should already be apparent from material covered in the preceding three sections. The geographical redistribution of plant types predicted in Sec. 7.4 and reported in Sec. 9.4, for example, will obviously lead to a similar redistribution of animal species which depend upon them for sustenance. The nutrient/carbon balance adjustments to atmospheric CO_2 enrichment discussed with respect to foraging insects in Sec. 8.8 will also have implications for larger herbivores.[55,56] And the numerous soil improvements described in Sec. 7.6 and 8.10 will be felt up and down various food chains. Indeed, the whole host of both direct and indirect effects of atmospheric CO_2 enrichment on the plant life of the planet -- both terrestrial and aquatic -- will impact the animal life of the planet through the cascading phenomena of herbivory[57,58] and predation.[59,60] Especially important in this regard could be the effects of atmospheric CO_2 enrichment on the production of secondary plant chemicals which act as attractants, repellants and toxins.[61] To work through all of the many potential ramifications of these interactions,[62] however, is beyond the scope of this book and would be highly speculative. Consequently, as with direct effects of CO_2 on animal life, the detailed elucidation of indirect effects will have to await the acquisition of a whole new data base. And these second generation research results will not be forthcoming until we have thoroughly defined the primary response characteristics of Earth's plant life to atmospheric CO_2 enrichment.

11 WITHER THE WORLD?

11.1 Forecasting Earth's Future

Every year since 1984, Lester R. Brown and his colleagues at the Worldwatch Institute have published a "State of the World" report,[1] wherein they give the Earth "a physical examination, checking its vital signs." In this enterprise they perform a valuable service, alerting the world community of concerned citizens to many of the unfortunate environmental consequences which result from the pressures of human population growth. They widely miss the mark in one important area, however, and that is their assessment of the significance of the rapidly rising CO_2 content of Earth's atmosphere. Giving no recognition at all to the many biological benefits of this phenomenon, they dwell entirely on perceived negative aspects of the CO_2 greenhouse effect, which the media have helped to make part and parcel of the public psyche. In their most recent report, for example, they state that

> climate change, like no other issue, calls the whole notion of human progress into question. The benefits of newer technologies, more efficient economies, and improved political systems could be overwhelmed by uncontrolled global warming. Some warming is in-avoidable. But unless trends are reversed, tragic changes could occur in just the next two decades. The challenge is to act be-fore it is too late -- which means before the scientific evidence is conclusive. The longer society waits, the more radical and draconian the needed responses will be.

Quite to the contrary, however, it is very possible that the CO_2 enrichment of the atmosphere which we are currently experiencing may well be a blessing in disguise, as suggested by the many studies reviewed in the preceding four sections. Hence, in an attempt to provide some balance to the type of treatment generally received by CO_2 in most analyses of this type, this final section provides a broad synthesis of all that has been presented to this point, concluding with an optimistic projection of what the net climatic and biological consequences of the rapidly rising CO_2 concentration of our atmosphere may portend for Earth's future.

11.2 Future Climate

In mankind's insatiable quest for knowledge of things to come, the subject of climate ranks high indeed. Like the weather, nearly everyone talks about it; but unlike the weather of the past, we may truly be doing something about the weather of the future.[2]

In the words of Michael McElroy:[3]

> Humans are now a force for change on a global scale. Our presence is evident from pole to pole, from the depths of the

ocean to the heights of the stratosphere. [As a result,] we face an immediate and important challenge: to understand and predict the consequences of our actions and to bring this knowledge to bear on policies so as to preserve the viability of the planet for ourselves and for generations yet unborn.

Within this context, it is clear that model predictions of a catastrophic CO_2 greenhouse effect currently hold center stage, commanding the attention of all the world.[4-10] But as Andrew McIntyre[11] has indicated, current climate models have a number of imposing limitations: "(1) the atmosphere cannot, as yet, be directly coupled dynamically with the ocean, (2) there is no one model that manages successfully to mimic all the meteorological processes and (3) none of the high-resolution GCMs can evolve through time like the forces that mold climate change." Indeed, as Thiele and Schiffer[12] have noted, "even the most comprehensive global climate models greatly oversimplify or misrepresent key climatic processes."

Because of these problems, and the host of others described in Sec. 3, the ability of state-of-the-art GCMs to correctly predict future climate is essentially nil, as recently reported by Reid Bryson:[13]

A statement of what the climate is going to be in the year AD 2050 is a 62-years' forecast. Do the models which are used as a basis for the forecast have a demonstrated capability of making 63-years' forecasts? No. A 6.3 year forecast? No. A 0.63 year forecast? No. Have they successfully simulated the climatic variation of the past century-and-a-half? No. Is the handling of the radiative transfer in the model tested against reality? No. Do the models take into account aerosols in the atmosphere? Rarely: they are marvels of mathematics and computer science, but rather crude imitators of reality.

Clearly, then, there is no rational basis for believing that the predictions of such models should bear any resemblance whatsoever to the future condition of Earth's climate. Yet the standard CO_2/trace gas greenhouse effect scenario is still described as "one of the best-established theories in atmospheric science."[14] And if anyone ventures to question the popular paradigm -- even if it be at a scientific conference and the speaker is the illustrious Mikhail Budyko -- it is often received as "swearing in the church."[15]

Nevertheless, the truly critical thinkers of our day are not dissuaded; and they continue to speak out.[16] Wallace Broecker,[17] for example, has also sized up the situation rather bleakly:

It is my feeling that we are in no better a position to make these predictions than are medical students when asked when and where cancer will strike a specific person... I am convinced that the basic architecture of these models denies the possibility for key interactions which occur in the real system [and] that at present no meaningful way exists to incorporate these interactions into models... As we have only recently become aware of the complexity of the linkages which tie together the ocean, atmosphere, ice, and terrestrial vegetation, we have not even begun to formulate means

by which these linkages might be modeled. Indeed, reliable modeling may never be possible.

Although our knowledge of Earth's climate system is thus rudimentary
indeed, and our ability to theoretically model it almost nil, those of us who
fully appreciate the enormity of the task ahead would by no means completely
wash our hands of GCM-type models. Rather, we would take the position
recently articulated by Bryson:[13]

> Please note that I do not say that the GCM-type models are
> [totally] useless or wrong, when they are the most sophisticated
> known way of considering the bulk of atmospheric physics simul-
> taneously over the whole interacting globe... I simply wish to
> point out that there is much more which needs to be done before
> they are sufficiently reliable for climate forecasting.

Consequently, although humanity is indeed beginning to impact the
environment on a planetary scale, "wise intervention," in the words of
McElroy,[3] "must presume a base of knowledge sufficient to permit the
comprehensive advanced assessment of the impact [and] it is clear that the
current state of global geoscience is wholly inadequate to such a demanding
task." Hence, to say that "it is not important to resolve the scientific
issues before taking action," as the U.S. Environmental Protection Agency has
recently done,[18] would appear to be the height of folly; and as a means to
avoid the likely "iatrogenic error" of such actions, Lovelock[19] has suggested
that putative planetary doctors should be required to take a Hippocratic-type
oath, "an oath to prevent the overzealous from applying a cure that would do
more harm than good."

11.3 The Empirical Greenhouse

In light of the current inability of state-of-the-art GCM-type models to
make valid climate predictions, we are left with empirical projection as the
only viable alternative for assessing the climatic course of the Earth in
response to CO_2/trace gas forcing. And within this context, our most
rational alternative would appear to be the comparative planetary climatology
relationship of Fig. 4.1(2).

Now for a simple 300 to 600 ppm doubling of the atmospheric CO_2
concentration, Fig. 4.1(2) suggests that the mean surface air temperature of
the Earth will rise by about 0.4°C. But since other trace gases are expected
to have produced a greenhouse warming equivalent to that of CO_2 by the time
that a CO_2 concentration of 600 ppm is achieved (Sec. 2.4), this figure should
probably be increased to 0.8°C. And assuming that this "doubling effect" of
non-CO_2 trace gas emissions holds true for CO_2 concentration increases above
600 ppm (which may actually be a somewhat conservative assumption), an
increase in atmospheric CO_2 concentration to 1,000 ppm will likely mean an

ultimate mean surface air temperature rise above the nominal 300 ppm base value of approximately 1.6°C (or slightly more, if non-CO_2 trace gas emissions significantly outstrip CO_2 emissions in the years ahead).

So when could such temperature increases be expected? Turning again to empirical projection as the technique most likely to produce reliable results, extension of the linear portion of the CO_2/population relationship of Fig. 1.3 to atmospheric CO_2 concentrations of 600 and 1,000 ppm suggests that these CO_2 concentrations will likely be reached when world population reaches 19.5 and 42.5 billion people, respectively. But when will that be?

From the population vs. time curve of Fig. 1.1, it is clear that world population has been growing at a phenomenal rate over the past several decades; but from that graph it is difficult to make the needed projections. Hence, the post-1900 population/time data of Table 1.1 are replotted in the format of Fig. 11.1, where it can be seen that a rather stable and readily projectable relationship has been maintained for the past forty years. And this relationship suggests that a world population of 19.5 billion people will likely be reached by AD 2059 and that a population of 42.5 billion people should be reached by AD 2100.

To these dates, however, must be added the lengths of time required for climate equilibrium to be achieved; and here there is no prior empirical relationship with which to work. Consequently, we are left with little choice but to refer to the climate model projections reported in Sec. 2.5, which range from a few decades to as long as a century or more for a 300 to 600 ppm doubling of the atmospheric CO_2 concentration. Appropriately scaling these estimates along the entire timeline required for 1,000-ppm CO_2 equilibrium warming to be realized, we thus get the family of curves depicted in Fig. 11.2 for 300 to 600 ppm lag-times of 0, 50, 100 and 200 years.

In viewing this graph, we note the following. First of all, the several curves all end abruptly at the point of achievement of equilibrium warming for a 1,000 ppm atmospheric CO_2 concentration. And this CO_2 concentration was achieved by following the world population vs. time curve defined over the past forty years right up to a world population of 42.5 billion people, which scenario suggests further phenomenal growth beyond this point as well. However, Earth's maximum "carrying capacity" for humankind has been estimated to be somewhere in the range of 40 to 50 billion people.[20] Hence, although a maximum warming of about 1.6°C may ultimately be expected, it will probably not be achieved until a much later date than what is suggested by the curves of Fig. 11.2. Thus, if a 50-year time lag is characteristic of the Earth's climate system, the type of population growth adjustment needed to stabilize the number of people on the planet near a value of 40 billion would lead to a true warming curve more like that depicted for a 200-year time lag.

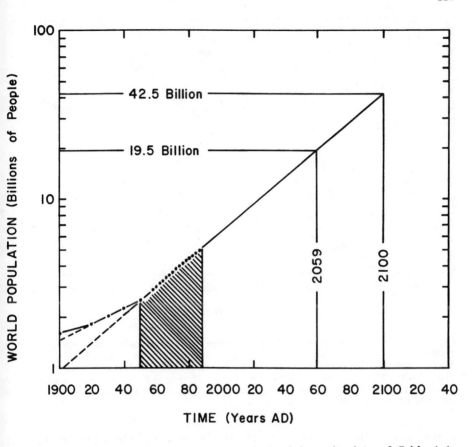

Fig. 11.1 World population vs. time, as derived from the data of Table 1.1. The hatched area highlights the period of the past 40 years, during which time the relationship used to predict the times of occurrance of world poplulations of 19.5 (corresponding to 600 ppm CO_2) and 42.5 (corresponding to 1,000 ppm CO_2) billion people was developed.

But even the extreme results of Fig. 11.2 suggest that both the magnitude and rate of warming predicted by current GCMs are way out of line with reality. Hence, most -- if not all -- of the "doom and gloom" scenarios of the past decade or so will probably never be realized. In fact, the warming predicted in Fig. 11.2 may be just what our planet needs, for according to Bryson,[13] the next glaciation of the Earth should begin "in about 400 plus or minus 200 years." And the impetus for warming depicted in Fig. 11.2 is of about the right magnitude and destined to appear at about the right time and at about the right rate to effectively counterbalance the expected impetus for cooling.

In terms of McElroy's plea[3] to preserve the viability of the Earth for both ourselves "and for generations yet unborn," mankind's flooding of the

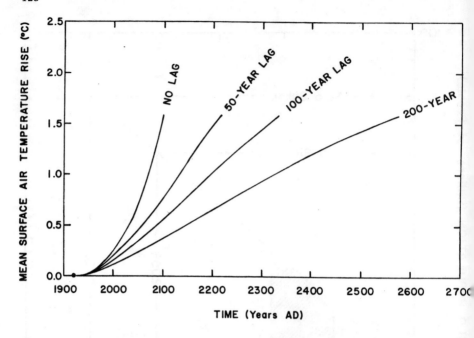

Fig. 11.2 Potential scenarios of mean global warming due to projected increases in CO_2 and other trace gases (assumed to be equivalent to CO_2 in their combined greenhouse effect) based upon the population projections of Fig. 11.1, the CO_2 vs. population curve of Fig. 1.3, and the comparative planetary climatology relationship of Fig. 4.1, for time lags to equilibrium warming of 0, 50, 100, and 200 years.

atmosphere with CO_2 may thus be the only realistic means we have of accomplishing that task. And the planetary elixir appears to be being administered in truly Gaian fashion; for as Lovelock[19] has written:

> Geophysiology reminds us that all ecosystems are interconnected [and] we can be altruistic and selfish simultaneously in a kind of unconscious enlightened self-interest. We most certainly are not a cancer of the Earth, nor is the Earth some mechanical contraption needing the services of a mechanic... [Rather,] we need a general practitioner of planetary medicine.

And in response to his plaintive query -- "Is there a doctor out there?"-- we can probably answer "yes." In fact, to savage a famous quotation, we could probably also say that we have met the doctor, and he is us!

11.4 A Biology of Bounty

In terms of the plant life of the planet, a 300 to 1,000 ppm increase in the CO_2 content of the atmosphere should produce about a tripling of the water use efficiency of almost all of the world's vegetation (Sec. 7.3), as a result of the direct effects of atmospheric CO_2 enrichment upon the primary lant processes of photosynthesis (Sec. 7.1) and transpiration (Sec. 7.2). nd to this phenomenal response must be added the effects of a number of econdary amplifying processes and synergistic feedbacks.

First of all, if the planet does warm (which it may not, due to the ompeting effect of the likely concurrent tendency towards glaciation), the igher air temperatures should enhance plant growth still more (Sec. 7.8.7). econdly, many plants will greatly expand their ranges with augmented water se efficiencies (Sec. 7.4), stabilizing the soil and protecting it from rosion (Sec. 7.6). And with greater above-ground productivity, there will e greater below-ground productivity (Sec. 7.8.3). Plant roots will grow eeper to utilize previously untapped nutrients (Sec. 7.5 and 8.10). More rganic matter will be returned to the soil (Sec. 7.6) Microbial populations ill rise (Sec. 8.10). Earthworms will increase in number (Sec. 7.6). Soil-orming processes will be enhanced (Sec. 8.10). Water (Sec. 8.2), salinity (Sec. 8.5), nutrient (Sec. 8.10), and air pollution (Sec. 8.6) stresses will e relieved. High (Sec. 8.3) and low (Sec. 8.4) temperature extremes should e better tolerated. Many plant diseases may be reduced (Sec. 8.8). And there will be a greater supply of forage and habitat available for supporting a vastly increased animal population of all types.

Considered in their totality, these many interrelated phenomena could readily double, triple or even quadruple the primary direct effects of atmospheric CO_2 enrichment upon plant growth and development on a global basis, especially when one considers the potential for vegetative expansion over the large portions of the planet which are presently barren and the possibility for increases in oceanic productivity. Hence, it is by no means inconceivable that the vitality of the biosphere, or the totality of all life processes on planet Earth, could rise by a full order of magnitude or more over the next few centuries; for as Boyd Strain[21] has recently noted, "we are moving from a carbon-starved to a carbon-fertilized world." Indeed, we seem destined to experience what no member of our species has ever before encountered, but something with which the earlier inhabitants of the planet were well acquainted.

11.5 Return to Eden?

The Earth was once considerably more vegetated, and the biosphere considerably more vigorous, than it is now. Throughout the Phanerozoic, for example, when the oxygen content of the atmosphere has been relatively stable at approximately 21 percent by volume,[19] life has generally flourished from pole to pole.[22] Throughout the Quaternary, however, periodic glaciations have regularly confined the Earth's terrestrial life forms to equatorial and middle latitudes, except for a number of geologically-brief interludes suc[h] as the one we have about concluded, i.e., the Holocene. And as noted in Sec 1.3, the CO_2 content of Earth's atmosphere during the dominant glacial epoch[s] of the Quaternary has often dropped so low as to be a bona fide threat to th[e] continued existence of the planet's predominant C_3 plants.

In light of these facts, it is easy to see that the past two million year[s] of Earth's history have been the exception to the rule by which the biospher[e] has governed itself over the past 600 million years. One possible cause o[f] this digression may have been the increase in uplift rates experienced in man[y] mountainous parts of the world over the past three to five millennia.[23-27] The resultant raised topography, for example, has recently been suggested t[o] have enhanced the meridionality of Northern Hemisphere atmospheric waves and[ ] caused a greater southward penetration of polar air masses over east-centra[l] North America and Europe, possibly triggering the development of Quaternary[ ] glaciations.[28] It has also been suggested that the uplifted terrain may have[ ] produced greater continental weathering rates and contributed to a general reduction of atmospheric CO_2,[29] via the mechanisms described in Sec. 1.2.

These two criteria, i.e., near-persistent glaciation and ultra-low[ ] atmospheric CO_2 concentrations,[30-32] thus set the past two millennia apart from the previous 600. And although the biosphere has responded to this life-threatening (for C_3 plants) perturbation by developing a new (C_4) mode of photosynthesis which can effectively cope with the anomalously low CO_2 concentrations of the dominant Quaternary environmental regime,[33,34] a more satisfying solution would be a return to pre-Quaternary CO_2 concentrations, as there appears to have been little variation in the primary photosynthetic reaction center throughout geologic time.[35] Such a turnabout, for instance, would preserve the Earth's predominant C_3 plants without sacrificing either C_4 or CAM species (which, although much less responsive to atmospheric CO_2 enrichment than C_3 plants, still function better at atmospheric CO_2 concentrations higher than those of today[36-43]), while at the same time helping to prevent further glaciations of the planet, as described in Sec. 11.3. As a bonus, such conditions would also amplify nearly all of the life-processes described in the preceding subsection, where nearly every species wins and practically none loses. This seeming to be the case, it would thus appear that man's inadvertent flooding of the atmosphere with CO_2 is a most fortunate and desirable phenomenon indeed.

11.6 The "Dark Side" of Gaian Man

That one aspect of the impact of humanity on the global environment may be beneficial does not, of course, imply that all of man's actions must

ecessarily result in good, nor that even the net effect of all of his
ctivities will be benign. And perhaps nowhere has this fact been made more
oundantly clear than in Time magazine's historic departure from tradition in
aking Endangered Earth its "Planet of the Year" and the lead cover story for
989.

"The Death of Birth" by Eugene Linden,[44] for example, chronicles man's
elentless destruction of tropical forests,[45,46] which contain some 50 to 80
ercent of Earth's plant and animal species and which are being felled at a
ate that is roughly equivalent to the area of one football field per
econd.[47,48] So great is the desecration of priceless habitat that it is
stimated that over the next three decades man will drive an average of 100
pecies to extinction every single day,[49] this in contrast to the estimated
xtinction rate of one species per year over the past 600 millennia.[50-52] In
he words of E.O. Wilson,[47] "the extinctions ongoing worldwide promise to be
t least as great as the mass extinction that occurred at the end of the age
f dinosaurs," while ecologist Norman Myers[44] refers to the phenomenon as
the greatest single setback to life's abundance and diversity since the
irst flickerings of life almost 4 billion years ago." Indeed, at present
ates of habitat destruction, we could well see the demise of one-quarter or
ore of all extant species over the next several decades.[53-56]

"A Stinking Mess" is the title of John Langone's[57] Time report on the
roblem of human waste, which is threatening land, water and air alike.
articularly perplexing, he notes, are the ozone-layer-destroying
hlorofluorocarbons and radioactive wastes, each of which topics merits a
eport of its own.[58,59] And driving everything else, of course, is the ever-
ncreasing mass of humanity.[60] According to recent estimates, for example,
uman beings currently use 20 to 40% of all solar energy that is converted to
rganic matter by terrestrial vegetation.[61,62] In the words of Brown and
aurer:[63]

> Never before in the history of the earth has a single species been
> so widely distributed and monopolized such a large fraction of
> the energetic resources. An ever-diminishing remainder of these
> limited resources is now being divided among the millions of other
> species. The consequences are predictable: contraction of geogra-
> phic ranges, reduction of population sizes, and increased probabil-
> ity of extinction for most wild species ... and the loss of bio-
> logical diversity at all scales from local to global.

The one major fault of Time's otherwise laudable issue is its all-too-
tandard treatment of the CO_2 greenhouse effect:[64] "We're talking about rates
f climate change perhaps 100 times faster than at any other time in human
istory... The possible consequences are so scary that it is only prudent
or governments to slow the buildup of CO_2 through preventive measures...
he answer is a tax on CO_2 emissions." Or as Carl Sagan[65] has more

explicitly phrased it, the answer is "taxing and phasing out fossil fuels. Indeed, there are many who believe, as stated in an unsigned editorial in the 9 March 1989 issue of Nature, that the Montreal Protocol to reduce CF emissions "is chiefly to be valued because it is a precedent for the much more constraining convention there will have to be if ever it becomes necessary to restrict by international agreement the emission of carbon dioxide so as to avoid the now-familiar greenhouse effect."

In light of the material presented in the preceding sections, such statements as these are clearly not only unfounded but contrary to what i actually known and counterproductive to our future well-being. Indeed, most of them may generally be attributed to what Peter Myers[66] of the Natur Conservancy has called "data-free speculation," and they form the basis fo what has led Kenneth E.F. Watt[12] to dub the standard CO_2 greenhouse effec scenario "the laugh of the century."

Perhaps even more distressing, however, is what is left unsaid Throughout the entire spate of Time articles, there is not the slightest hin that atmospheric CO_2 enrichment has any positive effects whatsoever. Yet the hundreds of studies cited in this book conclusively demonstrate tha increasing the CO_2 content of the air leads to many beneficial biologica consequences. In fact, the several climatic analyses presented herei suggest that even in this area the effects of atmospheric CO_2 enrichment ma well be positive.

Time's reporting team's failure to acknowledge these facts is especially serious for two different reasons. First of all, as they correctly indicate,[67] "problems of agriculture are likely to be critical in the next century, as growing populations ... put even more pressure on a badly strained food-supply system." Consequently, it would be like cutting our own throats -- or, more properly,[3] the throats of "generations yet unborn" -- to attempt to thwart the very phenomenon (the steadily rising atmospheric CO_2 concentration) which has the proven ability to dramatically boost crop yields, enhance plant water use efficiencies, and give us the edge we need in our fight against world hunger. Not even genetic engineering, for example can presume to solve the problem of insufficient agricultural water supplies in many high-birthrate countries; for in spite of all that man might do for them, as Time quotes plant physiologist Anthony Hall,[67] "you can't grow plants without water." Thus, the plus-600 ppm atmospheric CO_2 concentrations of the 2060s and beyond, which will enable plants to utilize water twice as efficiently as they did in the early 1900s, may well provide our only hope of properly feeding the peoples of the world at that future time without destroying what remains of the planet's natural ecosystems. Indeed, the revitalization of degraded lands via the many positive phenomena set in otion by atmospheric CO_2 enrichment may well be the single most important eterrent to the further destruction of tropical forests.[68]

The second danger of ignoring the biological benefits of Earth's steadily ising atmospheric CO_2 concentration is equally real but somewhat sychological at the same time: psychological, because as Senator Albert ore[69] points out in his contribution, the solutions to many of the problems reated by the "dark side" of man's activities are "almost unimaginably ifficult," and to carry them out, "changes in human thinking of a magnitude omparable to the changes that brought about the abolition of slavery must ake place in one generation." Hence, unless there is a glimmer of hope, ome indication that there is at least one powerful force operating in our ehalf, we may not be able to muster the resolve required to tackle the roblems about which we truly can do something.

And here is where the very real danger of ignoring the facts comes into lay; for as Lemonick[64] has correctly written, "humanity's contribution to he greenhouse effect [i.e., increasing atmospheric CO_2] comes from so many asic activities that man cannot realistically expect to stop the process." 'hus, with an incomplete or improper understanding of the issue, we could asily squander our limited resources on an unwinnable battle, which even if t were winnable would be to our detriment, leading to the loss of other more mportant battles which are crucial to the winning of the war. Else, as uggested by Toufexis,[60] "we may be well on the way to producing a subhuman ind of race," which will devastate the biosphere to a greater extent than ny cataclysm the Earth has ever experienced.[52]

1.7 In Conclusion

In concluding this treatise on the current transitory status of planet arth, we ought not to forget our roots, that the soil "is the medium of rimaeval importance, and most organisms in terrestrial ecosystems still live here."[70] Indeed, soil is where the action is, for the soil microcosm upports all other levels of the terrestrial trophic hierarchy; and in the words of P.W. Price,[70] "it is no leap of faith to declare that micro-organisms have dominated the biosphere and still do [for] the nesting of taxa in their trophic relations within the confines of formerly evolved groups, and the great species richness of earlier taxa, places most of ecological interactions squarely among small organisms, most of which are in, or closely associated with, the soil."

But of the ultimate effects of atmospheric CO_2 enrichment upon this hidden world, we know precious little. From bits and pieces of information, it is possible to construct a scenario of vastly beneficial consequences resulting from the accelerated addition of organic matter to the soil system, which would seem to be almost assured in terms of what we have learned of basic growth responses of higher plants. Nevertheless, there have been practically no studies of the whole sequence of events, from initial atmospheric enrichment with CO_2 to ultimate synergistic soil feedback effects. Hence, such conclusions must remain somewhat tentative until such time as definitive experiments are conducted in these areas, which should surely receive high priority in the research agendas of major funding agencies.

By the same token, the world of the microscopic should not be neglected in our quest to unravel the mystery of climate; for if the evolution of plants and animals has depended entirely on micro-organisms,[71-73] why not climate also? Indeed, the currently popular phytoplankton-DMS-cloud reflectivity relationship discussed in Sec. 3.3 and 5.9 is likely but one of a whole host of such relationships which we have barely begun to appreciate. Here, too, a major research effort should be mounted.

"It is often difficult," Lovelock[19] has remarked, "to recognize the larger entity of which we are a part;" and this has been a major failure of climate research to date. Now, however, with the awakening of the scientific community to the vast interconnectedness of the physical, chemical and biological components of the Earth's surface matter -- its organized complexity[74] and richness of relationships[75] -- we are on the verge of embarking upon a new age of research, as the Earth itself enters a new age of existence. But as we begin this exciting endeavor, we must be sure, as the leadership of the American Meteorological Society and the University Corporation for Atmospheric Research[76] have recently cautioned, that our actions are based "on sound scientific conclusions that are themselves derived from careful, long-term measurements and on prediction models whose accuracy has been thoroughly tested and verified."

To this end, the government of the United States has recently inaugurated an ambitious research program,[77] designed "to establish the scientific basis for national and international policy making related to natural and human-induced changes in the global earth system," realizing that "rational response strategies and sound policy can only be built upon reliable information, predictions, and assessments of the complex phenomena of the global earth system" and that "responding to these changes without a strong scientific basis could be futile and costly."

Focusing on four key questions -- (1) What forces initiate global change? (2) How does the earth system respond to changes in forcing functions? (3) How has earth's environment changed in the past? and (4) How well can global change be predicted? -- and emphasizing seven major interdisciplinary science elements -- biogeochemical dynamics, ecological systems and dynamics, climate and hydrologic dynamics, human interactions, earth system history, solid earth processes and solar influences -- the program is well conceived. It is now up to us to follow through and obtain the needed answers.

REFERENCES

1. Holford, I. (1977) The Guinness Book of Weather Facts and Feats. Guinness Superlatives Ltd., Enfield, UK.
2. Rosenberg, N.J., Ed. (1978) North American Droughts. Westview Press, Boulder, CO.
3. Schulz, E.F., Koelzer, V.A. and Mahmood, K., Eds. (1973) Floods and Droughts. Water Resources Pub., Fort Collins, CO.
4. Hoyt, W.G. and Langbein, W.B. (1955) Floods. Princeton Univ. Press, Princeton, NJ.
5. Sears, P.B. (1980) Deserts on the March. Oklahoma Press, Norman, OK.
6. Eckholm, E.P. (1987) Spreading Deserts -- The Hand of Man. Worldwatch Institute, Washington, DC.
7. Glantz, M.H., Ed. (1977) Desertification: Environmental Degradation in and Around Arid Lands. Westview Press, Boulder, CO.
8. Hare, F.K. (1985) Climate Variations, Drought and Desertification. World Meteorol. Org., Geneva, Switzerland.
9. Johnson, A.H. and Siccama, T.G. (1983) Acid deposition and forest decline. Environ. Sci. Tech. 17, 294A-305A.
10. Hornbeck, J.W. and Smith, R.B. (1985) Documentation of red spruce growth decline. Can. J. For. Res. 15, 1199-1201.
11. Schutt, P. and Cowling, E.B. (1985) Waldsterben, a general decline of forests in central Europe: Symptoms, development and possible causes. Plant Dis. 69, 548-558.
12. Hinrichsen, D. (1987) The forest decline enigma. BioScience 37, 542-546.
13. Krogstad, E.J., Balakrishnan, S., Mukhopadhyay, D.K. Rajamani, V. and Hanson, G.N. (1989) Plate tectonics 2.5 billion years ago: Evidence at Kolar, south India. Science 243, 1337-1340.
14. Glen, W. (1982) The Road to Jaramillo: Critical Years of the Revolution in Earth Science. Stanford Univ. Press, Stanford, CA.
15. Cox, A. and Hart, R.B. (1986) Plate Tectonics: How it Works. Blackwell Sci. Pub., Palo Alto, CA.
16. Anderson, D.L. (1989) Where on Earth is the crust? Phys. Today 42(3), 38-46.
17. Hsu, K.J., Ed. (1982) Mountain Building Processes. Academic Press, New York, NY.
18. Schaer, J.-P. and Rodgers, J., Eds. (1987) The Anatomy of Mountain Ranges. Princeton Univ. Press, Princeton, NJ.
19. Huges, N.F., Ed. (1973) Organisms and Continents Through Time: Methods of Assessing Relationships Between Past and Present Biologic Distributions and the Positions of Continents. The Palaeontological Assoc., London, UK.
20. Brenchley, P.J., Ed. (1984) Fossils and Climate. John Wiley & Sons, New York, NY.
21. Crowley, T.J. and North, G.R. (1988) Abrupt climate change and extinction events in Earth history. Science 240, 996-1002.
22. Hays, J.D., Imbrie, J. and Shackleton, M.J. (1976) Variations in the Earth's orbit: Pacemaker of the ice ages. Science 194, 1121-1132.
23. Imbrie, J. and Imbrie, K.P. (1979) Ice Ages: Solving the Mystery. Enslow Pub., Short Hills, NJ.
24. Tooley, M.J. and Shennan, I., Eds. (1987) Sea-Level Changes. Basil Blackwell Ltd., Oxford, UK.
25. Ronov, A.B., Khain, V.E., Balukovsky, A.N. and Seslavinsky, K.B. (1980) Quantitative analysis of Phanerozoic sedimentation. Sediment. Geol. 25, 311-325.
26. Nitecki, M.H., Ed. (1984) Extinctions. Univ. Chicago Press, Chicago, IL.
27. Elliott, D.K., Ed. (1986) Dynamics of Extinction. Johy Wiley and Sons, New York, NY.
28. Schneider, S.H. and Londer, R. (1984) The Coevolution of Climate and

REFERENCES

Life. Sierra Club Books, San Francisco, CA.

29. Nance, R.D., Worsley, T.R. and Moody, J.B. (1988) The supercontinent cycle. Sci. Amer. 259(1), 72-79.

30. Lovelock, J.E. and Margulis, L. (1974) Atmospheric homeostasis by and for the biosphere: the Gaia hypothesis. Tellus 26, 2-10.

31. Lovelock, J.E. (1979) Gaia: A New Look at Life on Earth. Oxford Univ. Press, Oxford, UK.

32. Lovelock, J.E. (1988) The Ages of Gaia: A Biography of Our Living Earth. W.W. Norton & Co., New York, NY.

33. Sattaur, O. (1987) Cuckoo in the nest. New Sci. 116(1592/1593), 16-18.

34. Kerr, R.A. (1988) No longer willful, Gaia becomes respectable. Science 240, 393-395.

35. Lindley, D. (1988) Is the Earth alive or dead? Nature 332, 483-484.

36. Pearce, F. (1988) Gaia: A revolution comes of age. New Sci. 117(1604), 32-33.

37. Lovelock, J.E. (1972) Gaia as seen through the atmosphere. Atmos. Environ. 6, 579-580.

38. Holland, H.D. (1984) The Chemical Evolution of the Atmosphere and Oceans. Princeton Univ. Press, Princeton, NJ.

39. Margulis, L. and Lovelock, J.E. (1974) Biological modulation of the Earth's atmosphere. Icarus 21, 471-489.

40. Barnes, I., Irwin, W.P. and White, D.E. (1978) Global distribution of carbon dioxide discharges and major zones of seismicity. U.S. Geol. Surv. Water Res. Inv., Open-file Rept., pp. 78-89.

41. Javoy, M., Pineau, F. and Allegre, C.J. (1982) Carbon geodynamic cycle. Nature 300, 171-173.

42. Gerlach, T.M. (1989) CO_2 from magma-chamber degassing. Nature 337, 124.

43. Volk, T. (1987) Feedbacks between weathering and atmospheric CO_2 over the last 100 million years. Amer. J. Sci. 287, 763-779.

44. Kasting, J.F. Toon, O.B. and Pollack, J.B. (1988) How climate evolved on the terrestrial planets. Sci. Amer. 258(2), 90-97.

45. Sundquist, E.T. and Broecker, W.S., Eds. (1985) The Carbon Cycle and Atmospheric CO_2: Natural Variations, Archean to Present. Amer. Geophys. Union, Washington, DC.

46. Lemon, E.R., Ed. (1983) CO_2 and Plants: The Response of Plants to Rising Levels of Atmospheric Carbon Dioxide. Westview Press, Boulder, CO.

47. Dahlman, R.C., Strain, B.R. and Rogers, H.H. (1985) Research on the response of vegetation to elevated atmospheric carbon dioxide. J. Environ. Qual. 14, 1-8.

48. Strain, B.R. and Cure, J.D., Eds. (1985) Direct Effects of Increasing Carbon Dioxide on Vegetation. U.S. Dept. Energy, Washington, DC.

49. National Research Council, U.S. (1979) Carbon Dioxide and Climate: A Scientific Assessment. National Academy Press, Washington, DC.

50. National Research Council, U.S. (1982) Carbon Dioxide and Climate: A Second Assessment. National Academy Press, Washington, DC.

51. National Research Council, U.S. (1983) Changing Climate. National Academy Press, Washington, DC.

52. MacCracken, M.C. and Luther, F.M., Eds. (1985) Projecting the Climatic Effects of Increasing Carbon Dioxide. U.S. Dept. Energy, Washington, DC.

53. Idso, S.B. (1982) Carbon Dioxide: Friend or Foe? IBR Press, Tempe, AZ.

54. Idso, S.B. (1982) The anthropic principle, Gaia, carbon dioxide and the human condition. Spec. Sci. Tech. 5, 455-459.

55. Keeling, C.D. (1988) Numeric Data Package NDP001. Carbon Dioxide Information Center, Oak Ridge, TN.

56. Sagan, C. and Mullen, G. (1972) Earth and Mars: Evolution of atmospheres and surface temperatures. Science 177, 52-56.

57. Owen, T., Cess, R.D. and Ramanathan, V. (1979) Enhanced CO_2 greenhouse

to compensate for reduced solar luminosity on early Earth. Nature 277, 640-642.

58. Schwarzschild, M., Howard, R. and Harm, R. (1957) Inhomogeneous stellar models. V. A solar model with convective envelope and inhomogeneous interior. Astrophys. J. 125, 233-241.

59. Ezer, D. and Cameron, A.G.W. (1965) A study of solar evolution. Can. J. Phys. 43, 1497-1517.

60. Bahcall, J.N. and Shaviv, G. (1968) Solar models and neutrino fluxes. Astrophys. J. 153, 113-126.

61. Iben, I. (1969) The Ci^{37} solar neutrino experiment and the solar helium abundance. Ann. Phys. (New York) 54, 164-203.

62. Newman, M.J. and Rood, R.T. (1977) Implication of solar evolution for the Earth's early history. Science 198, 1035-1037.

63. Schopf, J.W. and Barghourn, E.S. (1967) Alga-like fossils from the early Precambrian of South Africa. Science 156, 507-512.

64. Knauth, L.P. and Epstein, S. (1976) Hydrogen and oxygen isotope ratios in modular and bedded cherts. Geochim. Cosmochim. Acta 40, 1095-1108.

65. Schopf, J.W. (1978) The evolution of the earliest cells. Sci. Amer. 239(3), 110-138.

66. Lowe, D.R. (1980) Archean sedimentation. Ann. Rev. Earth Planetary Sci. 8, 145-167.

67. Wigley, T.M.L. and Brimblecombe, P. (1981) Carbon dioxide, ammonia and the origin of life. Nature 291, 213-215.

68. Woese, C.R. (1984) The Origin of Life. Carolina Biological Supply Co., Burlington, IA.

69. Schidlowski, M. (1988) A 3,800-million-year isotopic record of life from carbon in sedimentary rocks. Nature 333, 313-318.

70. Zahnle, K.J., Kasting, J.F. and Pollack, J.B. (1988) Evolution of a steam atmosphere during Earth's accretion. Icarus 74, 62-97.

71. Hart, M.H. (1978) The evolution of the atmosphere of the Earth. Icarus 33, 23-39.

72. Walker, J.C.G. (1985) Carbon dioxide on the early Earth. Origins Life 16, 117-127.

73. Kasting, J.F. and Ackerman, T.P. (1986) Climatic consequences of very high carbon dioxide levels in Earth's early atmosphere. Science 234, 1383-1385.

74. Siever, R. (1968) Sedimentological consequences of steady-state ocean-atmosphere. Sedimentol. 11, 5-29.

75. Broecker, W.S. (1971) A kinetic model for the chemical composition of sea water. Quat. Res. 1, 188-207.

76. Garrels, R.M. and Mackenzie, F.T. (1971) Evolution of Sedimentary Rocks. W.W. Norton, New York, NY.

77. Walker, J.C.G. (1977) Evolution of the Atmosphere. Macmillan, New York, NY.

78. Holland, H.D. (1978) The Chemistry of the Atmosphere and Oceans. Interscience, New York, NY.

79. Alekseyev, V.V., Zaytsev, S.I., Lyamin, M.Ya. and Kiseleva, S.V. (1987) Experimental modeling of CO_2 washout from the atmosphere by precipitation. Izvest. Atmos. Ocean. Phys. 23, 785-788.

80. Berner, R.A. and Lasaga, A.C. (1989) Modeling the geochemical carbon cycle. Sci. Amer. 260(3), 74-81.

81. Marshall, H.G., Walker, J.C.G. and Kuhn, W.R. (1988) Long-term climate change and the geochemnical cycle of carbon. J. Geophys. Res. 93, 791-801.

82. Volk, T. (1989) Sensitivity of climate and atmospheric CO_2 to deep-ocean and shallow-ocean carbonate burial. Nature 337, 637-640.

83. Walker, J.C.G., Hays, P.B. and Kasting, J.F. (1981) A negative feedback mechanism for the long-term stabilization of Earth's surface temperature. J. Geophys. Res. 86, 9776-9782.

84. Berner, R.A., Lasaga, A.C. and Garrels, R.M. (1983) The carbonate-silicate geochemical cycle and its effect on atmospheric carbon dioxide over the past 100 million years. Amer. J. Sci. 283, 641-683.
85. Bolin, B., Degens, E.T., Kempe, S. and Ketner, P. (1979) The Global Carbon Cycle. SCOPE 13. Wiley, New York, NY.
86. Buyanovsky, G.A. and Wagner, G.H. (1983) Annual cycles of carbon dioxide level in soil air. Soil Sci. Soc. Amer. J. 47, 1139-1145.
87. Holland, H.D., Lazar, B. and McCaffrey, M. (1986) Evolution of atmosphere and oceans. Nature 320, 27-33.
88. Lovelock, J.E. and Whitfield, M. (1982) Life span of the biosphere. Nature 296, 561-563.
89. Broecker, W.S. (1982) Glacial to interglacial changes in ocean chemistry. Prog. Oceanog. 11, 151-197.
90. Broecker, W.S. (1982) Ocean chemistry during glacial time. Geochim. Cosmochim. Acta 46, 1689-1705.
91. McElroy, M.B. (1983) Marine biological controls on atmospheric CO_2 and climate. Nature 302, 328-329.
92. Knox, F. and McElroy, M.B. (1984) Changes in atmospheric CO_2: Influence of the marine biota at high latitude. J. Geophys. Res. 89, 4629-4637.
93. Lovelock, J.E. and Watson, A.J. (1982) The regulation of carbon dioxide and climate: Gaia or geochemistry. Planet. Space Sci. 30, 795-802.
94. Ittekkot, V. (1988) Global trends in the nature of organic matter in river suspensions. Nature 332, 436-438.
95. DesMarais, D.J., Cronin, S., Nguyen, H. and D'Amelio, E. (1988) Isotopes and stromatolites as a key to understanding the history of the carbon cycle. Trans. Amer. Geophys. Union 69, 1084.
96. Salisbury, F.B. and Ross, C.W. (1978) Plant Physiology. Wadsworth Pub. Co., Belmont, CA.
97. Berner, W., Stauffer, B. and Oeschger, H. (1979) Past atmospheric composition and climate, gas parameters measured in ice cores. Nature 275, 53-55.
98. Delmas, R.J., Ascencio, J.-M. and Legrand, M. (1980) Polar ice evidence that atmospheric CO_2 20,000 yr BP was 50% of present. Nature 284, 155-157.
99. Neftel, A., Oeschger, H., Schwander, J., Stauffer, B. and Zumbrunn, R. (1982) Ice core sample measurements give atmospheric CO_2 content during the past 40,000 yr. Nature 295, 220-223.
100. Barnola, J.M., Raynaud, D., Korotkevich, Y.S. and Lorius, C. (1987) Vostok ice core provides 160,000-year record of atmospheric CO_2. Nature 329, 408-414.
101. Neftel, A., Oeschger, H., Staffelbach, T. and Stauffer, B. (1988) CO_2 record in the Byrd ice core 50,000-5,000 years BP. Nature 331, 609-611.
102. Fifield, R. (1988) Frozen assets of the ice cores. New Sci. 118(1608), 28-29.
103. Suarez, M.J. and Held, I.M. (1976) Modelling climatic response to orbital parameter variations. Nature 263, 46-47.
104. Pollard, D. (1978) An investigation of the astronomical theory of the ice ages using a simple climate-ice sheet model. Nature 272, 233-235.
105. Schneider, S.H. and Thompson, S.L. (1979) Ice ages and orbital variations: Some simple theory modeling. Quat. Res. 12, 188-203.
106. Berger, A. (1978) Long-term variations of daily insolation and Quaternary climatic changes. J. Atmos. Sci. 35, 2362-2367.
107. Budyko, M.I. (1982) The Earth's Climate: Past and Future. Academic Press, New York, NY.
108. Maranto, G. (1986) Are we close to the road's end? Discover 7(1), 28-50.
109. Mills, H.B. (1959) The importance of being nourished. Trans. Illinois State Acad. Sci. 52, 3-12.

REFERENCES

110. Westing, A.H. (1981) A note on how many humans that have ever lived. BioScience 31, 523-524.

111. Bogue, D.J. (1987) Population. In Encyclopedia Americana, Grolier, Inc., Danbury, CT, Vol. 22, pp.402-408.

112. Keepin, W., Mintzer, I. and Kristoferson, L. (1986) Emission of CO_2 into the atmosphere: The rate of release of CO_2 as a function of future energy developments. In Bolin, B., Doos, B.R., Jager, J. and Warrick, R.A. (eds) The Greenhouse Effect, Climatic Change, and Ecosystems. SCOPE 29. John Wiley & Sons, New York, NY, pp. 35-91.

113. Baes, C.F., Jr., Goeller, H.E., Olson, J.S. and Rotty, R.M. (1976) The Global Carbon Dioxide Problem. Oak Ridge Nat. Lab., Oak Ridge, TN.

114. Melillo, J.M., Palm, C.A., Houghton, R.A., Woodwell, G.M. and Myers, N. (1985) A comparison of two recent estimates of disturbance in tropical forests. Environ. Conserv. 12, 37-40.

115. Lugo, A.E. (1988) The future of the forest: Ecosystem rehabilitation in the tropics. Environment 30(7), 16-20, 41-45.

116. Stuiver, M. (1978) Atmospheric carbon dioxide and carbon reservoir changes. Science 199, 253-258.

117. Wilson, A.T. (1978) Pioneer agriculture explosion and CO_2 levels in the atmosphere. Nature 273, 40-41.

118. Woodwell, G.M. (1987) The carbon dioxide question. Sci. Amer. 238(1), 34-43.

119. Woodwell, G.M., Hobbie, J.E., Houghton, R.A., Melillo, J.M., Moore, B., Peterson, B.J. and Shaver, G.R. (1983) Global deforestation: Contribution to atmospheric carbon dioxide. Science 222, 1081-1086.

120. Trabalka, J.R., Ed. (1985) Atmospheric Carbon Dioxide and the Global Carbon Cycle. U.S. Dept. Energy, Washington, DC.

121. Neftel, A., Moor, E., Oeschger, H. and Stauffer, B. (1985) Evidence from polar ice cores for the increase in atmospheric CO_2 in the past two centuries. Nature 315, 45-47.

122. Friedli, H., Lotscher, H., Oeschger, H., Siegenthaler, U. and Stauffer, B. (1986) Ice core record of the $^{13}C/^{12}C$ ratio of atmospheric CO_2 in the past two centuries. Nature 324, 237-238.

123. Raynaud, D. and Barnola, J.M. (1985) An Antarctic ice core reveals atmospheric CO_2 variations over the past few centuries. Nature 315, 309-311.

124. Pearman, G.I., Etheridge, D., de Silva, F. and Fraser, P.J. (1986) Evidence of changing concentrations of atmospheric CO_2, N_2O and CH_4 from air bubbles in Antarctic ice. Nature 320, 248-250.

125. Schulze, F. (1871) Tagliche Beobachtungen uber den Kohlensauregehalt der Atmosphare zu Rostok vom 18 October 1868 bis 31 Juli 1871. Landwirtscha. Vers. Stn. 14, 366-388.

126. Reiset, J.A. (1882) Recherches sur la proportion de l'acide carbonique dans l'air. Ann. Chim. Phys. 26, 145-221.

127. Muntz, A. and Aubin, E. (1886) Mission Scientifique du Cap Horn, 1882-1883. Recherches sur la Constitution Chimique de l'Atmosphere. Tome III. Les Ministeres de la Marine et de l'Instruction publique, Paris, France.

128. Petermann, A. and Graftiau, J. (1892-1893) Recherches sur la composition de l'atmosphere. Premiere partie. Acide carbonique contenu dans l'air atmospherique. Bruxelles Mem. Couronn. 47, 1-79.

129. Letts, E.A. and Blake, R.F. (1900) The carbonic anhydride of the atmosphere. Roy. Dublin Soc. Rep. 9, 107-270.

130. Brown, H.T. and Escombe, F. (1905) On the variations in the amount of carbon dioxide in the air of Kew during the years 1893-1901. Proc. Roy. Soc. London, Biol. 76, 118-121.

131. Fraser, P.J., Elliott, W.P. and Waterman, L.S. (1983) Atmospheric CO_2 record from direct chemical measurements during the 19th century. In Trabalka, J.R. and Reichle, D.E., Eds. The Changing Carbon Cycle: A

Global Analysis. Springer-Verlag, New York, NY, pp. 66-88.

132. Newell, N.D. and Marcus, L. (1987) Carbon dioxide and people. Palaios 2, 101-103.

133. Gammon, R.H., Sundquist, E.T. and Fraser, P.J. (1985) History of carbon dioxide in the atmosphere. In: Trabalka, J.R. (ed) Atmospheric Carbon Dioxide and the Global Carbon Cycle. U.S. Dept. Energy, Washington, DC, pp. 25-62.

134. Foerster, H., Mora, P.M. and Amiot, L.W. (1960) Doomsday: Friday, 13 November, A.D. 2026. Science 132, 1291-1295.

135. Umpleby, S.A. (1987) World population: Still ahead of schedule. Science 237, 1555-1556.

136. Broecker, W.S. (1987) Unpleasant surprises in the greenhouse? Nature 328, 123-126.

137. Cure, J.D. and Acock, B. (1986) Crop response to carbon dioxide doubling: A literature survey. Agric. For. Meteorol. 38, 127-145.

138. Kimball, B.A. (1983) Carbon dioxide and agricultural yield: An assemblage and analysis of 430 prior observations. Agron. J. 75, 779-788.

139. Kimball, B.A. (1983) Carbon Dioxide and Agricultural Yield: An Assemblage and Analysis of 770 Prior Observations. U.S. Water Conserv. Lab., Phoenix, AZ.

140. Kimball, B.A. and Idso, S.B. (1983) Increasing atmospheric CO_2: Effects on crop yield, water use, and climate. Agric. Water Management 7, 55-73.

141. Wong, S.C. (1980) Effects of elevated partial pressure of CO_2 on rate of CO_2 assimilation and water use efficiency in plants. In Pearman, G.I. (ed) Carbon Dioxide and Climate: Australian Research. Aust. Acad. Sci., Canberra, Australia, pp. 159-166.

142. Valle, R., Mishoe, J.W., Jones, J.W. and Allen, L.H., Jr. (1985) Transpiration rate and water use efficiency of soybean leaves adapted to different CO_2 environments. Crop Sci. 25, 477-482.

143. Rogers, H.H., Thomas, J.F. and Bingham, G.E. (1983) Response of agronomic and forest species to elevated atmospheric carbon dioxide. Science 220, 428-429.

144. Idso, S.B. and Quinn, J.A. (1983) Vegetational Redistribution in Arizona and New Mexico in Response to a Doubling of the Atmospheric CO_2 Concentration. Lab. of Climatol., Ariz. State Univ., Tempe, AZ.

145. Idso, S.B. (1984) What if increases in atmospheric CO_2 have an inverse greenhouse effect? I. Energy balance considerations related to surface albedo. J. Climatol. 4, 399-409.

146. Idso, S.B. (1986) Industrial age leading to the greening of the Earth? Nature 320, 22.

147. Williams, J., Ed. (1978) Carbon Dioxide, Climate and Society. Pergamon press, Oxford, UK.

148. Kellogg, W. and Schware, R. (1981) Climate Change and Society. Westview Press, Boulder, CO.

149. Barth, M.C. and Titus, J.G., Eds. (1984) Greenhouse Effect and Sea Level Rise: A Challenage for this Generation. Van Nostrand Reinhold, New York, NY.

150. Meier, M.F., Aubrey, D.G., Bentley, C.R., Broecker, W.S., Hansen, J.E., Peltier, W.R. and Somerville, R.C.J. (1985) Glaciers, Ice Sheets, and Sea Level: Effect of a CO_2-Induced Climatic Change. U.S. Dept. Energy, Washington, DC.

151. Robin, G. deQ. (1986) Changing sea level: Projecting the rise in sea level caused by warming of the atmosphere. In Bolin, B., Doos, B.R., Jager, J. and Warrick, R.A. (eds) The Greenhouse Effect, Climatic Change, and Ecosystems. SCOPE 29. John Wiley & Sons, New York, NY, pp. 323-359.

152. Emanuel, K.A. (1987) The dependence of hurricane intensity on climate.

Nature 326, 483-485.

153. Anonymous (1988) Hurricane Gilbert matures in the greenhouse. New Sci. 119(1631), 22.

154. Joyce, C. (1988) Underwater methane could fuel warming. New Sci. 120(1643), 9.

155. Shine, K.P. (1988) Comment on "Southern Hemisphere temperature trends: A possible greenhouse effect?" Geophys. Res. Lett. 15, 843-844.

156. Raloff, J. (1988) Ozone hole of 1988: Weak and eccentric. Sci. News 134, 260.

157. Wigley, T.M.L., Jones, P.D. and Kelly, P.M. (1980) Scenario for a warm, high-CO_2 world. Nature 283, 17-21.

158. Kellogg, W.W. (1982) Precipitation trends on a warmer Earth. In Reck, R.A. and Hummel, J.R. (eds) AIP Conf. Proc. No. 82: Interpretation of Climate and Photochemical Models, Ozone and Temperature Measurements. Amer. Institute Physics, New York, NY, pp. 35-46.

159. Manabe, S. and Wetherald, R.T. (1986) Reduction in summer soil wetness induced by an increase in atmospheric carbon dioxide. Science 232, 626-628.

160. Miller, P.R. and O'Brian, M.J. (1952) Plant disease forecasting. Bot. Rev. 18, 547-601.

161. Gregory, P.H. (1961) The Microbiology of the Atmosphere. Interscience, New York, NY.

162. Wallin, J.R. (1967) Ground level climate in relation to forecasting plant diseases. In Shaw, R.H. (ed) Ground Level Climatology. Amer. Assoc. Adv. Sci., Washington, DC, pp. 149-163.

163. Wellington, W.G. (1957) The synoptic approach to studies of insects and climate. Ann. Rev. Entomol. 2, 143-162.

164. Rainey, R.C. (1973) Airborne pests and the atmospheric environment. Weather 28, 224-239.

165. Waggoner, P.E. (1984) Agriculture and carbon dioxide. Amer. Sci. 72, 179-184.

166. Decker. W.S., Jones, V. and Achutuni, R. (1985) The impact of CO_2-induced climate change on U.S. agriculture. In White, M.R. (ed) Characterization of Information Requirements for Studies of CO_2 Effects: Water Resources, Agriculture, Fisheries, Forests and Human Health. U.S. Dept. Energy, Washington, DC, pp. 69-93.

167. Warrick, R.A., Gifford, R.M. and Parry, M.L. (1986) CO_2, climatic change and agriculture. In Bolin, B., Doos, B.R., Jager, J. and Warrick, R.A. (eds) The Greenhouse Effect, Climatic Change, and Ecosystems. SCOPE 29. John Wiley & Sons, New York, NY, pp. 393-473.

168. Parry, M.L. et al., Eds. (1988) The Impact of Climatic Variations on Agriculture. Vol. 1. Assessment in Cool Temperate and Cold Regions. Reidel, Dordrecht, Holland.

169. Smit, B., Ludlow, L. and Brklacich, M. (1988) Implications of a global climatic warming for agriculture: A review and appraisal. J. Environ. Qual. 17, 519-527.

170. Wilks, D.S. (1988) Estimating the consequences of CO_2-induced climatic change on North American grain agriculture using general circulation model information. Clim. Change 13, 19-42.

171. Riebsame, W.E. (1988) Adjusting water resources management to climate change. Clim. Change 13, 69-97.

172. Cohen, S.J. and Allsopp, T.R. (1988) The potential impacts of a scenario of CO_2-induced climatic change on Ontario, Canada. J. Climate 1, 669-681.

173. Revelle, R.R. and Waggoner, P.E. (1983) Effects of a carbon dioxide-induced climatic change on water supplies in the western United States. In National Research Council, U.S. Changing Climate. National Academy Press, Washington, DC, pp. 419-432.

174. Callaway, J.M. and Currie, J.W. (1985) Water resource systems and

REFERENCES   SECTION 1

      chanages in climate and vegetation. In White, M.R. (ed)
      Characterization of Information Requirements for Studies of CO₂
      Effects: Water Resources, Agriculture, Fisheries, Forests and Human
      Health. U.S. Dept. Energy, Washington, DC, pp. 23-67.
175. Cohen, S.J. (1987) Projected increases in municipal water use in the
      Great Lakes due to CO_2-induced climate change. Water Res. Bull. 23,
      91-101.
176. Marchand, D., Sanderson, M., Howe, D. and Alpaugh, C. (1988) Climatic
      change and Great Lakes levels -- The impact on shipping. Clim. Change
      12, 107-133.
177. Sibley, T.H. and Strickland, R.M. (1985) Fisheries: Some relationships
      to climate change and marine environmental factors. In White, M.R.
      (ed) Characterization of Information Requirements for Studies of CO₂
      Effects: Water Resources, Agriculture, Fisheries, Forests and Human
      Health. U.S. Dept. Energy, Washington, DC, pp. 95-143.
178. Tangley, L. (1988) Preparing for climate change. BioScience 38(1), 14-
      18.
179. Meisner, J.D. (1988) Assessing impacts of climatic warming on freshwater
      fishes in North America. Climatol. Bull. 22, 48-49.
180. Hepting, G.H. (1963) Climate and forest diseases. Ann. Rev. Phytopath.
      1, 31-50.
181. Solomon, A.M. and West, D.C. (1985) Potential responses of forests to
      CO_2-induced climate change. In White, M.R. (ed) Characterization of
      Information Requirements for Studies of CO₂ Effects: Water Resources,
      Agriculture, Fisheries, Forests and Human Health. U.S. Dept. Energy,
      Washington, DC, pp. 145-169.
182. Shugart, H.H., Antonovsky, M.Ya., Jarvis, P.G. and Sandford, A.P. (1986)
      CO₂, climatic change and forest ecosystems. In Bolin, B., Doos, B.R.,
      Jager, J. and Warrick, R.A. (eds) The Greenhouse Effect, Climatic
      Change, and Ecosystems. SCOPE 29. John Wiley & Sons, New York, NY, pp.
      475-521.
183. Shands, W.E. and Hoffman, J.S., Eds. (1987) The Greenhouse Effect,
      Climate Change, and U.S. Forests. Conservation Foundation, Washington,
      DC.
184. Hamburg, S.P. and Cogbill, C.V. (1988) Historical decline of red spruce
      populations and climatic warming. Nature 331, 428-431.
185. Pastor, J. and Post, W.M. (1988) Response of northern forests to CO₂-
      induced climate change. Nature 334, 55-58.
186. Clark, J.S. (1988) Effect of climate change on fire regimes in
      northwestern Minnesota. Nature 334, 233-235.
187. Roberts, L. (1989) How fast can trees migrate? Science 243, 735-737.
188. White, M.R. and Hertz-Picciotto, I. (1985) Human health: Analysis of
      climate related to health. In White, M.R. (ed) Characterization of
      Information Requirements for Studies of CO₂ Effects: Water Resources,
      Agriculture, Fisheries, Forests and Human Health. U.S. Dept. Energy,
      Washington, DC, pp. 171-206.
189. McLean, D.M (1978) A terminal Mesozoic "greenhouse": Lessons from the
      past. Science 201, 401-406.
190. Peters, R.L. and Darling, J.D.S. (1985) The greenhouse effect and nature
      reserves. BioScience 35, 707-717.
191. McLean, D.M. (1986) Embryogenesis dysfunction in the Pleistocene/
      Holocene transition mammalian extinctions, dwarfing, and skeletal
      abnormality. In McDonald, J.N. and Bird, S.O. (eds) The Quaternary of
      Virginia -- A Symposium Volume. Virginia Div. of Mineral Resources,
      Pub. No. 75, pp. 105-120.
192. Joyce, C. (1988) Global warming could wipe out wildlife. New Sci.
      117(1598), 29.
193. Allen, R. (1980) The impact of CO₂ on world climate. Environ. 22(10),
      6-13, 37-38.

REFERENCES   SECTION

194. Seidel, S. and Keyes, D. (1983) Can We Delay a Greenhouse Warming? U.S
     Environ. Protection Agency, Washington, DC.
195. Bach, W. (1984) Our Threatened Climate. D. Reidel, Dordrecht, Holland.
196. Lemonic, M.D. (1987) The heat is on. Time 130(16), 58-67.
197. Schneider, S.H. (1988) The whole Earth dialogue. Issues Sci. Tech. 4(3)
     93-99.
198. Revkin, A.C. (1988) Endless summer: Living with the greenhouse effect
     Discover 9(10), 50-61.
199. Glantz, M.H., Ed. (1988) Societal Responses to Regional Climati‹
     Change: Forecasting by Analogy. Westview Press, Boulder, CO.
200. Revelle, R.R. and Suess, H.E. (1957) Carbon dioxide exchange betwee‹
     atmosphere and ocean and the question of an increase of atmospheri‹
     CO_2 during the past decades. Tellus 9, 18-27.
201. Idso, S.B. (1986) Environmental effects of atmospheric CO_2 enrichment
     Good news for the biosphere. CO_2/Clim. Dial. 1(2), 16-28.
202. Pain, S. (1988) No escape from the global greenhouse. New Sci
     120(1638), 38-43.
203. McGourty, C. (1988) Global warming becomes an international political
     issue. Nature 336, 194.
204. Koomanoff, F.A., Duzheng, Y., Jianping, Z., Riches, M.R., Wang, W.-C.
     and Shiyan, T. (1988) The United States' Department of Energy and the
     People's Republic of China's Chinese Academy of Sciences joint
     research on the greenhouse effect. Bull. Amer. Meteorol. Soc. 69,
     1301-1308.
205. Schneider, S.H. (1989) The greenhouse effect: Science and policy.
     Science 243, 771-781.
206. Abrahamson, D.E., Ed. (1989) The Challenge of Global Warming. Island
     Press, Covelo, CA.

REFERENCES   SECTION 2

1. Schneider, S.H. (1987) Climate modeling. Sci. Amer. 256(5), 72-80.
2. Rind, D., Rosenzweig, A. and Rosenzweig, C. (1988) Modelling the future:
   A joint venture. Nature 334, 483-486.
3. Barnett, T.P. and Schlesinger, M.E. (1987) Detecting changes in global
   climate induced by greenhouse gases. J. Geophys. Res. 92, 14,772-
   14,780.
4. Idso, S.B. (1980) The climatological significance of a doubling of
   earth's atmospheric carbon dioxide concentration. Science 207, 1462-
   1463.
5. Idso, S.B. (1984) An empirical evaluation of earth's surface air temperature response to radiative forcing, including feedback, as applied to
   the CO_2-climate problem. Arch. Meteorol. Geophys. Bioclim. Ser. B 34,
   1-19.
6. MacCracken, M.C. and Luther, F.M. (1985) Executive summary. In MacCracken, M.C. and Luther, F.M. (eds) Projecting the Climatic Effects of
   Increasing Carbon Dioxide, U.S. Dept. Energy, Washington, DC, pp.
   xvii-xxv.
7. Schneider, S.H. (1988) The whole Earth dialogue. Issues Sci. Tech. 4(3),
   93-99.
8. North, G.R., Cahalan, R.F. and Coakley, J.A. (1981) Energy balance climate models. Rev. Geophys. Space Phys. 19, 91-121.
9. Ramanathan, V. and Coakley, J.A. (1978) Climate modeling through radiative-convective models. Rev. Geophys. Space Phys. 16, 465-489.
10. Saltzman, B. (1978) A survey of statistical-dynamical models of the ter-
    restrial climate. Adv. Geophys. 20, 183-303.
11. Hansen, J.E., Russell, G., Rind, D., Stone, P., Lacis, A., Lebedeff,

EFERENCES

S., Ruedy, R. and Travis, L. (1983) Efficient three-dimensional global models for climate studies: Models I and II. Mon. Wea. Rev. 111, 609-662.

12. Simmons, A.J. and Bengtsson, L. (1984) Atmospheric general circulation models: Their design and use for climate studies. In Houghton, J.T. (ed) The Global Climate, Cambridge Univ. Press, Cambridge, MA, pp. 37-62.

13. Schneider, S.H. and Dickinson, R.E. (1974) Climate modeling. Rev. Geophys. Space Phys. 12, 447-493.

14. Schlesinger, M.E. (1983) Simulating CO_2-induced climatic change with mathematical climate models: Capabilities, limitations and prospects. Proceedings, Carbon Dioxide Research Conference: Carbon Dioxide, Science and Consensus. U.S. Dept. Energy, Washington, DC, pp. III3-III139.

15. Dickinson, R.E. (1986) The climate system and modeling of future climate. In Bolin, B., Doos, B.A., Jager, J. and Warrick, R.A. (eds) The Greenhouse Effect, Climate Change, and Ecosystems. John Wiley & Sons, New York, NY, pp. 206-270.

16. Harrington, J.B. (1987) Climatic change: A review of causes. Can. J. For. Res. 17, 1313-1339.

17. Schlesinger, M.E. (1984) Climate model simulations of CO_2-induced climatic change. Adv. Geophys. 26, 141-235.

18. Birchfield, G.E., Weertman, J. and Lunde, A.T. (1982) A model study of the role of high latitude topography in the climate response to orbital insolation anomalies. J. Atmos. Sci. 39, 71-87.

19. Barron, E.J., Thompson, S.L. and Hay, W.W. (1984) Continental distribution as a forcing factor for global-scale temperature. Nature 310, 574-575.

20. Crowley, T.J., Mengel, J.G. and Short, D.A. (1987) Gondwanaland's seasonal cycle. Nature 329, 803-807.

21. Barron, E.J. (1987) Explaining glacial periods. Nature 329, 764-765.

22. Sellers, W.D. (1969) A climate model based on the energy balance of the earth-atmosphere system. J. Appl. Meteorol. 8, 392-400.

23. Lian, M.S. and Cess, R.D. (1977) Energy balance climate models: A reappraisal of ice-albedo feedback. J. Atmos. Sci. 34, 1058-1062.

24. Wang, W.-C. and Stone, P.H. (1980) Effect of ice-albedo feedback on global sensitivity in a one-dimensional radiative-convective climate model. J. Atmos. Sci. 37, 545-552.

25. Hansen, J., Lacis, A., Rind, D., Russell, G., Stone, P., Fung, I., Ruedy, R. and Lerner, J. (1984) Climate sensitivity analysis of feedback mechanisms. In Hansen, J.E. and Takahashi, T. (eds) Climate Processes and Climate Sensitivity. Amer. Geophys. Union, Washington, DC., pp. 130-163.

26. Shukla, J. and Mintz, Y. (1982) Influence of land-surface evapotranspiration on the earth's climate. Science 215, 1498-1501.

27. Dickinson, R.E. (1983) Land surface processes and climate-surface albedos and energy balance. Adv. Geophys. 25, 305-353.

28. Henderson-Sellers, A. and Gornitz, V. (1984) Possible climatic impacts of land cover transformations, with particular emphasis on tropical deforestation. Clim. Change 6, 231-257.

29. Mintz, Y. (1984) The sensitivity of numerically-simulated climates to land-surface boundary conditions. In Houghton, J.T. (ed) The Global Climate. Cambridge Univ. Press, Cambridge, MA, pp. 79-106.

30. Kiehl, J. T. and Ramanathan, V. (1982) Radiative heating due to increased CO_2: The role of H_2O continuum absorption in the 12-18 μm region. J. Appl. Meteorol. 39, 2923-2926.

31. Idso, S.B. (1984) What if increases in atmospheric CO_2 have an inverse greenhouse effect? I. Energy balance considerations related to surface albedo. J. Climatol. 4, 399-409.

146

REFERENCES

32. Manabe, S. and Wetherald, R.T. (1975) The effects of doubling the CO₂ concentration on the climate of a general circulation model. J. Atmos Sci. 32, 3-15.

33. Washington, W.M. and Meehl, G.A. (1983) General circulation model experiments on the climatic effects due to a doubling and quadrupling of carbon dioxide concentration. J. Geophys. Res. 88, 6600-6610.

34. Namias, J. (1980) Some concomitant regional anomalies associated with hemispherically averaged temperature variations. J. Geophys. Res. 85 1585-1590.

35. Hughes, T. (1973) Is the West Antarctic ice sheet disintegrating? J Geophys. Res. 78, 7884-7910.

36. Weertman, J. (1974) Stability of the junction of an ice sheet and an ice shelf. J. Glaciol. 13, 3-11.

37. Mercer, J.H. (1978) West Antarctic ice sheet and CO_2 greenhouse effect A threat of disaster. Nature 271, 321-325.

38. Saari, M.R., Yuen, D.A. and Schubert, G. (1987) Climatic warming and basal melting of large ice sheets: Possible implications for Eas Antarctica. Geophys. Res. Lett. 14, 33-36.

39. Van Der Veen, C.J. (1987) Ice Sheets and the CO_2 problem. Surv. Geophys 9, 1-42.

40. Kopec, R.J. (1971) Global climate change and the impact of a maximum sea level on coastal settlement. J. Geogr. 70, 541-550.

41. Hoffman, J. (1983) Projecting Future Sea Level: Methodology, Estimates to the Year 2100, and Research Needs. U.S. Environmental Protection Agency, Washington, DC.

42. Seidel, S. and Keyes, D. (1983) Can We Delay a Greenhouse Warming? U.S. Environmental Protection Agency, Washington, DC.

43. Giese, G.S. (1987) Losing coastal upland to relative sea-level rise: 3 Scenarios for Massachusetts. Oceanus 30(3), 17-22.

44. Barnett, T.P. (1982) Recent changes in sea level and their possible causes. Clim. Change 5, 15-38.

45. Gornitz, V., Lebedeff, S. and Hansen, J. (1982) Global sea level trend in the past century. Science 215, 1611-1614.

46. Revelle, R.R. (1983) Probable future changes in sea level resulting from increased atmospheric carbon dioxide. In National Research Council, U.S. (eds) Changing Climate. National Academy Press, Washington, DC, pp. 433-448.

47. Wigley, T.M.L. and Raper, S.C.B. (1987) Thermal expansion of sea water associated with global warming. Nature 330, 127-131.

48. Rind, D. (1988) The doubled CO_2 climate and the sensitivity of the modeled hydrologic cycle. J. Geophys. Res. 93, 5385-5412.

49. Schlesinger, M.E. and Mitchell, J.F.B. (1985) Model projections of the equilibrium climatic response to increased carbon dioxide. In MacCracken, M.C. and Luther, F.M. (eds) Projecting the Climatic Effects of Increasing Carbon Dioxide. U.S. Dept. Energy, Washington, DC, pp. 81-147.

50. Manabe, S., Wetherald, R.T. and Stouffer, R.J. (1981) Summer dryness due to an increase of atmospheric CO_2 concentration. Clim. Change 3, 347-386.

51. Mitchell, J.F.B. and Lupton, G. (1984) A 4 x CO_2 integration with prescribed changes in sea surface temperatures. Prog. Biometeorol. 3, 353-374.

52. Manabe, S. and Wetherald, R.T. (1987) Large-scale changes of soil wetness induced by an increase in atmospheric carbon dioxide. J. Atmos. Sci. 44, 1211-1235.

53. Wilson, C.A. and Mitchell, J.F.B. (1987) A doubled CO_2 climate sensitivity experiment with a global climate model including a simple ocean. J. Geophys. Res. 92, 13,315-13,343.

54. Kellogg, W.W. and Zhao, Z.-c. (1988) Sensitivity of soil moisture to

doubling of carbon dioxide in climate model experiments. Part I: North America. J. Climate 1, 348-366.

5. Zhao, Z.-c. and Kellogg, W.W. (1988) Sensitivity of soil moisture to doubling of carbon dioxide in climate model experiments. Part II: The Asian monsoon region. J. Climate 1, 367-378.

6. Manabe, S. and Wetherald, R.T. (1986) Reduction in summer soil wetness induced by an increase in atmospheric carbon dioxide. Science 232, 626-628.

7. Gleick, P.H. (1987) The development and testing of a water balance model for climate impact assessment: Modeling the Sacramento Basin. Water Resources Res. 23, 1049-1061.

8. Gleick, P.H. (1987) Regional hydrologic consequences of increases in atmospheric CO_2 and other trace gases. Clim. Change 10, 137-161.

9. Bultot, F., Dupriez, G.L. and Gellens, D. (1988) Estimated annual regime of energy-balance components, evapotranspiration and soil moisture for a drainage basin in the case of a CO_2 doubling. Clim. Change 12, 39-56.

0. Wigley, T.M.L., Briffa, K.R. and Jones, P.D. (1984) Predicting plant productivity and water resources. Nature 312, 102-103.

1. Wigley, T.M.L. and Jones, P.D. (1985) Influence of precipitation changes and direct CO_2 effects on streamflow. Nature 314, 149-152.

2. Revelle, R.R. and Waggoner, P.E. (1983) Effects of a carbon dioxide-induced climatic change on water supplies in the western United States. In National Research Council, U.S. (eds) Changing Climate. National Academy Press, Washington, DC, pp. 419-432.

3. Kimball, B.A. and Idso, S.B. (1983) Increasing atmospheric CO_2: Effects on crop yield, water use and climate. Agric. Water Manage. 7, 55-73.

4. Aston, A.R. (1984) The effect of doubling atmospheric CO_2 on streamflow: A simulation. J. Hydrol. 67, 273-280.

5. Idso, S.B. and Brazel, A.J. (1984) Rising atmospheric carbon dioxide concentrations may increase streamflow. Nature 312, 51-53.

6. Washington, W.M. and Meehl, G.A. (1984) Seasonal cycle experiment on the climate sensitivity due to a doubling of CO_2 with an atmospheric general circulation model coupled to a simple mixed-layer ocean model. J. Geophys. Res. 89, 9475-9503.

7. Rind, D. and Lebedeff, S. (1984) Potential Climatic Impacts of Increasing Atmospheric CO_2 with Emphasis on Water Availability and Hydrology in the United States. Environmental Protection Agency, Washington, DC.

8. Meehl, G.A. and Washington, W.M. (1985) Tropical response to increased CO_2 in a GCM with a simple mixed layer ocean: Similarities to an observed Pacific warm event. Mon. Wea. Rev. 114, 667-674.

9. MacCracken, M.C., Schlesinger, M.E. and Riches, M.R. (1986) Atmospheric carbon dioxide and summer soil wetness. Science 234, 659-660.

0. Mitchell, J.F.B. and Warrilow, D.A. (1987) Summer dryness in northern mid-latitudes due to increased CO_2. Nature 330, 238-240.

1. Revelle, R.R. and Waggoner, P.E., Eds. (1989) Climatic Variability, Climate Change, and U.S. Water Resources. John Wiley & Sons, New York, NY.

72. Revkin, A.C. (1988) Endless summer: Living with the greenhouse effect. Discover 9(10), 50-61.

73. Dracup, J.A. and Kendall, D.R. (1989) Climatic change and its impact on floods and droughts. In Revelle, R.R. and Waggoner, P.E. (eds) Climatic Variability, Climate Change, and U.S. Water Resources. John Wiley & Sons, New York, NY, in press.

74. Ramanathan, V., Cicerone, R.J., Singh, H.B. and Kiehl, J.T. (1985) Trace gas trends and their potential role in climate change. J. Geophys. Res. 90, 5547-5566.

75. Wang, W-C. and Molnar, G. (1985) A model study of greenhouse effects due

to increasing atmospheric CH_4, N_2O, CF_2Cl_2, and $CFCl_3$. J. Geophys Res. 90, 12,971-12,980.

76. Ramanathan, V., Callis, L., Cess, R., Hansen, J., Isaksen, I., Kuhn, W. Lacis, A., Luther, F., Mahlman, J., Reck, R. and Schlesinger, M (1987) Climate-chemical interactions and effects of changin atmospheric trace gases. Rev. Geophys. 25, 1441-1482.

77. Rogers, J.D. and Stephens, R.D. (1988) Absolute infrared intensitie for F-113 and F-114 and an assessment of their greenhouse warmin potential relative to other chloroflourocarbons. J. Geophys. Res.93 2423-2428.

78. Khalil, M.A.K. and Rasmussen, R.A. (1983) Increase and seasonal cycle in the atmospheric concentration of nitrous oxide (N_2O) Tellus 35B 161-169.

79. Rasmussen, R.A. and Khalil, M.A.K. (1986) Atmospheric trace gases Trends and distributions over the last decade. Science 232, 1623-1624

80. Prinn, R., Cunnold, D., Rasmussen, R., Simmonds, P., Alyea, F. Crawford, A., Fraser, P. and Rosen, R. (1987) Atmospheric trends i methylchloroform and the global average for the hydroxyl radical Science 238, 945-950.

81. Blake, D.R. and Rowland, F.S. (1988) Continuing worldwide increase i tropospheric methane, 1978-1987. Science 239, 1129-1131.

82. Craig, H. and Chou, C.C. (1982) Methane: The record in polar ice cores Geophys. Res. Lett. 9, 1221-1224.

83. Rasmussen, R.A. and Khalil, M.A.K. (1984) Atmospheric methane in th recent and ancient atmospheres: Concentrations, trends, an interhemispheric gradient. J. Geophys. Res. 89, 11,599-11,605.

84. Stauffer, B., Fischer, G., Neftel, A. and Oeschger, H. (1985) Increas of atmospheric methane recorded in Antarctic ice core. Science 229 1386-1388.

85. Pearman, G.I. and Fraser, P.J. (1988) Sources of increased methane Nature 332, 489-490.

86. Stauffer, B., Lochbronner, E., Oeschger, H. and Schwander, J. (1988) Methane concentration in the glacial atmosphere was only half that o the preindustrial Holocene. Nature 332, 812-814.

87. Raynaud, D., Chappellaz, J., Barnola, J.M., Korotkevich, Y.S. an Lorius, C. (1988) Climatic and CH_4 cycle implications of glacial-interglacial CH_4 change in the Vostok ice core. Nature 333, 655-657.

88. Craig, H., Chou, C.C., Welhan, J.A., Stevens, C.M. and Engelkemeir, A. (1988) The isotopic composition of methane in polar ice cores. Science 242, 1535-1539.

89. Kerr, R.A. (1983) Trace gases could double climate warming. Science 220 1364-1365.

90. Dickinson, R.E. and Cicerone, R.J. (1986) Future global warming from atmospheric trace gases. Nature 319, 109-115.

91. Bruhl, C. and Crutzen, P.J. (1988) Scenarios of possible changes in atmospheric temperatures and ozone concentrations due to man's activities, estimated with a one-dimensional coupled photochemical climate model. Clim. Dynam. 2, 173-203.

92. Vupputuri, R.K.R. (1988) Potential effects of anthropogenic trace gas emissions on atmospheric ozone and temperature structure and surface climate. Atmos. Environ. 22, 2809-2818.

93. Madden, R.A. and Ramanathan, V. (1980) Detecting climatic change due to increasing carbon dioxide. Science 209, 763-768.

94. Ellsaesser, H.W. (1985) Do the recorded data of the past century indicate a CO_2 warming? Proceedings: Third Conference on Climate Variations and Symposium on Contemporary Climate: 1850-2100. Amer. Meteorol. Soc., Boston, MA, pp. 87-88.

95. Parker, D.E. (1985) On the detection of temperature changes induced by increasing atmospheric carbon dioxide. Quart. J. Roy. Meteorol. Soc.

111, 587-601.

96. Barnett, T.P. (1986) Detection of changes in the global troposphere temperature field induced by greenhouse gases. J. Geophys. Res. 91, 6659-6667.

97. Solow, A.R. (1987) Testing for climate change: An application of the two-phase regression model. J. Clim. Appl. Meteorol. 26, 1401-1405.

98. Hansen, J. and Lebedeff, S. (1988) Global surface air temperatures: Update through 1987. Geophys. Res. Lett. 15, 323-326.

99. Monastersky, R. (1988) Has the greenhouse taken effect? Sci. News 133, 282.

00. Jones, P.D., Raper, S.C.B., Bradley, R.S., Diaz, H.F., Kelly, P.M. and Wigley, T.M.L. (1986) Northern Hemispheric surface temperature variations: 1851-1984. J. Clim. Appl. Meteorol. 25, 161-179.

01. Jones, P.D., Raper, S.C.B. and Wigley, T.M.L. (1986) Southern Hemispheric surface air temperature variations: 1851-1984. J. Clim. Appl. Meteorol. 25, 1213-1230.

02. Jones, P.D., Wigley, T.M.L. and Wright, P.B. (1986) Global temperature variations between 1861 and 1984. Nature 322, 430-434.

03. Ellsaesser, H.W., MacCracken, M.C., Walton, J.J. and Grotch, S.L. (1986) Global climatic trends as revealed by the recorded data. Rev. Geophys. 24, 745-792.

04. Hansen, J. and Lebedeff, S. (1987) Global trends of measured surface air temperature. J. Geophys. Res. 25, 13,345-13,372.

05. Labitzke, K., Brasseur, G., Naujokat, B. and De Rudder, A. (1986) Long-term temperature trends in the stratosphere: Possible influence of anthropogenic gases. Geophys. Res. Lett. 13, 52-55.

06. Karoly, D.J. (1987) Southern Hemisphere temperature trends: A possible greenhouse gas effect? Geophys. Res. Lett. 14, 1139-1141.

07. Kerr, R.A. (1988) Is the greenhouse here? Science 239, 559-561.

08. Jones, P.D., Wigley, T.M.L., Folland, C.K., Parker, D.E., Angell, J.K., Lebedeff, S. and Hansen, J.E. (1988) Evidence for global warming in the past decade. Nature 332, 790.

09. Hansen, J., Fung, I., Lacis, A., Rind, D., Lebedeff, S., Ruedy, R., Russell, G. and Stone, P. (1988) Global climate changes as forecast by Goddard Institute for Space Studies three-dimensional model. J. Geophys. Res., 93, 9341-9364.

110. Bradley, R.S., Diaz, H.F., Eischeid, J.K., Jones, P.D., Kelly, P.M. and Goodess, C.M. (1987) Precipitation fluctuations over Northern Hemisphere land areas since the mid-19th century. Science 237, 171-175.

111. Danard, M.B. and Murty, T.S. (1988) Some recent trends in precipitation in western Canada and their possible link to rising carbon dioxide. Atmos.-Ocean 26, 139-145.

112. Schneider, S.H. and Thompson, S.L. (1981) Atmospheric CO_2 and climate: Importance of the transient response. J. Geophys. Res. 86, 3135- 3147.

113. North, G.R., Mengle, J.G. and Short, D.A. (1984) On the transient response patterns of climate to time dependent concentrations of atmospheric CO_2. In Hansen, J.E. and Takahashi, T. (eds) Geophys. Monogr. 29: Climate Processes and Climate Sensitivity. Amer. Geophys. Union, Washington, DC, pp. 164-170.

114. Hoffert, M.I. and Flannery, B.P. (1985) Model projections of the time-dependent response to increasing carbon dioxide. In MacCracken, M.C. and Luther, F.M. (eds) Projecting the Climatic Effects of Increasing Carbon Dioxide. U.S. Dept. Energy, Washington, DC, pp. 149-190.

115. Cess, R.D. and Goldenberg, S.D. (1981) The effect of ocean heat capacity upon global warming due to increasing carbon dioxide. J. Geophys. Res. 86, 498-602.

116. Bryan, K., Komro, F.G., Manabe, S. and Spelman, M.J. (1982) Transient climate response to increasing atmospheric carbon dioxide. Science

215, 56-58.

117. Wigley, T.M.L. and Schlesinger, M.E. (1985) Analytical solution for the effect of increasing CO_2 on global mean temperature. Nature 315, 649-652.

118. Ramanathan, V. (1988) The greenhouse theory of climate change: A test by an inadvertent global experiment. Science 240, 293-299.

119. Schlesinger, M.E. and Jiang, X. (1988) The transport of CO_2-induced warming into the ocean: An analysis of simulations by the OSÚ coupled atmosphere-ocean general circulation model. Clim. Dynam. 3, 1-17.

120. Bryan, K., Manabe, S. and Spelman, M.J. (1988) Interhemispheric asymmetry in the transient response of a coupled ocean-atmosphere model to a CO_2 forcing. J. Phys. Oceanogr. 18, 851-867.

121. Lebedeff, S.A. (1988) Analytic solution of the box diffusion model for a global ocean. J. Geophys. Res. 93, 14,243-14,255.

122. Wigley, T.M.L. (1987) The effect of model structure on projections of greenhouse-gas-induced climatic change. Geophys. Res. Lett. 14, 1135-1138.

123. Wigley, T.M.L. and Jones, P.D. (1981) Detecting CO_2-induced climatic change. Nature 292, 205-208.

124. Bell, T.L. (1982) Optimal weighting of data to detect climatic change: Application to the carbon dioxide problem. J. Geophys. Res. 87, 11,161-11,170.

125. Farman, J.C., Gardiner, B.G. and Shanklin, J.D. (1985) Large losses of total ozone in the Antarctica reveal seasonal ClO_x/NO_x interaction. Nature 315, 207-210.

126. Rodgers, C. (1988) Global ozone trends reassessed. Nature 332, 201.

127. Urbach, F. (1969) The Biologic Effects of Ultraviolet Radiation. Pergamon Press, New York, NY.

128. Newton, J.W., Tyler, D.D. and Slodki, M.E. (1979) Effect of ultraviolet-B (280-320 nm) radiation on blue-green algae (cyanobacteria), possible indicators of stratospheric ozone depletion. Appl. Environ. Microbiol. 37, 1137-1141.

129. Tevini, M. and Iwanzik, W. (1983) Inhibition of photosynthetic activity by UV-B radiation in radish seedlings. Physiol. Plant. 58, 395-400.

130. Worrest, R.C. (1983) Impact of solar ultraviolet-B radiation (290-320 nm) upon marine microalgae. Physiol. Plant. 58, 428-434.

131. Wellmann, E., Schneider-Ziebert, U. and Beggs, C.J. (1984) UV-B inhibition of phytochrome-mediated anthocyanin formation in Sinapis alba L. cotyledons. Plant Physiol. 75, 997-1000.

132. Hader, D.P., Watanabe, M. and Furuya, M. (1986) Inhibition of motility in the cyanobacterium, Phormidium unicatum, by solar and monochromatic UV radiation. Plant Cell Physiol. 27, 887-894.

133. National Research Council, U.S. (1984) Causes and Effects of Changes in Stratospheric Ozone: An Update. National Academy Press, Washington, DC.

134. Urbach, F. (1984) Ultraviolet radiation and skin cancer. In Smith, K.C. (ed) Topics in Photomedicine. Plenum Press, New York, NY, pp. 39-142.

135. Beardsley, T. (1986) Deaths foretold. Nature 324, 102.

136. Doughty, M.J. and Cullen, A.P. (1989) Long-term effects of a single dose of ultraviolet-B on albino rabbit cornea -- I. in vivo analyses. Photochem. Photobiol. 49, 185-196.

137. Shine, K.P. (1988) Comment on "Southern Hemisphere temperature trends: A possible greenhouse effect?" Geophys. Res. Lett. 15, 843-844.

138. Singer, S.F. (1988) Does the Antarctic ozone hole have a future? Trans. Amer. Geophys. Union 69, 1588.

139. Solomon, S., Garcia, R.R., Rowland, F.S. and Wuebbles, D.J. (1986) On the depletion of Antarctic ozone. Nature 321, 755-758.

140. Levi, B.G. (1988) Ozone depletion at the poles: The hole story emerges. Phys. Today 41(7), 17-21.

141. Hofmann, D.J. (1989) Direct ozone depletion in springtime Antarctic lower stratospheric clouds. Nature 337, 447-449.
142. Molina, M.J., Tso, T.-L., Molina, L.T. and Wang, F. C.-Y. (1987) Antarctic stratospheric chemistry of chlorine nitrate, hydrogen chloride, and ice: Release of active chlorine. Science 238, 1253-1257.
143. Hanson, D. and Mauersberger, K. (1988) Solubility and equilibrium vapor pressures of HCl dissolved in polar stratospheric cloud materials: Ice and the trihydrate of nitric acid. Geophys. Res. Lett. 15, 1507-1510.
144. Rowland, F.S. (1989) Chlorofluorocarbons and the depletion of stratospheric ozone. Amer. Sci. 77, 36-45.
145. Kawahira, K. and Hirooka, T. (1989) Interannual temperature changes in the Antarctic lower stratosphere -- A relation to the ozone hole. Geophys. Res. Lett. 16, 41-44.

1. Kellogg, W.W. and Mead, M., Eds. (1977) Fogarty International Center Proceedings No. 39: The Atmosphere Endangered and Endangering. National Institutes of Health, Washington, DC.
2. Marchetti, C. (1977) On geoengineering and the CO_2 problem. Clim. Change 1, 59-68.
3. Seidel, S. and Keyes, D. (1983) Can We Delay a Greenhouse Warming? U.S. Environmental Protection Agency, Washington, DC.
4. Advisory Group on Greenhouse Gases (1988) World climate programme. WMO Bull. 37, 111-112.
5. Idso, S.B. (1984) Through a glass darkly. New Sci. 102, 34.
6. Reck, R.A. (1982) Confidence in climate models including those with suspended particles. In Reck, R.A. and Hummel, J.R. (eds) AIP Conf. Proc. No. 82: Interpretation of Climate and Photochemical Models, Ozone and Temperature Measurements. American Institute of Physics, New York, NY, pp. 23-33.
7. Choudhury, B. and Kukla, G. (1979) Impact of CO_2 on cooling of snow and water surfaces. Nature 280, 668-676.
8. Hansen, J., Johnson, D., Lacis, A., Lebedeff, S., Lee, P., Rind, D. and Russell, G. (1981) Climate impact of increasing atmospheric carbon dioxide. Science 213, 957-966.
9. Thiele, O. and Schiffer, R.A., Eds. (1985) Understanding Climate: A Strategy for Climate Modeling and Predictability Research. NASA, Washington, DC.
10. Kellogg, W.W. (1980) Modeling future climate. Ambio 9, 216-221.
11. Newell, R.E. and Dopplick, R.G. (1979) Questions concerning the possible influence of anthropogenic CO_2 on atmospheric temperature. J. Appl. Meteorol. 18, 822-825.
12. Idso, S.B. (1980) The climatological significance of a doubling of earth's atmospheric carbon dioxide concentration. Science 207, 1462-1463.
13. Idso, S.B. (1981) Carbon dioxide -- An alternative view. New Sci. 92, 444-446.
14. Idso, S.B. (1982) Temperature limitation by evaporation in hot climates and the greenhouse effects of water vapor and carbon dioxide. Agric. Meteorol. 27, 105-109.
15. Idso, S.B. (1983) Carbon dioxide and global temperature: What the data show. J. Environ. Qual. 12, 159-163.
16. Idso, S.B. (1984) An empirical evaluation of Earth's surface air temperature response to radiative forcing, including feedback, as applied to the CO_2-climate problem. Archiv. Meteorol. Geophys. Bioclim. Ser. B 34, 1-19.

17. Ellsaesser, H.W. (1984) The climatic effect of CO_2: A different view. Atmos. Environ. 18, 431-434.

18. Essex, C. (1986) Trace gases and the problem of false invariants in climate models. Climatol. Bull. 20, 19-25.

19. Schneider, S.H., Kellogg, W.W., Ramanathan, V.; Leovy, C.B.; Idso, S.B. (1980) Carbon dioxide and climate. Science 210, 6-8.

20. Watts, R.G. (1980) Discussion of "Questions concerniung the possible influence of anthropogenic CO_2 on atmospheric temperature." J. Appl. Meteorol. 19, 494-495.

21. Newell, R.E. and Dopplick, T.G. (1981) Reply to Robert G. Watts' "Discussion of 'Questions concerning the possible influence of anthropogenic CO_2 on atmospheric temperature.'" J. Appl. Meteorol. 20, 114-117.

22. National Research Council, U.S. (1982) Carbon Dioxide and Climate: A Second Assessment. National Academy Press, Washington, DC.

23. Idso, S.B. (1982) Carbon Dioxide: Friend or Foe? IBR Press, Tempe, AZ.

24. Smagorinsky, J. (1983) Effects on climate. In National Research Council, U.S. (eds) Changing Climate. National Academy Press, Washington, DC, pp. 266-284.

25. Schneider, S.H. (1984) "Natural experiments" and CO_2-induced climate change: The controversy drags on -- An editorial. Clim. Change 6, 317-321.

26. Cess, R.D. and Potter, G.L. (1984) A commentary on the recent CO_2-climate controversy. Clim. Change 6, 365-376.

27. Webster, P.J. (1984) The carbon dioxide/climate controversy: Some personal comments on two recent publications. Clim. Change 6, 377-390.

28. Kiehl, J.T. (1984) Reply to Dr. Idso, Reply to Dr. Gribbin. Clim. Change 6, 395-396.

29. Idso, S.B. (1986) My response to the concluding 1984 issue of Climatic Change dealing with the CO_2/climate controversy. CO_2/Clim. Dial. 1(1), 7-66. Available from the Laboratory of Climatology, Arizona State University, Tempe, AZ 85287.

30. Idso, S.B. (1987) A clarification of my position on the CO_2/climate connection. Clim. Change 10, 81-86.

31. Luther, F.M. and Cess, R.D. (1985) Appendix B: Review of the recent carbon dioxide-climate controversy. In MacCracken, M.C. and Luther, F.M. (eds) Projecting the Climatic Effects of Increasing Carbon Dioxide. U.S. Dept. Energy, Washington, DC, pp. 321-335.

32. Idso, S.B. (1986) My response to Appendix B -- "Review of the recent carbon dioxide-climate controversy" -- of the DOE-sponsored state-of-the-art volume: The potential climatic effects of increasing carbon dioxide. CO_2/Clim. Dial. 1(2), 29-37. Available from the Laboratory of Climatology, Arizona State University, Tempe, AZ 85287.

33. Potter, G.L., Kiehl, J.T. and Cess, R.D. (1987) A clarification of certain issues related to the CO_2-climate problem. Clim. Change 10, 87-95.

34. Idso, S.B. (1987) Carbon dioxide and climate: The legacy of logic. CO_2/Clim. Dial. 2(1), 9-17. Available from the Laboratory of Climatology, Arizona State University, Tempe, AZ 85287.

35. Idso, S.B. (1988) Me and the modelers: Perhaps not so different after all. Clim. Change 12, 93.

36. Schlesinger, M.E. and Mitchell, J.F.B. (1987) Climate model simulations of the equilibrium climatic response to increased carbon dioxide. Rev. Geophys. 25, 760-798.

37. Williamson, D.L. (1988) The effect of vertical finite difference approximations on simulations with the NCAR community climate model. J. Climate 1, 40-58.

38. Grotch, S.L. (1988) Regional Intercomparisons of General Circulation Model Predictions and Historical Climate Data. U.S. Dept. Energy,

REFERENCES

Washington, DC.
39. Straus, D.M. and Shukla, J. (1988) A comparison of a GCM simulation of the seasonal cycle of the atmosphere with observations. Part I: Mean fields and the annual harmonic. Atmos.-Ocean 26, 541-574.
40. Straus, D.M. and Shukla, J. (1988) A comparison of a GCM simulation of the seasonal cycle of the atmosphere with observations. Part II: Stationary waves and transient fluctuations. Atmos.-Ocean 26, 575-607.
41. Weickmann, K.M. and Chervin, R.M. (1988) The observed and simulated atmospheric seasonal cycle. Part I: Global wind field modes. J. Climate 1, 265-289.
42. Gutowski, W.J., Gutzler, D.S., Portman, D. and Wang, W.-C. (1988) Surface Energy Balance of Three General Circulation Models: Current Climate and Response to Increasing Atmospheric CO_2. U.S. Dept. Energy, Washington, DC.
43. Weare, B. (1988) A comparison of radiation variables calculated in the UCLA general circulation model to observations. J. Climate 1, 485-499.
44. Geleyn, J.F., Hensen, A. and Preuss, H.J. (1982) A comparison of model generated radiation fields with satellite measurements. Contrib. Atmos. Phys. 55, 253-286.
45. Charlock, T.P. and Ramanathan, V. (1985) The albedo field and cloud radiative forcing produced by a general circulation model with internally generated cloud optics. J. Atmos. Sci. 42, 1408-1429.
46. Paltridge, G.W. (1987) Comparison of NIMBUS 7 wide-angle earth radiation budget measurements with the output of four cloud-generating numerical climate models. J. Geophys. Res. 92, 4097-4106.
47. Ramanathan, V. (1987) The role of earth radiation budget studies in climate and general circulation modeling. J. Geophys. Res. 92, 4075-4095.
48. Neeman, B.U., Ohring, G. and Joseph, J.H. (1988) The Milankovitch theory and climate sensitivity. 1. Equilibrium climate model solutions for the present surface conditions. J. Geophys. Res. 93, 11,153-11,174.
49. Boer, G.J. and Lazare, M. (1988) Some results concerning the effect of horizontal resolution and gravity-wave drag on simulated climate. J. Climate 1, 789-806.
50. Manabe, S., Hahn, D.G. and Holloway, J.L. (1979) Climate simulations with GFDL spectral models of the atmosphere. Report of the JOC study conference on climate models: Performance, intercomparison and sensitivity studies. GARP Pub. Ser. No. 22 1, 41-94.
51. Dyson, J.F. (1985) The effect of resolution and diffusion on the simulated climate. Research Activities in Atmospheric and Oceanic Modelling, Report No. 8. World Meteorol. Org., Geneva, Switzerland.
52. Palmer, T.N., Shutts, G.L. and Swinbank, R. (1986) Alleviation of a systematic westerly bias in general circulation and numerical weather prediction models through an orographic gravity wave drag parameterization. Quart. J. Roy. Meteorol. Soc. 112, 1001-1039.
53. Cess, R.D. and Potter, G.L. (1988) A methodology for understanding and intercomparing atmospheric climate feedback processes in general circulation models. J. Geophys. Res. 93, 8305-8314.
54. Bryan, K., Manabe, S. and Spelman, M.J. (1988) Interhemispheric asymmetry in the transient response to a coupled ocean-atmosphere model to a CO_2 forcing. J. Phys. Oceanogr. 18, 851-867.
55. Hansen, J., Fung, I., Lacis, A., Rind, D., Lebedeff, S., Ruedy, R., Russell, G. and Stone. P. (1988) Global climate changes as forecast by Goddard Institute for Space Studies three-dimensional model. J. Geophys. Res. 93, 9341-9364.
56. Ramanathan, V. (1988) The greenhouse theory of climate change: A test by an inadvertent global experiment. Science 240, 293-299.
57. Covey, C. (1988) Atmospheric and oceanic heat transport: Simulations

versus observations. <u>Clim</u>. <u>Change</u> 13, 149-159.

58. Wang, W.-C., Molnar, G. and Mitchell, T.P. (1984) Effects of dynamical heat fluxes on model climate sensitivity. <u>J</u>. <u>Geophys</u>. <u>Res</u>. 89, 4699-4711.

59. Spelman, M.J. and Manabe, S. (1984) Influence of oceanic heat transport upon the sensitivity of a model climate. <u>J</u>. <u>Geophys</u>. <u>Res</u>. 89, 571-586.

60. Henderson-Sellers, B. (1987) Modelling sea surface temperature rise resulting from increasing atmospheric carbon dioxide concentrations. <u>Clim</u>. <u>Change</u> 11, 349-359.

61. Rosati, A. and Miyakoda, K. (1988) A general circulation model for upper ocean simulation. <u>J</u>. <u>Phys</u>. <u>Oceanog</u>. 18, 1601-1626.

62. Watts, R.G. (1985) Global climate variation due to fluctuations in the rate of deep water formation. <u>J</u>. <u>Geophys</u>. <u>Res</u>. 90, 8067-8070.

63. Boyle, E.A. and Keigwin, L.D. (1984) Deep circulation of the North Atlantic over the last 200,000 years: Geochemical evidence. <u>Science</u> 218, 784-787.

64. Roemmich, D. and Wunsch, C. (1984) Apparent changes in the climatic state of the deep North Atlantic Ocean. <u>Nature</u> 307, 447-450.

65. Duplessy, J.-C. and Shackleton, N.J. (1985) Response of global deep-water circulation to earth's climatic change 135,000-109,000 years ago. <u>Nature</u> 316, 500-507.

66. Labeyrie, L.D., Duplessy, J.C. and Blanc, P.L. (1987) Variations in mode of formation and temperature of oceanic deep waters over the past 125,000 years. <u>Nature</u> 327, 477-482.

67. Whitehead, J.A. (1989) Giant ocean cataracts. <u>Sci</u>. <u>Amer</u>. 260(2), 50-57.

68. Kerr, R.A. (1989) Did the ocean once run backwards. <u>Science</u> 243, 740.

69. Shaffer, G. (1989) A model of biogeochemical cycling of phosphorus, nitrogen, oxygen, and sulphur in the ocean: One step toward a global climate model. <u>J</u>. <u>Geophys</u>. <u>Res</u>. 94, 1979-2004.

70. Kepkay, P.E. and Johnson, B.D. (1989) Coagulation on bubbles allows microbial respiration of oceanic dissolved organic carbon. <u>Nature</u> 338, 63-65.

71. Hibler, W.D., III. (1984) The role of sea ice dynamics in modeling CO_2 increases. <u>In</u> Hansen, J.E. and Takahashi, T. (eds) <u>Geophys</u>. <u>Monogr</u>. <u>29</u>: <u>Climate</u> <u>Processes</u> <u>and</u> <u>Climate</u> <u>Sensitivity</u>. Amer. Geophys. Union, Washington, DC, pp. 238-253.

72. Ledley, T.A. (1988) A coupled energy balance climate-sea ice model: Impact of sea ice and leads on climate. <u>J</u>. <u>Geophgys</u>. <u>Res</u>. 93, 15,919-15,932.

73. Curry, J.A., Radke, L.F., Brock, C.A. and Ebert, E.E. (1988) Arctic ice-crystal haze. <u>Proc</u>.: <u>Symposium</u> <u>on</u> <u>the</u> <u>Role</u> <u>of</u> <u>Clouds</u> <u>in</u> <u>Atmospheric</u> <u>Chemistry</u> <u>and</u> <u>Global</u> <u>Climate</u>. Amer. Meteorol. Soc., Boston, MA, pp. 114-117.

74. Hoff, R.M. and Leaitch, W.R. (1988) Ground-based cirrus clouds in the Arctic. <u>Proc</u>.: <u>Symposium</u> <u>on</u> <u>the</u> <u>Role</u> <u>of</u> <u>Clouds</u> <u>in</u> <u>Atmospheric</u> <u>Chemistry</u> <u>and</u> <u>Global</u> <u>Climate</u>. Amer. Meteorol. Soc., Boston, MA, pp. 324-329.

75. Sperber, K.R., Hameed, S., Gates, W.L. and Potter, G.L. (1987) Southern oscillation simulated in a global climate model. <u>Nature</u> 329, 140-142.

76. Dickinson, R.E., Ed. (1987) <u>The</u> <u>Geophysiology</u> <u>of</u> <u>Amazonia</u>. <u>Vegetation</u> <u>and</u> <u>Climate</u> <u>Interactions</u>. John Wiley & Sons, New York, NY.

77. Henderson-Sellers, A., Dickinson, R.A. and Wilson, M.F. (1988) Tropical deforestation: Important processes for climate models. <u>Clim</u>. <u>Change</u> 13, 43-67.

78. Charney, J.G. (1975) Dynamics of deserts and drought in the Sahel. <u>Quart</u>. <u>J</u>. <u>Roy</u>. <u>Meteorol</u>. <u>Soc</u>. 101, 193-202.

79. Shukla, J. and Mintz, Y. (1982) Influence of land surface evapo-transpiration on the Earth's climate. <u>Science</u> 215, 1498-1501.

80. Sud, Y.C., Chou, M.D., Sellers, P.J., Minz, Y., Walker, G.K., Smith,

W.E. and Molod, A. (1988) Winter and summer simulations using the GLA GCM with and without the simple biosphere model (SiB). Trans. Amer. Geophys. Union 69, 1057.

81. Sato, N. and Sellers, P.J. (1988) Effects of implementing the simple biosphere model in a general circulation model. Trans. Amer. Geophys. Union 69, 1057.

82. Delworth, T.L. and Manabe, S. (1988) The influence of potential evaporation on the variabilities of simulated soil wetness and climate. J. Climate 1, 523-547.

83. Abramopoulos, F., Rosenzweig, C. and Choudhury, B. (1988) Improved ground hydrology calculations for global climate models (GCMs): Soil water movement and evapotranspiration. J. Climate 1, 921-941.

84. Somerville, R.C.J. (1985) Clouds and climate regulation. Nature 315, 713-714.

85. Crane, R.G. and Barry, R.G. (1984) The influence of clouds on climate with a focus on high latitude interactions. J. Climatol. 4, 71-93.

86. Henderson-Sellers, A. (1987) Climate is a cloudy issue. New Sci. 115(1570), 37-39.

87. Slingo, A. and Slingo, J.M. (1988) The response of a general circulation model to cloud longwave radiative forcing. I: Introduction and initial experiments. Quart. J. Roy. Meteorol. Soc. 114, 1027-1062.

88. Del Genio, A.D. and Yao, M.-S. (1988) Sensitivity of a global climate model to the specification of convective updraft and downdraft mass fluxes. J. Atmos. Sci. 45, 2641-2668.

89. Washington, W.M. and Meehl, G.A. (1984) Seasonal cycle experiment on the climate sensitivity due to a doubling of CO_2 with an atmospheric general circulation model coupled to a simple mixed-layer ocean model. J. Geophys. Res. 89, 9475-9503.

90. Hansen, J., Lacis, A., Rind, D., Russell, G., Stone, P., Fung, I., Ruedy, R. and Lerner, J. (1984) Climate sensitivity: Analysis of feedback mechanisms. In Hansen, J. and Takahashi, T. (eds) Geophys. Monogr. Ser. 29: Climate Processes and Climate Sensitivity. Amer. Geophys. Union, Washington, DC, pp. 130-163.

91. Wetherald, R.T. and Manabe, S. (1986) An investigation of cloud cover change in response to thermal forcing. Clim. Change 8, 5-23.

92. Wilson, C.A. and Mitchell, J.F.B. (1987) A doubled CO_2 climate sensitivity experiment with a global climate model including a simple ocean. J. Geophys. Res. 92, 13,315-13,343.

93. Wetherald, R.T. and Manabe, S. (1988) Cloud feedback processes in a general circulation model. J. Atmos. Sci. 45, 1397-1415.

94. Anonymous (1986) Clouds may offest "greenhouse effect." Bull. Amer. Meteorol. Soc. 67, 1425.

95. Ramanathan, V., Cess, R.D., Harrison, E.F., Minnis, P., Berkstrom, B.R., Ahmed, E. and Hartmann, D. (1989) Cloud-radiative forcing and climate: Results from the Earth Radiation Budget Experiment. Science 243, 57-63.

96. Nullet, D. and Ekern, P.C. (1988) Temperature and insolation trends in Hawaii. Theoret. Appl. Climatol. 39, 90-92.

97. Nullet, D. (1987) Sources of energy for evaporation on tropical islands. Phys. Geogr. 8, 36-45.

98. Webster, P.J. and Stephens, G.L. (1984) Cloud-radiation interaction and the climate problem. In Houghton, J.T. (ed) The Global Climate. Cambridge Univ. Press, Cambridge, UK, pp. 63-78.

99. Platt, C.M.R. and Harshvardhan (1988) Temperature dependence of cirrus extinction: Implications for climate feedback. J. Geophys. Res. 93, 11,051-11,058.

100. Ackerman, S.A. and Cox, S.K. (1988) Shortwave radiative parameterization of large atmospheric aerosols: Dust and water clouds. J. Geophys. Res. 93, 11,063-11,073.

REFERENCES   SECTION 3

101. Boville, B.A. and Cheng, X. (1988) Upper boundary effects in a general
     circulation model. J. Atmos. Sci. 34, 2591-2606.
102. Delage, Y. (1988) The position of the lowest levels in the boundary
     layer of atmospheric circulation models. Atmos.-Ocean 26, 329-340.
103. Bengtsson, L. and Shukla, J. (1988) Integration of space and in situ
     observations to study global climate change. Bull. Amer. Meteorol.
     Soc. 69, 1130-1143.
104. Skatskii, V.I. (1965) Some results from experimental study of the
     liquid-water content in cumulus clouds. Izv. Atmos. Oceanic Phys. 1,
     833-844.
105. Voyt, F.Y. and Mazin, I.P. (1972) Liquid water content in cumulus
     clouds. Izv. Atmos. Oceanic Phys. 8, 1166-1176.
106. Petukhov, V.K., Feygelson, Ye.M. and Manuylova, N.I. (1975) The
     regulating role of clouds in the heat effects of anthropogenic
     aerosols and carbon dioxide. Izv. Atmos. Oceanic Phys. 11, 802-808.
107. Paltridge, G.W. (1980) Cloud-radiation feedback to climate. Quart. J.
     Roy. Meteorol. Soc. 106, 895-899.
108. Charlock, T.P. (1981) Cloud optics as a possible stabilizing factor in
     climate change. J. Atmos. Sci. 38, 661-663.
109. Charlock, T.P. (1982) Cloud optical feedback and climate stability in a
     radiative-convective model. Tellus 34, 245-254.
110. Roeckner, E. (1988) A GCM analysis of the cloud optical depth feedback.
     Proc.: Symposium on the Role of Clouds in Atmospheric Chemistry and
     Global Climate. Amer. Meteorol. Soc., Boston, MA, pp. 67-68.
111. Betts, A.K. and Harshvardhan (1987) Thermodynamic constraint on the
     cloud liquid water feedback in climate models. J. Geophys. Res. 92,
     8483-8485.
112. Roeckner, E., Schlese, U., Biercamp, J. and Loewe, P. (1987) Cloud
     optical depth feedbacks and climate modeling. Nature 329, 138-140.
113. Somerville, R.C.J. and Remer, L.A. (1984) Cloud optical thickness
     feedbacks in the CO_2 climate problem. J. Geophys. Res. 89, 9668-9672.
114. Feigelson, E.M. (1978) Preliminary radiation model of a cloudy
     atmosphere, 1, Structure of clouds and solar radiation. Beitr. Phys.
     Atmos. 51, 203-229.
115. Schlesinger, M.E. (1988) Negative or positive cloud optical depth
     feedback? Nature 335, 303-304.
116. Roeckner, E. (1988) Reply. Nature 335, 304.
117. Somerville, R.C.J. and Iacobellis, S. (1988) Climate stability and cloud
     optical thickness feedbacks. Proc.: Symposium on the Role of Clouds in
     Atmospheric Chemistry and Global Climate. Amer. Meteorol. Soc.,
     Boston, MA, pp. 60-62.
118. Molnar, G. and Wang, W.-C. (1988) Effects of cloud optical property
     feedbacks on the greenhouse effect of CO_2 and other trace gases.
     Proc.: Symposium on the Role of Clouds in Atmospheric Chemistry and
     Global Climate. Amer. Meteorol. Soc., Boston, MA, pp. 63-66.
119. Hookings, G.A. (1965) Precipitation maintained downdrafts. J. Appl.
     Meteorol. 4, 190-195.
120. Yao, M.-S. and Stone, P.H. (1987) Development of a two-dimensional
     zonally averaged statistical-dynamical model. Part I: The
     parameterization of moist convection and its role in the general
     circulation. J. Atmos. Sci. 44, 65-82.
121. Hartman, D.L., Hendon, H.H. and Houze, R.A. (1984) Some implications of
     the mesoscale circulations in tropical cloud clusters for large-scale
     dynamics and climate. J. Atmos. Sci. 41, 113-121.
122. Lau, K.M. and Peng, L. (1987) Origin of low-frequency (intraseasonal)
     oscillations in the tropical atmosphere. Part I: Basic theory. J.
     Atmos. Sci. 44, 950-972.
123. Chao, W.C. (1987) On the origin of the tropical intraseasonal
     oscillation. J. Atmos. Sci. 44, 1940-1949.

REFERENCES

124. Neelin, J.D., Held, I.M. and Cook, K.H. (1987) Evaporation-wind feedback and low frequency variability in the tropical atmosphere. _J_. _Atmos_. _Sci_. 44, 2341-2348.

125. Zipser, E.J. (1977) Mesoscale and convective-scale downdrafts as distinct components of squall-line circulation. _Mon_. _Wea_. _Rev_. 105, 1568-1589.

126. Johnson, R.H. (1984) Partitioning tropical heat and moisture budgets into cumulus and mesoscale components: Implications for cumulus parameterization. _Mon_. _Wea_. _Rev_. 112, 1590-1601.

127. Shaw, G.E. (1983) Bio-controlled thermostasis involving the sulfur cycle. _Clim_. _Change_ 5, 297-303.

128. Shaw, G.E. (1987) Aerosols as climate regulators: A climate-biosphere linkage? _Atmos_. _Environ_. 21, 985-986.

129. Charlson, R.J., Lovelock, J.E., Andreae, M.O. and Warren, S.G. (1987) Oceanic phytoplankton, atmospheric sulfur, cloud albedo and climate. _Nature_ 326, 655-661.

130. Twomey, S.A. (1977) _Atmospheric_ _Aerosols_. Elsevier, Amsterdam, The Netherlands.

131. Bonsang, B., Nguyen, B.C., Gaudry, A. and Lambert, G. (1980) Sulfate enhancement in marine aerosols owing to biogenic sulfur compounds. _J_. _Geophys_. _Res_. 85, 7410-7416.

132. Shaw, G.E. (1988) Antarctic aerosols: A review. _Rev_. _Geophys_. 26, 89-112.

133. Andreae, M.O. (1980) The production of methylated sulfur compounds by marine phytoplankton. _In_ Trudinger, P.A., Walter, M.R. and Ralph, B.J. (eds) _Biogeochemistry_ _of_ _Ancient_ _and_ _Modern_ _Environments_. Springer-Verlag, New York, NY, pp. 253-259.

134. Bates, T.S., Charlson, R.J. and Gammon, R.H. (1987) Evidence for the climatic role of marine biogenic sulphur. _Nature_ 329, 319-321.

135. Charlson, R.J. and Bates, T.S. (1988) The role of the sulfur cycle in cloud microphysics, cloud albedo, and climate. _Proc_.: _Symposium_ _on_ _the_ _Role_ _of_ _Clouds_ _in_ _Atmospheric_ _Chemistry_ _and_ _Global_ _Climate_. Amer. Meteorol. Soc., Boston, MA, pp. 1-3.

136. Eppley, R.W. (1972) Temperature and phytoplankton growth in the sea. _Fish_. _Bull_. 70, 1063-1085.

137. Goldman, J.C. and Carpenter, E.J. (1974) A kinetic approach to the effect of temperature on algal growth. _Limnol_. _Oceanogr_. 19, 756-766.

138. Rhea, G.-Y. and Gotham, I.J. (1981) The effect of environmental factors on phytoplankton growth: Temperature and the interactions of temperature with nutrient limitation. _Linmol_. _Oceanogr_. 26, 635-648.

139. Platt, T. and Sathyendranath, S. (1988) Oceanic primary production: Estimation by remote sensing at local and regional scales. _Science_ 241, 1613-1620.

140. Sakshaug, E. (1988) Light and temperature as controlling factors of phytoplankton growth rate in temperate and polar regions. _Trans_. _Amer_. _Geophys_. _Union_ 69, 1081.

141. Vairavamurthy, A., Andreae, M.O. and Iverson, R.L. (1985) Biosynthesis of dimethylsulfide and dimethylpropiothetin by _Hymenomonas_ _carterae_ in relation to sulfur source and salinity variations. _Limnol_. _Oceanogr_. 30, 59-70.

142. Caldeira, K. (1989) Evolutionary pressures on planktonic production of atmospheric sulphur. _Nature_ 337, 732-734.

143. Nguyen, B.C., Belviso, S., Mihalopoulos, N., Gostan, J. and Nival, P. (1988) Dimethyl sulfide production during natural phytoplanktonic blooms. _Marine_ _Chem_. 24, 133-141.

144. Dacey, J.W.H. and Wakeham, S.G. (1988) Oceanic dimethylsulfide: Production during zooplankton grazing on phytoplankton. _Science_ 233, 1314-1316.

145. Turner, S.M., Malin, G., Liss, P.S., Harbour, D.S. and Holligan, P.M.

(1988) The seasonal variation of dimethyl sulfide and dimethylsulfoniopropionate concentrations in nearshore waters. Limnol. Oceanogr. 33, 364-375.

146. Nguyen, B.C., Gaudry, A., Bonsang, B. and Lambert, G. (1978) Reevaluation of the role of dimethyl sulphide in the sulfur budget. Nature 275, 637-639.

147. Andreae, M.O. and Raemdonck, H. (1983) Dimethyl sulfide in the surface ocean and the marine atmosphere: A global view. Science 221, 744-747.

148. Wakeham, S.G., Howes, B.L. and Dacey, J.W.H. (1984) Dimethyl sulfide in a stratified coastal salt pond. Nature 310, 770-772.

149. Hatakeyama, S.D., Okuda, M. and Akimoto, H. (1982) Formation of sulfur dioxide and methane sulfonic acid in the photo-oxidation of dimethylsulfide in the air. Geophys. Res. Lett. 9, 583-586.

150. Saltzman, E.S., Savoie, D.L., Zika, R.G. and Prospero, J.M. (1983) Methane-sulfonic acid in the marine atmosphere. J. Geophys. Res. 88, 10,897-10,902.

151. Grosjean, D. (1984) Photo-oxidation of methylsulfide, ethyl-sulfide and methanethiol. Environ. Sci. Technol. 18, 460-468.

152. Saxena, V.K., Curtin, T.B. and Parungo, F.P. (1985) Aerosol formation by wave action over Ross Sea, Antarctica. J. Rech. Atmos. 19, 213-224.

153. Andreae, M.O. (1986) The ocean as a source of atmospheric sulphur compounds. In Buat-Menard, P. (ed) The Role of Air-Sea Exchange in Geochemical Cycling. Reidel, Hingham, MA, pp. 331-362.

154. Clark, A.D., Ahlquist, N.C. and Covert, D.S. (1987) The Pacific marine aerosol: Evidence for natural acid sulfates. J. Geophys. Res. 92, 4179-4190.

155. Andreae, M.O., Berresheim, H., Andreae, T.W., Kritz, M.A., Bates, T.S. and Merril, J.T. (1988) Vertical distribution of dimethylsulfide, sulfur dioxide, aerosol ions and radon over the northeast Pacific Ocean. J. Atmos. Chem. 6, 149-173.

156. Covert, D.S. (1988) North Pacific marine background aerosol: Average ammonium to sulfate molar ratio equals 1. J. Geophys. Res. 93, 8455-8458.

157. Kreidenweis, S.M. and Seinfeld, J.H. (1988) Nucleation of sulfuric acid-water and methanesulfonic acid-water solution particles: Implications for the atmospheric chemistry of organosulfur species. Atmos. Environ. 22, 283-296.

158. Elliott, W.P. and Egami, R. (1975) CCN measurements over the ocean. J. Atmos. Sci. 32, 371-374.

159. Hoppel, W.A. (1979) Measurement of the size distribution and CCN supersaturation spectrum of submicron aerosols over the ocean. J. Atmos. Sci. 36, 2006-2015.

160. Saxena, V.K. (1983) Evidence of the biogenic nuclei involvement in Antarctic coastal clouds. J. Phys. Chem. 87, 4130.

161. Conover, J.H. (1966) Anomalous cloud lines. J. Atmos. Sci. 23, 778-785.

162. Twomey, S.A., Howell, H.B. and Wojceichowski, T.A. (1968) Comments on "Anomalous cloud lines." J. Atmos. Sci. 25, 333-334.

163. Coakley, J.A., Bernstein, R.L. and Durkee, P.A. (1987) Effect of ship-stack effluents on cloud reflectivity. Science 237, 1020-1022.

164. Scorer, R.A. (1987) Ship trails. Atmos. Environ. 21, 1417-1425.

165. Twomey, S.A. and Warner, J. (1967) Comparison of measurements of cloud droplets and cloud nuclei. J. Atmos. Sci. 24, 702-703.

166. Warner, J. and Twomey, S.A. (1967) The production of cloud nuclei by cane fires and the effect on cloud droplet concentration. J. Atmos. Sci. 24, 704-706.

167. Hunt, G.E. (1973) Radiative properties of terrestrial clouds at visible and infrared thermal window wavelengths. Quart. J. Roy. Meteorol. Soc. 99, 346-369.

168. Hudson, J.D. (1983) Effects of CCN concentrations on stratus clouds. J.

Atmos. Sci. 40, 480-486.

169. Noonkester, V.R. (1984) Droplet spectra observed in marine stratus cloud layers. J. Atmos. Sci. 41, 829-845.

170. Durkee, P.A. (1988) Observations of aerosol-cloud interactions in satellite-detected visible and near-infrared radiance. Proc.: Symposium on the Role of Clouds in Atmospheric Chemistry and Global Climate. Amer. Meteorol. Soc., Boston, MA, pp. 157-160.

171. Lovelock, J.E. (1988) The Ages of Gaia: A Biography of our Living Earth. W.W. Norton & Co., New York, NY.

172. Went, F.W. (1966) On the nature of Aitken condensation nuclei. Tellus 18, 549-555.

173. Mooney, H.A., Vitousek, P.M. and Matson, P.A. (1987) Exchange of materials between terrestrial ecosystems and the atmosphere. Science 238, 926-932.

174. Duce, R.A., Mohnen, V.A., Zimmerman, P.R., Grosjean, D., Cautreels, W., Chatfield, R., Jaenicke, R., Ogsen, J.A., Pillizzari, E.D. and Wallace, G.T. (1983) Organic material in the global troposphere. Rev. Geophys. Space Phys. 21, 921-952.

175. Roosen, R.G. and Angione, R.J. (1984) Atmospheric transmission and climate: Results from Smithsonian measurements. Bull. Amer. Meteorol. Soc. 65, 950-957.

176. Meszaros, E. (1988) On the possible role of the biosphere in the control of atmospheric clouds and precipitation. Atmos. Environ. 22, 423-424.

177. Idso, S.B. (1986) Industrial age leading to the greening of the Earth? Nature 320, 22.

178. Albrecht, B.A. (1988) Modulation of boundary layer cloudiness by precipitation processes. Proc.: Symposium on the Role of Clouds in Atmospheric Chemistry and Global Climate. Amer. Meteorol. Soc., Boston, MA, pp. 9-13.

179. Squires, P. (1958) The microstructure and colloidal stability of warm clouds. Part II: The causes of the variations in microstructure. Tellus 10, 262-271.

180. Twomey, S. and Wojciechowski, T.A. (1969) Observations of the variation of cloud nuclei. J. Atmos. Sci. 26, 684-688.

181. Ludlam, F.H. (1980) Clouds and Storms. Penn. State Univ. Press, University Park, PA.

182. Brost, R.A., Lenschow, D.H. and Wyngaard, J.C. (1982) Marine stratocumulus layers. Part II: Turbulence budgets. J. Atmos. Sci. 39, 818-836.

183. Nicholls, S. (1984) The dynamics of stratocumulus: Aircraft observations and comparisons with a mixed layer model. Quart. J. Roy. Meteorol. Soc. 110, 783-820.

184. Radke, L.F. (1988) Airborne observations of cloud microphysics modified by anthropogenic forcing. Proc.: Symposium on the Role of Clouds in Atmospheric Chemistry and Global Climate. Amer. Meteorol. Soc., Boston, MA, pp. 310-315.

185. SCEP (1971) Study of Critical Environmental Problems. MIT Press, Cambridge, MA.

186. SMIC (1971) Study of Man's Impact on Climate. MIT Press, Cambridge, MA.

187. Twomey, S.A., Piepgrass, M. and Wolfe, T.L. (1984) An assessment of the impact of pollution on global cloud albedo. Tellus 36B, 356-366.

188. Countess, R.J., Wolfe, G.T. and Cadle, S.H. (1980) The Denver winter aerosol: A comprehensive chemical characterization. J. Air Poll. Contr. Assoc. 30, 1194-1200.

189. Twohy, C.H., Clarke, A.D., Warren, S.G., Radke, L.F. and Charlson, R.J. (1988) Measurements of the light-absorbing material inside cloud droplets and its effect on cloud albedo. Proc.: Symposium on the Role of Clouds in Atmospheric Chemistry and Global Climate. Amer. Meteorol. Soc., Boston, MA, pp. 215-220.

REFERENCES   SECTION

190. Hobbs, P.V., Radke, L.F. and Shumway, S.E. (1970) Cloud condensatio
     nuclei from industrial sources and their apparent influence o
     precipitation in Washington state. J. Atmos. Sci. 27, 81-89.
191. Braham, R. (1974) Cloud physics of urban weather modification:
     preliminary report. Bull. Amer. Meteorol. Soc. 55, 100-106.
192. Schmidt, M. (1974) Verlauf der Kondensationskernkonzentration an zwe
     suddeutschen Stationen. Meteorol. Rundsch. 27, 151-153.
193. Twomey, S., Davidson, K.A. and Seton, K.J. (1978) Results of 5 years
     observations of cloud nucleus concentrations at Robertson, New Soutl
     Wales. J. Atmos. Sci. 35, 650-656.
194. Cobb, W.E. and Wells, H.J. (1970) The electrical conductivity of oceani
     air and its correlation to global pollution. J. Atmos. Sci. 27, 814-
     819.
195. Neeman, B.U., Ohring, G. and Joseph, J.H. (1988) The Milankovitch theory
     and climate sensitivity. 2. Interaction between the Northern
     Hemisphere ice sheets and the climate system. J. Geophys. Res. 93
     11,175-11,191.
196. Winograd, I.J., Szabo, B.J., Coplen, T.B. and Riggs, A.C. (1988) A
     250,000-year climatic record from Great Basin vein calcite
     Implications for Milankovitch theory. Science 242, 1275-1280.
197. Imbrie, J. and Imbrie, J.Z. (1980) Modelling the climatic response to
     orbital variations. Science 207, 943-953.
198. LeTreut, H. and Ghil, M. (1983) Orbital forcing, climatic interactions,
     and glaciation cycles. J. Geophys. Res. 88, 5167-5190.
199. Saltzman, B., Sutera, A. and Hansen, A.R. (1984) Earth-orbital
     eccentricity variations and climatic change. In Berger, A., Imbrie,
     J., Hays, J., Kukla, G. and Saltzman, B. (eds) Milankovitch and
     Climate. D. Reidel, Hingham, MA, pp. 615-636.
200. Lindzen, R.S. (1986) A simple model for 100 K-year oscillations in
     glaciation. J. Atmos. Sci. 43, 986-996.
201. Woillard, G. (1979) Abrupt end of the last interglacial in north-east
     France. Nature 281, 558-562.
202. Flohn, H. (1984) A possible mechanism of abrupt climatic changes. In
     Morner, N.-A. and Karlen, W. (eds) Climatic Changes on a Yearly to
     Millenial Basis. D. Reidel, Amsterdam, Holland, pp. 521-531.
203. Broecker, W.S., Peteet, D.M. and Rind, D. (1985) Does the ocean-
     atmosphere system have more than one stable mode of operation? Nature
     315, 21-26.
204. Jacoby, G.C., Jr., Cook, E.R. and Ulan, L.D. (1985) Reconstructed summer
     degree days in central Asia and north-western Canada since 1524. Quat.
     Res. 23, 18-26.
205. Payette, S., Filion, L., Gauthier, L. and Boutin, Y. (1985) Secular
     climate change in old growth tree-line vegetation of northern Quebec.
     Nature 315, 135-138.
206. Conkey, L.E. (1986) Red spruce tree-ring widths and densities in
     eastern North America as indicators of past climate. Quat. Res. 26,
     232-243.
207. Denton, G.H., Hughes, T.J. and Karlen, W. (1986) Global ice-sheet system
     interlocked by sea level. Quat. Res. 26, 3-26.
208. Kuhle, M. (1987) Subtropical mountain- and highland-glaciation as ice age
     triggers and the waning of the glacial periods in the Pleistocene.
     GeoJournal 14, 393-421.
209. Kerr, R.A. (1988) Linking earth, ocean, and air at the AGU. Science 239,
     259-260.
210. Broecker, W.S., Andree, M., Bonani, G., Wolfli, W., Oeschger, H. and
     Klas, M. (1988) Can the Greenland climatic jumps be identified in
     records from ocean and land? Quat. Res. 30, 1-6.
211. Stephens, G.L. and Webster, P.J. (1984) Cloud decoupling of the surface
     and planetary radiative budgets. J. Atmos. Sci. 41, 681-686.

REFERENCES   SECTION 3

12. Manabe, S. and Stouffer, R.J. (1988) Two stable equilibria of a coupled ocean-atmosphere model. J. Climate 1, 841-866.
13. Stromel, H. (1961) Thermohaline convection with two stable regimes of flow. Tellus 13, 224-230.
14. Rooth, C. (1982) Hydrology and ocean circulation. Prog. Oceanogr. 11, 131-149.
15. Bryan, F. (1986) High latitude salinity effects and interhemispheric thermohaline circulations. Nature 323, 301-304.
16. Twomey, S.A., Piepgrass, M. and Wolfe, T.L. (1985) Reply. Tellus 37B, 310-312.
17. Heintzenberg, J. and Ogren, J.A. (1985) An assessment of the impact of pollution on global cloud albedo: Comment. Tellus 37B, 308-309.
18. Byrne, G. (1989) Bagging the albatross. Science 243, 32.
19. Kerr, R.A. (1989) How to fix the clouds in greenhouse models. Science 243, 28-29.
20. Stephens, G.L. (1988) Aspects of cloud-climate feedback. Proc.: Symposium on the Role of Clouds in Atmospheric Chemistry and Global Climate. Amer. Meteorol. Soc., Boston, MA, pp. 30-34.

REFERENCES   SECTION 4

1. Broecker, W.S. (1987) Unpleasant surprises in the greenhouse? Nature 328, 123-126.
2. Meadows, D., Richardson, J. and Bruckmann, G. (1982) Groping in the Dark: The First Decade of Global Modelling. John Wiley & Sons, New York, NY.
3. Earth System Sciences Committee, NASA Advisory Council (1988) Earth System Science: A Closer View. National Aeronautics and Space Administration, Washington, DC.
4. National Research Council, U.S. (1982) Carbon Dioxide and Climate: A Second Assessment. National Academy Press, Washington, DC.
5. National Research Council, U.S. (1983) Changing Climate. National Academy Press, Washington, DC.
6. Oyama, Y.I., Carle, G.C., Woeller, F. and Pollack, J.B. (1979) Venus lower atmospheric composition: Analysis by gas chromatography. Science 203, 802-805.
7. Pollack, J.B., Toon, O.B. and Boese, R. (1980) Greenhouse models of Venus' high surface temperature, as constrained by Pioneer Venus measurements. J. Geophys. Res. 85, 8223-8231.
8. Kasting, J.F., Toon, O.B. and Pollack, J.B. (1988) How climate evolved on the terrestrial planets. Sci. Amer. 258(2), 90-97.
9. Pollack, J.B. (1979) Climate change on the terrestrial planets. Icarus 37, 479-553.
10. McKay, C. (1983) Section 6. Mars. In Smith, R.E. and West, G.S. (eds) Space and Planetary Environment Criteria Guidelines for Use in Space Vehicle Development, 1982 Revision (Volume 1). NASA, Marshall Space Flight Center, AL.
11. Idso, S.B. (1988) The CO_2 greenhouse effect on Mars, Earth, and Venus. Sci. Tot. Environ. 77, 291-294.
12. Lovelock, J.E. and Whitfield, M. (1982) Life span of the biosphere. Nature 296, 561-563.
13. Kasting, J.F. and Ackerman, T.P. (1986) Climatic consequences of very high carbon dioxide levels in earth's early atmosphere. Science 234, 1383-1385.
14. Durham, R. and Chamberlain, J.W. (1989) A comparative study of the early terrestrial atmospheres. Icarus 77, 59-66.
15. Cess, R.D., Ramanathan, V. and Owen, T. (1980) The martian paleoclimate

REFERENCES   SECTION

    and enhanced atmospheric carbon dioxide. Icarus 41, 159-165.

16. Toon, O.B., Pollack, J.B., Ward, W., Burns, J.A. and Bilski, K. (1980) The astronomical theory of climate change on Mars. Icarus 44, 552-607

17. Hoffert, M.I., Calegari, A.J., Hsieh, C.T. and Ziegler, W. (1981) Liqui water on Mars: An energy balance climate model for CO_2/H_2 atmospheres. Icarus 47, 112-129.

18. Ramanathan, V. (1981) The role of ocean-atmosphere interactions in th CO_2 climate problem. J. Atmos. Sci. 38, 918-930.

19. Idso, S.B. (1982) CO_2 and climate: Where is the water vapor feedback Archiv. Meteorol. Geophys. Bioclimatol. Ser. B 31, 325-329.

20. Ellsaesser, H.W. (1984) The climatic effect of CO_2: A different view Atmos. Environ. 18, 431-434.

21. Lindzen, R.S., Hou, A.Y. and Farrell, B.F. (1982) The role of convectiv model choice in calculating the climate impact of doubling CO_2. J Atmos. Sci. 39, 1189-1205.

22. Idso, S.B. (1982) An empirical evaluation of Earth's surface ai temperature response to an increase in atmospheric carbon dioxid concentration. In Reck, R.A. and Hummel, J.R. (eds) AIP Conf. Proc No. 82: Interpretation of Climate and Photochemical Models, Ozone an Temperature Measurements. Amer. Inst. Physics, New York, NY, pp. 119- 134.

23. Idso, S.B. (1982) A surface air temperature response function fo Earth's atmosphere. Boundary Layer Meteorol. 22, 227-232.

24. Cess, R.D. and Potter, G.L. (1984) A commentary on the recent CO_2 climate controversy. Clim. Change 6, 365-376.

25. Ellis, J.S., Vonder Haar, T.H., Levitus, S. and Oort, A.H. (1978) Th annual variation in the global heat balance of the earth. J. Geophys. Res. 83, 1958-1962.

26. Kiehl, J.T. and Ramanathan, V. (1982) Radiative heating due to increased CO_2: The role of H_2O continuum absorption in the 12-18 um region. J. Atmos. Sci. 39, 2923-2926.

27. Matthews, R.K. and Poore, R.Z. (1980) Tertiary ^{18}O record and glacio-eustatic sea-level fluctuations. Geology 8, 501-504.

28. Barron, E.J., Thompson, S.L. and Schneider, S.H. (1981) An ice-free Cretaceous? Results from climate model simulations. Science 212, 501-508.

29. Idso, S.B. (1981) A set of equations for full spectrum and 8- to 14-um and 10.5- to 12.5-um thermal radiation from cloudless skies. Water Resources Res. 17, 295-304.

30. Idso, S.B. (1989) An upper limit to the greenhouse effect of earth's atmosphere. Theoret. Appl. Climatol., in press.

31. Somerville, R.C.J. (1985) Clouds and climate regulation. Nature 315, 713-714.

32. Kimball, B.A., Idso, S.B. and Aase, J.K. (1982) A model of thermal radiation from partly cloudy and overcast skies. Water Resources Res. 18, 931-936.

33. Priestley, C.H.B. (1966) The limitation of temperature by evaporation in hot climates. Agric. Meteorol. 3, 241-246.

34. Priestley, C.H.B. and Taylor, R.J. (1972) On the assessment of surface heat flux and evaporation using large-scale parameters. Mon. Wea. Rev. 100, 81-92.

35. Newell, R.E., Vincent, D.G. and Boer, G.J. (1974) The General Circulation of the Tropical Atmosphere and Interactions with Extratropical Latitudes. Vol. 2. MIT Press, Cambridge, MA.

36. Newell, R.E., Navato, A.R. and Hsiung, J. (1978) Long-term global sea surface temperature fluctuations and their possible influence on atmospheric CO_2 concentrations. Pure Appl. Geophys. 116, 351-371.

37. Newell, R.E. and Dopplick, T.G. (1979) Questions concerning the possible influence of anthropogenic CO_2 on atmospheric temperature. J. Appl.

Meteorol. 18, 822-825.

38. Newell, R.E. and Dopplick, T.G. (1981) Reply to Robert G. Watt's "Discussion of 'Questions concerning the possible influence of anthropogenic CO_2 on atmospheric temperature.'" J. Appl. Meteorol. 20, 114-117.

39. Brutsaert, W. (1982) Evaporation into the Atmosphere. D. Reidel, Dordrecht, Holland.

40. Gadgil, S., Joseph, P.V. and Jose, N.V. (1984) Ocean-atmosphere coupling over monsoon regions. Nature 312, 141-143.

41. Graham, N.E. and Barnett, T.P. (1987) Sea surface temperature, surface wind divergence, and convection over tropical oceans. Science 238, 657-659.

42. Flannery, B.P, Callegari, A.J. and Hoffert, M.I. (1984) Energy balance models incorporating evaporative buffering of equatorial thermal response. In Hansen, J.E. and Takahashi, T. (eds) Geophys. Monogr. 29: Climate Processes and Climate Sensitivity. Amer. Geophys. Union, Washington, DC, pp. 108-117.

43. Idso, S.B., Clawson, K.L. and Anderson, M.G. (1986) Foliage temperature: Effects of environmental factors with implications for plant water stress assessment and the CO_2/climate connection. Water Resources Res. 22, 1702-1716.

44. Aase, J.K. and Idso, S.B. (1978) A comparison of two formula types for calculating long-wave radiation from the atmosphere. Water Resources Res. 14, 623-625.

45. Hatfield, J.L., Reginato, R.J. and Idso, S.B. (1983) Comparison of long-wave radiation calculation methods over the United States. Water Resources Res. 19, 285-288.

46. Idso, S.B. (1980) On the apparent incompatability of different atmospheric thermal radiation data sets. Quart. J. Roy. Meteorol. Soc. 106, 375-376.

47. Simpson, J.J. and Paulson, C.A. (1979) Mid-ocean observations of atmospheric radiation. Quart. J. Roy. Meteorol. Soc. 105, 487-502.

48. Arnfield, A.J. (1979) Evaluation of empirical expressions for the estimation of hourly and daily totals of atmospheric longwave emission under all sky conditions. Quart J. Roy. Meteorol. Soc. 105, 1041-1052.

49. Idso, S.B. (1981) On the systematic nature of diurnal patterns of differences between calculations and measurements of clear sky atmospheric thermal radiation. Quart. J. Roy. Meteorol. Soc. 107, 737-741.

50. Manabe, S. and Wetherald, R.T. (1967) Thermal equilibrium of the atmosphere with a given distribution of relative humidity. J. Atmos. Sci. 24, 241-259.

51. Hickcox, D.H. (1988) How hot can it get? Weatherwise 41, 157-159.

52. Idso, S.B. (1982) Carbon Dioxide: Friend or Foe? IBR Press, Tempe, AZ.

53. Idso, S.B. (1985) An upper limit to global surface air temperature. Archiv. Meteorol. Geophys. Bioclimatol. Ser. A 34, 141-144.

54. Manabe, S. and Stouffer, R.J. (1980) Sensitivity of a global climate model to an increase of CO_2-concentration in the atmosphere. J. Geophys. Res. 85, 5529-5554.

55. Spelman, M.J. and Manabe, S. (1984) Influence of oceanic heat transport upon the sensitivity of a model climate. J. Geophys. Res. 89, 571-586.

56. Cess, R.D. and Potter, G.L. (1988) A methodology for understanding and intercomparing atmospheric climate feedback processes in general circulation models. J. Geophys. Res. 93, 8305-8314.

57. Wilson, C.A. and Mitchell, J.F.B. (1987) A 2 x CO_2 climate sensitivity experiment with a global climate model including a simple ocean. J. Geophys. Res. 92, 13,315-13,343.

58. Hansen, J., Lacis, A., Rind, D., Russel, G., Stone, P., Fung, I. and Lerner, J. (1984) Climate sensitivity: Analysis of feedback

REFERENCES   SECTION ∢

      mechanisms. In Hansen, J.E. and Takahashi, T. (eds) Geophys. Monogr
      Ser., Vol. 29: Climate Processes and Climate Sensitivity. Amer
      Geophys. Union, Washington, DC, pp. 130-163.
59. Wetherald, R.T. and Manabe, S. (1986) An investigation of cloud cover
      change in response to thermal forcing. Clim. Change 8, 5-23.
60. Wetherald, R.T. and Manabe, S. (1975) The effect of changing the solar
      constant on the climate of a general circulation model. J. Atmos. Sci
      32, 2044-2059.
61. Potter, G.L. and Cess, R.D. (1984) Background tropospheric aerosols:
      Incorporation within a statistical-dynamical climate model. J
      Geophys. Res. 89, 9521-9526.
62. Washington, W.M. and Meehl, G.A. (1986) General circulation model CO_2
      sensitivity experiments: Snow-sea ice albedo parameterizations and
      globally averaged surface air temperature. Clim. Change 8, 321-341.
63. Landsberg, H.E. and Albert, J.M. (1971) The summer of 1816 and
      volcanism. Weatherwise 27, 63-66.
64. Self, S., Rampino, M.R. and Barbera, J.J. (1981) The possible effects of
      large 19th and 20th century volcanic eruptions on zonal and
      hemispheric surface temperatures. J. Volcanol. Geotherm. Res. 11, 41-
      60.
65. Baldwin, B., Pollack, J.B., Summers, A., Toon, O.B., Sagan, C. and Van
      Camp, W. (1976) Stratospheric aerosols and climatic change. Nature
      263, 551-555.
66. Pollack, J.B., Toon, O.B., Sagan, C., Summers, A., Baldwin, B. and Van
      Camp, W. (1976) Volcanic explosions and climatic change: A theoretical
      assessment. J. Geophys. Res. 81, 1071-1083.
67. Angell, J.K. and Korshover, J. (1985) Surface temperature changes
      following the six major volcanic episodes between 1780 and 1980. J.
      Clim. Appl. Meteorol. 24, 937-951.
68. Ellsaesser, H.W. (1986) Comments on "Surface temperature changes
      following the six major volcanic episodes between 1780 and 1980." J.
      Clim. Appl. Meteorol. 25, 1184-1185.
69. Self, S. and Rampino, M.R. (1988) The relationship between volcanic
      eruptions and climate change: Still a conundrum? Trans. Amer. Geophys.
      Union 69, 74-75, 85-86.
70. Sear, C.B., Kelly, P.M., Jones, P.D. and Goodess, C.M. (1987) Global
      surface temperature responses to major volcanic eruptions. Nature 330,
      365-367.
71. Hines, C.O. and Halevy, I. (1977) On the reality and nature of a certain
      sun-weather correlation. J. Atmos. Sci. 34, 382-404.
72. Haurwitz, M.W. and Brier, G.W. (1981) A critique of the superposed epoch
      analysis method: It's application to solar-weather relations. Mon.
      Wea. Rev. 109, 2074-2079.
73. Idso, S.B. (1989) On the stability of earth's climate. Theoret. Appl.
      Climatol. 39, 177-178.
74. Bradley, R.S. (1988) The explosive volcanic eruption signal in northern
      hemisphere continental temperature records. Clim. Change 12, 221-243.

REFERENCES   SECTION 5

1. Begley, S., Miller, M. and Hager, M. (1988) The endless summer?
      Newsweek 122(2), 18-20.
2. Anderson, I. (1988) Greenhouse warming grips American corn belt. New
      Sci. 118(1619), 35.
3. Kerr, R. (1989) 1988 ties for warmest year. Science 243, 891.
4. Sidney, H. (1988) The big dry. Time 132(1), 12-15.
5. Palca, J. (1988) Heated response to US drought. Nature 334, 92.

6. Wainger, L.A. (1988) Worst drought since '36 spurs climate research. Trans. Amer. Geophys. Union 69, 715.
7. Revkin, A.C. (1988) Endless summer: Living with the greenhouse effect. Discover 9(10), 50-61.
8. Schneider, S.H. (1988) The greenhouse effect and the U.S. summer of 1988: Cause and effect or a media event. Clim. Change 13, 113-115.
9. Shabecoff, P. (1988) Global warming has begun, expert tells senate. The New York Times 137(47,546), A1, A14.
10. Rifkin, J. (1988) The greenhouse doomsday scenario. The Washington Post 111(239), C3.
11. Ramirez, A. (1988) A warming world: What it will mean. Fortune 118(1), 102-107.
12. Twomey, S. (1988) Earth's radiation balance. Trans. Amer. Geophys. Union 69, 1045.
13. Sowell, T. (1988) The doom boom. The Washington Times 7(212), E1.
14. Joyce, C. (1988) America counts the cost of global warming. New Sci. 120(1636), 26.
15. Swinbanks, D. (1989) US Congress plans greenhouse legislation. Nature 338, 3.
16. Gore, A., Jr. (1986) A congressional perspective of the greenhouse effect. In Trabalka, J.R. and Reichle, D.E. (eds) The Changing Carbon Cycle: A Global Analysis. Springer-Verlag, New York, NY, pp. xxi-xxvi.
17. Anonymous. (1988) Toronto delegates call for a "law of the atmosphere." New Sci. 119(1620), 24.
18. White, G.E. (1988) Global warming: Uncertainty and action. Environment 30(6), i.
19. Dickman, S. (1988) Atmosphere may rise to the top of the West German agenda. Nature 334, 459.
20. Hare, F.K. (1988) Jumping the greenhouse gun? Nature 334, 646.
21. Pearman, G.I., Ed. (1988) Greenhouse. CSIRO, Canberra, Australia.
22. Gleick, P.H. (1988) The United States - Soviet "greenhouse/glasnost" teleconference. Ambio 17, 297-298.
23. Rosswall, T. (1988) The International Geosphere-Biosphere Programme: A study of global change (IGBP). Ambio 17, 299.
24. Crawford, M. (1988) Planning for climate change. Science 242, 510.
25. Anderson, A. (1988) US legislators attack greenhouse gases. Nature 335, 583.
26. Maddox, J. (1988) Jumping the greenhouse gun. Nature 334, 9.
27. Singer, S.F. (1988) Fact and fancy on greenhouse earth. Wall Street J. 212(42), 22.
28. Lorenz, E.N. (1984) Formulation of a low order model of a moist general circulation. J. Atmos. Sci. 41, 1933-1945.
29. Lorenz, E.N. (1986) The index cycle is alive and well. Namias Symposium, Lib. Cong. #86-50752, pp. 188-196.
30. Karl, T.R. (1988) Multi-year fluctuations of temperature and precipitation: The gray area of climate change. Clim. Change 12, 179-197.
31. Jager, J. (1988) Anticipating climatic change: Priorities for action. Environment 30(7), 12-15, 30-33.
32. Allman, W.F. (1988) Rediscovering planet earth. U.S. News World Rept. 105(17), 56-68.
33. Hecht, A.D. and Doos, B.R. (1988) Climate change, economic growth and energy policy: A recommended U.S. strategy for the coming decades / An editorial. Clim. Change 13, 1-3.
34. Pearce, F. (1986) How to stop the greenhouse effect. New Sci. 111(1526), 29-30.
35. Edmonds, J.A., Reilly, J., Trabalka, J.R., Reichle, D.E., Rind, D., Lebedeff, S., Palutikof, J.P., Wigley, T.M.L., Lough, J.M., Blasing, T.J., Solomon, A.M., Seidel, S., Keyes, D. and Steinberg, M. (1986)

REFERENCES   SECTION

Future Atmospheric Carbon Dioxide Scenarios and Limitation Strategies
      Noyes Publications, Park Ridge, NJ.
36. Mintzer, I.M. (1987) A Matter of Degrees: The Potential for Controlling
      the Greenhouse Effect. World Resources Institute, Washington, DC.
37. Kerr, R.A. (1988) Report urges greenhouse action now. Science 241, 23-
      24.
38. Nisbet, E.G. (1988) The business of planet management. Nature 333, 617.
39. Tangley, L. (1988) Greenhouse effect already here? Scientists call for
      action. BioScience 38, 538.
40. Woodwell, G.M. (1988) The global carbon cycle. Science 241, 1736-1737.
41. Booth, W. (1988) Johny Appleseed and the greenhouse. Science 242, 19-20.
42. Byrne, G. (1988) Let 100 million trees bloom. Science 242, 371.
43. Anderson, A. (1988) Swapping power for trees in the Americas. Nature
      335, 583.
44. Edmonds, J.A., Ashton, W.B., Cheng, H.C. and Steinberg, M. (1989) A
      Preliminary Analysis of U.S. CO$_2$ Emissions Reduction Potential from
      Energy Conservation and the Substitution of Natural Gas for Coal in
      the Period to 2010. U.S. Dept. Energy, Washington, DC.
45. Kellogg, W.W. (1988) Foreword to Climate Shocks: Natural and
      Anthropogenic by K.Ya. Kondratyev. John Wiley & Sons, New York, NY,
      pp. v-ix.
46. Kukla, G., Gavin, J. and Karl, T.R. (1986) Urban warming. J. Clim. Appl.
      Meteorol. 25, 1265-1270.
47. Tanner, W.F. (1988) TER-QUA climate meeting. Trans. Amer. Geophys. Union
      69, 1674.
48. Wood, F.B., Jr. (1988) Comment: On the need for validation of the Jones
      et al. temperature trends with respect to urban warming. Clim. Change
      12, 297-312.
49. Karl, T.R., Diaz, H.F. and Kukla, G. (1988) Urbanization: Its detection
      and effect in the United States climate record. J. Climate 1, 1099-
      1123.
50. Balling, R.C., Jr. and Idso, S.B. (1989) Historical temperature trends
      in the United States and the effect of urban population growth. J.
      Geophys. Res. 94, 3359-3363.
51. Mitchell, J.M., Jr. (1961) The temperature of cities. Weatherwise 14,
      224-227.
52. Oke, T.R. (1973) City size and the urban heat island. Atmos. Environ.
      7, 769-779.
53. Garstang, M., Tyson, P.D. and Emmitt, G.D. (1975) The structure of heat
      islands. Rev. Geophys. Space Phys. 13, 139-165.
54. Lowry, W.P. (1977) Empirical estimation of urban effects on climate: A
      problem analysis. J. Appl. Meteorol. 16, 124-135.
55. Oke, T.R. (1982) The energetic basis of the urban heat island. Quart. J.
      Roy. Meteorol. Soc. 108, 1-24.
56. Ellsaesser, H.W., MacCracken, M.C., Walton, J.J. and Grotch, S.L. (1986)
      Global climatic trends as revealed by the recorded data. Rev. Geophys.
      24, 745-792.
57. Landsberg, H.E. (1981) The Urban Climate. Academic Press, Orlando, FL.
58. Landsberg, H.E. (1981) City climate. In Landsberg, H.E. (ed) General
      Climatology: World Survey of Climatology Vol. 3. Elsevier, Amsterdam,
      Holland, pp. 302-312.
59. Cayan, D.R. and Douglas, A.V. (1984) Urban influences on surface
      temperatures in the southwestern United States during recent decades.
      J. Clim. Appl. Meteorol. 23, 1520-1530.
60. Karl, T.R. (1985) Perspective on climate change in North America during
      the twentieth century. Phys. Geogr. 6, 207-229.
61. Karl, T.R. and Quayle, R.G. (1988) Climatic change in fact and theory:
      Are we collecting the facts? Clim. Change 13, 5-17.
62. Feng, J.Z. and Petzold, D.E. (1988) Temperature trends through

REFERENCES

urbanization in metropolitan Washington, D.C., 1945-1979. Meteorol. Atmos. Phys. 38, 195-201.

63. Jones, P.D., Raper, S.C.B., Bradley, R.S., Diaz, H.F., Kelly, P.M. and Wigley, T.M.L. (1986) Northern hemispheric surface air temperature variations: 1851-1984. J. Clim. Appl. Meteorol. 25, 161-179.

64. Jones, P.D., Raper, S.C.B. and Wigley, T.M.L. (1986) Southern hemispheric surface air temperature variations: 1851-1984. J. Clim. Appl. Meteorol. 25, 1213-1230.

65. Jones, P.D., Wigley, T.M.L. and Wright, P.B. (1986) Global temperature variations between 1861 and 1984. Nature 322, 430-434.

66. Jones, P.D., Kelly, P.M., Goodess, C.M. and Karl, T. (1989) The effect of urban warming on the Northern Hemisphere temperature average. J. Climatol., in press.

67. Wigley, T.M.L. and Jones, P.D. (1988) Do large-area-average temperature series have an urban warming bias? Clim. Change 12, 313-319.

68. Hansen, J. and Lebedeff, S. (1987) Global trends of measured surface air temperature. J. Geophys. Res. 25, 13,345-13,372.

69. Quinlan, F.T., Karl, T.R. and Williams, C.N., Jr. (1987) United States Historical Climatology Network (HCN) Serial Temperature and Precipitation Data. NDP-019. Carbon Dioxide Information Center, Oak Ridge Nat. Lab., Oak Ridge, TN.

70. Karl, T.R., Kukla, G. and Gavin, J. (1986) Relationship between decreased temperature range and precipitation trends in the United States and Canada, 1941-80. J. Clim. Appl. Meteorol. 26, 1878-1886.

71. Karl, T.R. and Williams, C.N., Jr. (1987) An approach to adjusting climatological time series for discontinuous inhomogeneities. J. Clim. Appl. Meteorol. 246, 1744-1763.

72. Karl, T.R. and Jones, P.D. (1989) Urban bias in area-averaged surface air temperature trends. Bull. Amer. Meteorol. Soc. 70, 265-270.

73. Oke, T.R., Ed. (1984) Proc. Tech. Conf.: Urban Climatology and its Applications with Special Regard to Tropical Areas. World Meteorological Organization, Geneva, Switzerland.

74. Matson, M. (1978) Satellite detection of urban heat islands. Mon. Wea. Rev. 106, 1725-1734.

75. Weller, G. (1982) Geophysical Institute Annual Report: Urban Climates in Alaska. Univ. Alaska, Fairbanks, AK.

76. Jones, P.D. (1988) Hemispheric surface temperature variations: Recent trends and an update to 1987. J. Climate 1, 654-660.

77. Schlesinger, M.E. and Mitchell, J.F.B. (1987) Climate model simulations of the equilibrium climatic response to increased carbon dioxide. Rev. Geophys. 25, 760-798.

78. Lachenbruch, A.H. and Marshall, B.V. (1986) Changing climate: Geothermnal evidence from permafrost in the Alaskan Arctic. Science 234, 689-696.

79. Michaels, P.J., Sappington, D.E. and Stooksbury, D.E. (1988) Anthropogenic warming in north Alaska? J. Climate 1, 942-945.

80. MacCracken, M.C., Cess, R.D. and Potter, G.L. (1986) Climatic effects of anthropogenic Arctic aerosols: An illustration of climate feedback mechanisms with one- and two-dimensional climate models. J. Geophys. Res. 91, 14,445-14,450.

81. Harte, J. and Williams, J. (1988) Arctic aerosol and Arctic climate: Results from an energy budget model. Clim. Change 13, 161-189.

82. Gribbon, J. (1988) Britain shivers in the global greenhouse. New Sci. 118(1616), 42-43.

83. Barnett, T.P. (1988) Recent changes in the world's oceans: How well are they known? Trans. Amer. Geophys. Union 69, 1045.

84. Oort, A.H., Pan, Y.H., Reynolds, R.W. and Ropelewski, C.F. (1987) Historical trends in the surface temperature over the oceans based on the COADS. Clim. Dynam. 2, 29-38.

REFERENCES

85. Namias, J. (1959) Recent seasonal interactions between North Pacific waters and the overlying atmospheric circulation. J. Geophys. Res. 64, 631-646.
86. Bjerknes, J. (1966) A possible response of the atmospheric Hadley circulation to equatorial anomalies of ocean temperature. Tellus 18, 820-829.
87. Bjerknes, J. (1969) Atmospheric teleconnections from the equatorial Pacific. Mon. Wea. Rev. 97, 163-172.
88. Murakami, T. (1987) Intraseasonal atmospheric teleconnection patterns during the Northern Hemisphere summer. Mon. Wea. Rev. 115, 2133-2154.
89. Murakami, T. (1988) Intraseasonal atmospheric teleconnection patterns during the Northern Hemisphere winter. J. Climate 1, 117-131.
90. Rowntree, P.R. (1972) The influence of tropical east Pacific Ocean temperatures on the atmosphere. Quart. J. Roy. Meteorol. Soc. 98, 290-321.
91. White, W.B. and Walker, A.E. (1973) Meriodional atmospheric teleconnections over the North Pacific from 1950 to 1972. Mon. Wea. Rev. 101, 817-822.
92. Rogers, J.C. (1976) Sea surface temperature anomalies in the eastern north Pacific and associated wintertime fluctuations over North America, 1960-73. Mon. Wea. Rev. 104, 985-993.
93. Rowntree, P.R. (1976) Tropical forcing of atmospheric motions in a numerical model. Quart. J. Roy. Meteorol. Soc. 102, 583-605.
94. Trenberth, K.E. and Paolino, D.A. (1981) Characteristic patterns of variability of sea level pressure in the Northern Hemisphere. Mon. Wea. Rev. 109, 1169-1189.
95. Wallace, J.M. and Gutzler, D.S. (1981) Teleconnections in the geopotential height field during the Northern Hemisphere winter. Mon. Wea. Rev. 109, 784-812.
96. Barnett, T.P. and Preisendorfer, R.W. (1978) Multifield analogue prediction of short-term climate fluctuations using a climate state vector. J. Atmos. Sci. 35, 1771-1787.
97. Webster, P.J. (1981) Mechanisms determining the atmospheric response to sea surface temperature anomalies. J. Atmos. Sci. 38, 554-571.
98. Webster, P.J. (1982) Seasonality in the local and remote atmospheric response to sea surface temperature anomalies. J. Atmos. Sci. 39, 41-52.
99. Erickson, C.O. (1983) Hemispheric anomalies of 700 mb height and sea level pressure related to mean summer temperatures over the United States. Mon. Wea. Rev. 111, 545-561.
100. Kawamura, R. (1984) Relation between atmospheric circulation and dominant sea surface temperature anomaly patterns in the North Pacific during the northern winter. J. Meteorol. Soc. Japan 62, 910-916.
101. Kawamura, R. (1986) Seasonal dependency of atmosphere-ocean interaction over the North Pacific. J. Meteorol. Soc. Japan 64, 363-371.
102. Iwasaka, N., Hanawa, K. and Toba, Y. (1987) Analysis of SST anomalies in the North Pacific and their relation to 500 mb height anomalies over the Northern Hemisphere during 1969-1979. J. Meteorol. Soc. Japan 65, 103-114.
103. Park, C.-K. and Kung, E.C. (1988) Principal components of the North American summer temperature field and the antecedent oceanic and atmospheric conditions. J. Meteorol. Soc. Japan 66, 677-690.
104. Reifsnyder, W.E. (1989) A tale of ten fallacies: The skeptical inquirer's view of the carbon dioxide/climate controversy. Agric. For. Meteorol., in press.
105. Currie, R.G. (1983) Detection of 18.6-year nodal induced drought in the Patagonian Andes. Geophys. Res. Lett. 10, 1089-1092.
106. Currie, R.G. (1984) Evidence for 18.6 year lunar nodal drought in western North America during the past millennium. J. Geophys. Res. 89,

REFERENCES

1295-1308.
107. Currie, R.G. and Fairbridge, R.W. (1985) Periodic 18.6-year and cyclic 11-year induced drought and flood in northeastern China. Quat. Sci. Rev. 4, 109-134.
108. Borchert, J.R. (1971) The dust bowl of the 1970s. Ann. Assoc. Amer. Geogr. 61, 1-22.
109. Rosenberg, N.J., Ed. (1978) North American Droughts. Westview Press, Boulder, CO.
110. Currie, R.G. (1981) Evidence for 18.6 year signal in temperature and drought conditions in North America since A.D. 1800. J. Geophys. Res. 86, 11,055-11,064.
111. Russell, N. (1988) Dust gets in your eyes. New Sci. 119(1625), 61.
112. LaMarche, V.C., Jr. (1974) Paleoclimatic inferences from long tree-ring records. Science 183, 1043-1048.
113. Fritts, H.C. (1976) Tree Rings and Climate. Academic Press, Orlando, FL.
114. Fritts, H.C., Lofgren, G.R. and Gordon, G.A. (1979) Variations in climate since 1602 as reconstructed from tree rings. Quat. Res. 12, 18-46.
115. Duvick, D.N. and Blasing, T.J. (1981) A dendroclimatic reconstruction of annual precipitation amounts in Iowa since 1680. Water Resources Res. 17, 1183-1189.
116. Blasing, T.J. and Duvick, D.N. (1984) Reconstruction of precipitation history in the North American corn belt using tree rings. Nature 307, 143-145.
117. Stahle, D.W. and Hehr, J.G. (1984) Dendroclimatic relationships of post oak across a precipitation gradient in the south-central United States. Ann. Assoc. Amer. Geogr. 74, 561-573.
118. Cook, E.R. and Jacoby, G.C., Jr. (1983) Potomac River stream flow since 1730 as reconstructed by tree rings. J. Clim. Appl. Meteorol. 22, 1659-1672.
119. Phipps, R.L. (1983) Streamflow on the Occoquan River in Virginia as reconstructed from tree-ring series. Water Resources Bull. 19, 735-747.
120. Brinkmann, W.A.R. (1987) Water supplies to the Great Lakes reconstructed from tree rings. J. Clim. Appl. Meteorol. 26, 530-538.
121. Cook, E.R. and Jacoby, G.C., Jr. (1977) Tree-ring-drought relationships in the Hudson Valley, New York. Science 198, 399-401.
122. Stockton, C.W. and Meko, D.M. (1983) Drought recurrence in the Great Plains as reconstructed from long-term tree-ring records. J. Clim. Appl. Meteorol. 22, 17-29.
123. Stahle, D.W. and Cleaveland, M.K. (1988) Texas drought history reconstructed and analyzed from 1698 to 1980. J. Climate 1, 59-74.
124. Blasing, T.J., Stahle, D.W. and Duvick, D.N. (1988) Tree ring-based reconstruction of annual precipitation in the south-central United States from 1750 to 1980. Water Resources Res. 24, 163-171.
125. Kellogg, W.W. (1988) Climate change is already here to stay. Trans. Amer. Geophys. Union 69, 1067.
126. Nicolis, C. (1987) Long term climatic variability and chaotic dynamics. Tellus 39A, 1-9.
127. Parker, D.E. and Folland, C.K. (1988) The nature of climatic variability. Meteorol. Mag. 117, 201-210.
128. Salazar, J.M. and Nicolis, C. (1988) Self-generated aperiodic behavior in a simple climate model. Clim. Dynam. 3, 105-114.
129. Lorenz, E.N. (1976) Nondeterministic theories of climatic change. Quat Res. 6, 495-506.
130. Chervin, R.M. (1986) Interannual variability and seasonal climate predictability. J. Atmos. Sci. 43, 233-251.
131. Gordon, H.B. and Hunt, B.G. (1987) Interannual variability of the simulated hydrology in a climatic model -- Implications for drought.

Clim. Dynam. 1, 113-130.

132. Hunt, B.G. and Gordon, H.B. (1988) The problem of "naturally"-occurring drought. Clim. Dynam. 3, 19-33.

133. Palmer, T.N. and Brankovic, C. (1989) The 1988 US drought linked to anomalous sea surface temperature. Nature 338, 54-57.

134. Linden, E. (1988) Big chill for the greenhouse. Time 132(18), 90.

135. Kung, E.C. (1988) Principal-components forecast of the North American summer temperature and precipitation with antecedent oceanic conditions. Trans. Amer. Geophys. Union 69, 1067.

136. Trenberth, K.E. and Bramstator, G.W. (1988) Origins of the 1988 North American drought. Trans. Amer. Geophys. Union 69, 1067.

137. Namias, J. (1988) The 1988 summer drought over the great plains -- A classic example of air-sea-land interactions. Trans. Amer. Geophys. Union 69, 1067.

138. Trenberth, K.E. (1989) The wayward winds. Nat. Hist. 89(1), 44-45.

139. Rasmussen, E.M. and Carpenter, T.H. (1982) Variations in tropical sea surface temperature and surface wind fields associated with the Southern Oscillation/El Nino. Mon. Wea. Rev. 110, 354-384.

140. Philander, S.G.H. (1983) El Nino Southern Oscillation phenomena. Nature 302, 295-301.

141. Gill, A.E. and Rasmussen, E.M. (1983) The 1982-83 climate anomaly in the equatorial Pacific. Nature 306, 229-234.

142. Nicholls, N. (1988) El Nino-Southern Oscillation and rainfall variability. J. Climate 1, 418-421.

143. Kerr, R.A. (1988) La Nina's big chill replaces El Nino. Science 241, 1037-1038.

144. Bradley, R., Diaz, H., Kiladis, G. and Eischeid, J. (1987) ENSO signal in continental temperature and precipitation records. Nature 327, 497-501.

145. Shaw, H.R. and Moore, J.G. (1988) Magmatic heat and the El Nino cycle. Trans. Amer. Geophys. Union 69, 1553, 1564-1565.

146. Cann, J. (1989) Sea-floor volcano erupts ideas. Nature 337, 603-604.

147. Cook, E.R., Kablack, M.A. and Jacoby, G.C. (1988) The 1986 drought in the southeastern United States: How rare an event was it? J. Geophys. Res. 93, 14,257-14,260.

148. Heim, R.R., Jr. (1988) About that drought. Weatherwise 41, 266-271.

149. Karl, T.R., Heim, R., McNab, A.L., Knight, R.W., Quinlan, F.T. and Reek, T. (1988) The 1988 drought in the United States: Just another drought? Trans. Amer. Geophys. Union 69, 1067.

150. Trenberth, K.E., Branstator, G.W. and Arkin, P.A. (1988) Origins of the 1988 North American drought. Science 242, 1640-1645.

151. Hanson, K., Maul, G.A. and Karl, T.R. (1989) Are atmospheric "greenhouse" effects apparent in the climatic record of the contiguous U.S. (1895-1987)? Geophys. Res. Lett. 16, 49-52.

152. Namias, J. (1989) Cold waters and hot summers. Nature 338, 15-16.

153. Eden, P. (1988) Hurricane Gilbert. Weather 43, 446-448.

154. Lawrence, M.B. (1989) Return of the hurricanes. Weatherwise 42(1), 22-27.

155. Emanuel, K.A. (1987) The dependence of hurricane intensity on climate. Nature 326, 483-485.

156. Hobgood, J.S. and Cerveny, R.S. (1988) Ice-age hurricanes and tropical storms. Nature 333, 243-245.

157. AMS Council and UCAR Board of Trustees. (1988) The changing atmosphere- - Challenges and opportunities. Bull. Amer. Meteorol. Soc. 69, 1434-1440.

158. Mitchell, J.F.B., Wilson, C.A. and Cunnington, W.M. (1987) On CO_2 climate sensitivity and model dependence of results. Quart. J. Roy. Meteorol. Soc. 113, 293-322.

159. Sellers, W.D. and Liu, W. (1988) Temperature patterns and trends in the

upper troposphere and lower stratosphere. _J. Climate_ 1, 573-581.

160. Anthes, R.A. (1982) _Tropical Cyclones, Their Evolution, Structure and Effects_. Amer. Meteorol. Soc., Boston, MA.

161. Wigley, T.M.L. (1988) Events frozen in time. _Nature_ 335, 23.

162. Choudhury, B. and Kukla, G. (1979) Impact of CO_2 on cooling of snow and water surfaces. _Nature_ 280, 668-676.

163. Idso, S.B. (1983) Do increases in atmospheric CO_2 have a _cooling_ effect on surface air temperature? _Climatol. Bull_. 17, 22-26.

164. Idso, S.B. (1983) CO_2 as an _inverse_ greenhouse gas. _In_ Spiro, I.J. and Mollicone, R.A. (eds) _Infrared Technology IX. Proc. SPIE_ 430, 232-239.

165. Idso, S.B. (1984) What if increases in atmospheric CO_2 have an _inverse_ greenhouse effect? I. Energy balance considerations related to surface albedo. _J. Climatol._ 4, 399-409.

166. Thompson, L.G., Mosley-Thompson, E., Dansgaard, W. and Grootes, P.M. (1986) The Little Ice Age as recorded in the stratigraphy of the tropical Quelccalya ice cap. _Science_ 234, 361-364.

167. Groveman, B.S. and Landsberg, H.E. (1979) Simulated Northern Hemisphere temperature departures 1579-1880. _Geophys. Res. Lett._ 6, 767-769.

168. Friedli, H., Moor, E., Oeschger, H., Siegenthaler, U. and Stauffer, B. (1984) $^{13}C/^{12}C$ ratios in CO_2 extracted from Antarctic ice. _Geophys. Res. Lett._ 11, 1145-1148.

169. Friedli, H., Lotscher, H., Oeschger, H., Siegenthaler, U. and Stauffer, B. (1986) Ice core record of the $^{13}C/^{12}C$ ratio of atmospheric CO_2 in the past two centuries. _Nature_ 324, 237-238.

170. Oeschger, H. and Stauffer, B. (1986) Review of the history of atmospheric CO_2 recorded in ice cores. _In_ Trabalka, J.R. and Reichle, D.E. (eds) _The Changing Carbon Cycle: A Global Analysis_. Springer-Verlag, New York, NY, pp. 89-108.

171. Idso, S.B. (1988) Greenhouse warming or Little Ice Age demise: A critical problem for climatology. _Theoret. Appl. Climatol._ 39, 54-56.

172. Grove, J.M. (1988) _The Little Ice Age_. Methuen, London, UK.

173. Wilson, A.T. (1978) Pioneer agriculture explosion and CO_2 levels in the atmosphere. _Nature_ 273, 40-41.

174. Flohn, H. (1980) _Possible Climatic Consequences of a Man-Made Global Warming_. International Institute for Applied Systems Analysis, Laxenburg, Austria.

175. Nichols, H. (1974) Arctic North American paleoecology: The recent history of vegetation and climate deduced from pollen analysis. _In_ Ives, J.D. and Barry, R.G. (eds) _Arctic and Alpine Environments_. Methuen, London, UK, pp. 660-667.

176. Heusser, C.J. (1974) Vegetation and the climate of the southern Chilean lake district during and since the last interglaciation. _Quat. Res._ 4, 290-315.

177. Bartlein, P.J., Webb, T., III and Fleri, E. (1984) Holocene climatic change in the Northern Midwest: Pollen-derived estimates. _Quat. Res._ 22, 361-374.

178. Guiot, J. (1987) Late Quaternary climatic change in France estimated from multivariate pollen time series. _Quat. Res._ 28, 100-118.

179. Huntley, B. and Prentice, I.C. (1988) July temperatures in Europe from pollen data, 6000 years before present. _Science_ 241, 687-690.

180. Delmas, R.J., Ascencio, J.-M. and Legrand, M. (1980) Polar ice evidence that atmospheric CO_2 20,000 yr BP was 50% of present. _Nature_ 284, 155-157.

181. Neftel, A., Oeschger, H., Schwander, J., Stauffer, B. and Zumbrunn, R. (1982) Ice core sample measurements give atmospheric CO_2 content during the past 40,000 yr. _Nature_ 295, 220-223.

182. Neftel, A., Oeschger, H., Staffelbach, T. and Stauffer, B. (1988) CO_2 record in the Byrd ice core 50,000-5,000 years BP. _Nature_ 331, 609-611.

REFERENCES   SECTION 5

183. Shackleton, N.J., Hall, M.A., Line, J. and Shuxi, C. (1983) Carbon
     isotope data in core V19-30 confirm reduced carbon dioxide
     concentration in the ice age atmosphere. Nature 306, 319-322.
184. Bryson, R.A. (1989) Late Quaternary volcanic modulation of Milankovitch
     climate forcing. Theoret. Appl. Climatol. 39, 115-125.
185. Bryson, R.A. and Goodman, B.M. (1986) Milankovitch and global ice volume
     simulation. Theoret. Appl. Climatol. 37, 22-28.
186. Kondratyev, K. Ya. and Khvorostyanov, V.I. (1989) Modelling of cloud
     formation due to air-sea interactions in an energy-active zone.
     Boundary-Layer Meteorol. 46, 229-249.
187. Bjerknes, J. (1965) Tech. Note No. 66: Atmosphere-Ocean Interactions
     During the Little Ice Age (17th to 19th Centuries A.D.). World
     Meteorological Organization, Geneva, Switzerland, pp. 77-88.
188. Monin, A.S. and Shishkov, Yu. A. (1979) The History of Climate.
     Gidrometeoizdat, Leningrad, USSR.
189. Kerr, R.A. (1988) New ways to chill Earth. Science 241, 532-533.
190. Campbell, P. (1987) The Antarctic cornucopia. Nature 329, 387.
191. Sundquist, E.T. (1987) Ice core links CO_2 to climate. Nature 329, 389-
     390.
192. Lorius, C., Barkov, N.I., Jouzel, J., Korotkevich, Y.S., Kotlyakov, V.M.
     and Raynaud, D. (1988) Antarctic ice core: CO_2 and climatic change
     over the last climatic cycle. Trans. Amer. Geophys. Union 69, 681 &
     683-684.
193. Lorius, C., Jouzel, J., Ritz, C., Merlivat, L., Barkov, N.I.,
     Korotkevich, Y.S. and Kotlyakov, V.M. (1985) A 150,000 year climatic
     record from Antarctic ice. Nature 316, 591-596.
194. Jouzel, J., Lorius, C., Petit, J.R., Genthon, C., Barkov, N.I.,
     Kotlyakov, V.M. and Petrov, V.M. (1987) Vostok ice core: A continuous
     isotope temperature record over the last climatic cycle (160,000
     years). Nature 329, 403-408.
195. Barnola, J.M., Raynaud, D., Korotkevich, Y.S. and Lorius, C. (1987)
     Vostok ice core provides 160,000-year record of atmospheric CO_2.
     Nature 329, 408-414.
196. Genthon, C., Barnola, J.M., Raynaud, D., Lorius, C., Jouzel, J., Barkov,
     N.I., Korotkevich, Y.S. and Kotlyakov, V.M. (1987) Vostok ice core:
     Climatic response to CO_2 and orbital forcing changes over the last
     climatic cycle. Nature 329, 414-418.
197. Tangley, L. (1988) Preparing for climate change. BioScience 38, 14-18.
198. Idso, S.B. (1988) Carbon dioxide and climate in the Vostok ice core.
     Atmos. Environ. 22, 2341-2342.
199. Martin, J.H. and Fitzwater, S.E. (1988) Iron deficiency limits
     phytoplankton growth in the north-east Pacific subarctic. Nature 331,
     341-343.
200. Gribbin, J. (1988) The oceanic key to climatic change. New Sci.
     118(1613), 32-33.
201. Broecker, W.S. (1982) Ocean chemistry during glacial time Geochim.
     Cosmochim. Acta 46, 1689-1705.
202. Broecker, W.S. (1982) Glacial to interglacial changes in ocean
     chemistry. Prog. Oceanog. 11, 151-197.
203. Sarmiento, J.L. and Toggweiler, J.R. (1984) A new model for the role of
     the oceans in determining atmospheric P_{CO2}. Nature 308, 621-624.
204. Siegenthaler, U. and Wenk, T. (1984) Rapid atmospheric CO_2 variations
     and ocean circulation. Nature 308, 624-626.
205. Barber, R.T. (1988) Ocean primary productivity. Trans. Amer. Geophys.
     Union 69, 1045.
206. Knox, F. and McElroy, M.B. (1984) Changes in atmospheric CO_2: Influence
     of the marine biota at high latitude. J. Geophys. Res. 89, 4629-4637.
207. Martin, J.H., Gordon, M. and Fitzwater, S. (1988) Oceanic iron
     distributions in relation to phytoplankton productivity. Trans. Amer.

REFERENCES

Geophys. Union 69, 1045.

208. Taylor, R.S. (1964) Abundance of chemical elements in the continental crust: A new table. Geochim. Cosmochim. Acta 28, 1273-1285.

209. DeAngelis, M., Barkov, N.I. and Petrov, V.N. (1987) Aerosol concentrations over the last climatic cycle (160 kyr) from an Antarctic ice core. Nature 325, 318-321.

210. Legrand, M.R., Lorius, C., Barkov, N.I. and Petrov, V.N. (1988) Vostok (Antarctica) ice core: Atmospheric chemistry changes over the last climatic cycle (160,000 yr). Atmos. Environ. 22, 317-331.

211. Petit, J.-R., Briat, M. and Royer, A. (1981) Ice age aerosol content from East Antarctic ice core samples and past wind strength. Nature 293, 391-394.

212. CLIMAP Project Members. (1981) Seasonal reconstructions of the earth's surface at the last glacial maximum. In McIntyre, A. and Cline, R. (eds) Map Charts Ser. MC 36. Geol. Soc. Amer., Boulder, CO.

213. Parkin, D.W. and Shackleton, N.J. (1973) Trade wind and temperature correlations down a deep-sea core off the Sahara coast. Nature 245, 455-457.

214. Sarnthein, M., Tetzlaff, G., Koopman, B., Walter, K. and Pflaumann, U. (1981) Glacial and interglacial wind regimes over the east subtropical Atlantic and N.W. Africa. Nature 293, 193-196.

215. Roger, T. and Wilson, S. (1989) Carbon dioxide and climate in the Vostok ice core: Why does the system oscillate? Atmos. Environ. 22, 2637-2638.

216. Ram, M., Gayley, R.I. and Petit, J.-R. (1988) Insoluble particles in Antarctic ice: Background aerosol size distribution and diatom concentration. J. Geophys. Res. 93, 8378-8382.

217. DeAngelis, M., Jouzel, J., Lorius, C., Merlivat, R., Petit, J.-R. and Raynaud, D. (1984) Ice age data for climate modelling from an Antarctic (Dome C) ice core. In Berger, A.L. and Nicolis, C. (eds) New Perspectives in Climate Modelling. Elsevier Sci. Pub., New York, NY, pp. 23-45.

218. Gaudichet, A., De Angelis, M., Lefever, R., Petit, J.R., Korotkevitch, Y.S. and Petrov, V.N. (1988) Mineralogy of insoluble particles in the Vostok Antarctic ice core over the last climatic cycle (150 kyr). Geophys. Res. Lett. 15, 1471-1474.

219. Batifol, F., Boutron, C. and de Angelis, M. (1989) Changes in copper, zinc and cadmium concentration in Antarctic ice during the past 40,000 years. Nature 337, 544-546.

220. Pedersen, T.F. (1983) Increased productivity in the eastern equatorial Pacific during the last glacial maximum (19,000 to 14,000 yr B.P.). Geology 11, 16-19.

221. Muller, P.J., Erlenkeuser, H. and von Grafenstein, R. (1983) Glacial-interglacial cycles in oceanic productivity inferred from organic carbon contents in eastern North Atlantic sediment cores. In Thiede, J. and Suess, E. (eds) Coastal Upwelling: Its Sediment Record. Part B: Sedimentary Records of Ancient Coastal Upwelling. Plenum Press, New York, NY, pp. 365-389.

222. Morris, R.J., McCartney, M.J. and Weaver, P.P.E. (1984) Sapropelic deposits in a sediment from the Guinea Basin, South Atlantic. Nature 309, 611-614.

223. Lyle, M., Murray, D.W., Finney, B.P., Dymond, J., Robbins, J.M. and Brooksforce, K. (1988) The record of late Pleistocene biogenic sedimentation in the eastern tropical Pacific Ocean. Paleoceanography 3, 39-59.

224. Lyle, M. (1988) Climatically forced organic carbon burial in equatorial Atlantic and Pacific Oceans. Nature 335, 529-532.

225. Pedersen, T.F., Pickering, M., Vogel, J.S., Southon, J.N. and Nelson, D.E. (1988) The response of benthic foraminifera to productivity

cycles in the eastern equatorial Pacific: Faunal and geochemical constraints on glacial bottom water oxygen levels. Paleoceanography 3, 157-168.

226. Imbrie, J. and Kipp, N.G. (1971) A new micropaleontological method for quantitative paleoclimatology: Application to a late Pleistocene Caribbean core. In Turekian, K.K. (ed) The Late Cenozoic Glacial Ages. Yale Univ. Press, New Haven, CT, pp. 71-181.

227. Mix, A.C. (1989) Influence of productivity variations on long-term atmospheric CO_2. Nature 337, 541-544.

228. Monastersky, R. (1988) Devil's Hole fires ice age debate. Sci. News 134, 356.

229. Saigne, C. and Legrand, M. (1987) Measurements of methanesulphonic acid in Antarctic ice. Nature 330, 240-242.

230. Legrand, M. and Saigne, C. (1988) Formate, acetate and methanesulfonate measurements in Antarctic ice: Some geochemical implications. Atmos. Environ. 22, 1011-1017.

231. Legrand, M.R., Delmas, R.J. and Charlson, R.J. (1988) Climate forcing implications from Vostok ice-core sulphate data. Nature 334, 418-420.

232. Palmer, E. and Newton, C.W. (1969) Atmospheric Circulation Systems. Academic Press, New York, NY.

233. Lamb. H.H. (1985) Climatic History and the Future. Princeton Univ. Press, Princeton, NJ.

234. Ramanathan, V., Cess, R.D., Harrison, E.F., Minnis, P., Barkstrom, B.R., Ahmad, E. and Hartmann, D. (1989) Cloud-radiative forcing and climate: Results from the Earth Radiation Budget Experiment. Science 243, 57-63.

235. Harvey, L.D.D. (1988) Climatic impact of ice-age aerosols. Nature 334, 333-335.

236. Slingo, T. (1988) Can plankton control climate? Nature 336, 421.

237. Schwartz, S.E. (1988) Are global cloud albedo and climate controlled by marine phytoplankton? Nature 336, 441-445.

238. Schwartz, S.E. (1988) Control of global cloud albedo and climate by marine dimethylsulfide emissions: A test by anthropogenic sulfur dioxide emissions. Proc.: Symposium on the Role of Clouds in Atmospheric Chemistry and Global Climate. Amer. Meteorol. Soc., Boston, MA, pp. 4-8.

239. Jones, P.D. (1988) Hemispheric surface air temperature variations: Recent trends and an update to 1987. J. Climate 1, 654-660.

240. Keeling, C.D. (1982) The oceans and terrestrial biosphere as future sinks for fossil fuel CO_2. In Reck, R.A. and Hummel, J.R. (eds) AIP Conf. Proc. No. 82: Interpretation of Climate and Photochemical Models, Ozone and Temperature Measurements. Amer. Inst. Physics, New York, NY, pp. 47-82.

241. Surlyk, F. (1980) The Cretaceous-Tertiary boundary event. Nature 285, 187-188.

242. Rensberger, B. (1986) Death of dinosaurs. The true story? Sci. Digest 94(5), 28-35, 77-78.

243. Alvarez, L.W., Alvarez, W., Asaro, F. and Michel, H.V. (1980) Extraterrestrial cause for the Cretaceous-Tertiary extinction. Science 208, 1095-1107.

244. Alvarez, L.W. (1983) Experimental evidence that an asteroid impact led to the extinction of many species 65 million years ago. Proc. Natl. Acad. Sci. U.S.A. 80, 627-642.

245. Alvarez, W., Kauffman, E.G., Surlyk, F., Alvarez, L.W., Asaro, F. and Michel, H.V. (1984) Impact theory of mass extinctions and the invertebrate fossil record. Science 223, 1135-1141.

246. Hsu, K.J. (1980) Terrestrial catastrophe caused by cometary impact at the end of Cretaceous. Nature 285, 201-203.

247. Rampino, M. (1989) Dinosaurs, comets and volcanoes. New Sci. 121(1652),

54-58.

248. Toon, O.B., Pollack, J.B., Ackerman, T.P., Turco, R.P, McKay, C.P. and
     Liu, M.S. (1982) Evolution of an impact-generated dust cloud and its
     effects on the atmosphere. Geol. Soc. Amer. Spec. Pap. 190, 187-200.
249. Pollack, J.B., Toon, O.B., Ackerman, T.P., McKay, C.P. and Turco, R.P.
     (1983) Environmental effects of an inpact-generated dust cloud:
     Implications for the Cretaceous-Tertiary extinctions. Science 219,
     287-289.
250. Kyte, F.T., Zhou, Z. and Wasson, J.T. (1980) Siderophile-enriched
     sediments from the Cretaceous-Tertiary boundary. Nature 288, 651-656.
251. Orth, C.J., Gilmore, J.S., Knight, J.D., Pillmore, C.L., Tschudy, R.H.
     and Fassett, J.E. (1981) An iridium abundance anomaly at the
     palynological Cretaceous-Tertiary boundary in northern New Mexico.
     Science 214, 1341-1343.
252. Ganapathy, R. (1980) A major meteorite impact on the Earth 65 million
     years ago: Evidence from the Cretaceous-Tertiary boundary clay.
     Science 209, 921-923.
253. Smit, J. and Hertogen, J. (1980) An extraterrestrial event at the
     Cretaceous-Tertiary boundary. Nature 285, 198-200.
254. McHone, J.F., Nieman, R.A., Lewis, C.F. and Yates, A.M. (1989)
     Stishovite at the Cretaceous-Tertiary boundary, Raton, New Mexico.
     Science 243, 1182-1184.
255. Wolbach, W.S., Lewis, R.S. and Anders, E. (1985) Cretaceous extinctions:
     Evidence for wildfires and search for meteoritic material. Science
     230, 167-170.
256. Wolbach, W.S., Gilmour, I., Anders, E., Orth, C.J. and Brooks, R.S.
     (1988) Global fire at the Cretaceous-Tertiary boundary. Nature 334,
     665-669.
257. Venkatesan, M.I. and Dahl, J. (1989) Organic geochemical evidence for
     global fires at the Cretaceous/Tertiary boundary. Nature 338, 57-
     60.
258. Bourgeois, J., Hansen, T.A., Wiberg, P.L. and Kauffman, E.G. (1988) A
     tsunami deposit at the Cretaceous-Tertiary boundary in Texas. Science
     241, 567-570.
259. Macdougall, J.D. (1988) Seawater strontium isotopes, acid rain, and the
     Cretaceous-Tertiary boundary. Science 239, 485-487.
260. Officer, C.B. and Geieve, R.A.F. (1986) The impact of impacts and the
     nature of nature. Trans. Amer. Geophys. Union 67, 633, 637.
261. Hallam, A. (1987) End-Cretaceous mass extinction event: Argument for
     terrestrial causation. Science 238, 1237-1242.
262. Duncan, R.A. and Pyle, D.G. (1988) Rapid eruption of the Deccan flood
     basalts at the Cretaceous/Tertiary boundary. Nature 333, 841-843.
263. Courtillot, V., Feraud, G., Maluski, H., Vandamme, D., Moreau, M.G. and
     Iesse, J. (1988) Deccan flood basalts and the Cretaceous/Tertiary
     boundary. Nature 333, 843-846.
264. Loper, D.E. and McCartney, K. (1988) Shocked quartz found at the K/T
     boundary. Trans. Amer. Geophys. Union 69, 961, 971-972.
265. Henbest, N. (1988) How astronomers scored a knockout cometary blow. New
     Sci. 119(1630), 32-33.
266. Rice, A. (1989) Snowbird II: A dissenting view. Science 243, 875-876.
267. Paul, G.S. (1989) Giant meteor impacts and great eruptions: Dinosaur
     killers? BioScience 39, 162-172.
268. Hecht, J. (1988) Evolving theories for old extinctions. New Sci.
     120(1638), 28-30.
269. Kerr, R.A. (1988) Snowbird II: Clues to Earth's impact history. Science
     242, 1380-1382.
270. Garwin, L. (1988) Mass extinction: Of impacts and volcanoes. Nature
     336, 714-716.
271. Brenneke, J.C. and Anderson, T.F. (1977) Carbon isotope variations in

pelagic carbonates. *Trans*. *Amer*. *Geophys*. *Union* 58, 415.

272. Thierstein, H.R. and Berger, W.H. (1978) Injection events in ocean history. *Nature* 276, 461-466.

273. Hsu, K.J. and McKenzie, J.A. (1985) A "strangelove" ocean in the earliest Tertiary. *In* Sundquist, E.T. and Broecker, W.S. (eds) *The Carbon Cycle and Atmospheric* CO_2: *Natural Variations Archean to Present*. Amer. Geophys. Union, Washington, DC, pp. 487-492.

274. Arthur, M.A., Zachos, J.C. and Jones, D.S. (1987)   Primary productivity and the Cretaceous/Tertiary boundary event in the oceans. *Cret*. *Res*. 8, 43-54.

275. Perch-Nielsen, K., McKenzie, J. and He, Q. (1982)   Biostratigraphy and isotope stratigraphy and the "catastrophic" extinction of calcareous nanoplankton at the Cretaceous/Tertiary boundary. *Geol*. *Soc*. *Amer*. *Spec*. *Pap*. 190, 353-371.

276. Hsu, K.J., He, Q., McKenzie, J.A., Weissert, H., Perch-Nielsen, K., Oberhansli, H., Kelts, K., LaBrecque, J., Tauve, L., Krahenbuhl, U., Percival, S.F., Wright, R.K., Carman, M.F., Jr. and Schreiber, E. (1982) Mass mortality and its environmental and evolutionary consequences. *Science* 216, 249-256.

277. Smit, J. (1982) Extinction and evolution of planktonic foraminifera after a major impact at the Cretaceous/Tertiary boundary. *Geol*. *Soc*. *Amer*. *Spec*. *Pap*. 190, 329-352.

278. Hsu, K.J., McKenzie, J.A. and He, Q.X. (1982) Terminal Cretaceous environmental and evolutionary changes. *Geol*. *Soc*. *Amer*. *Spce*. *Pap*. 190, 317-328.

279. Thierstein, H.R. (1982) Terminal Cretaceous plankton extinctions: A critical assessment. *Geol*. *Soc*. *Amer*. *Spec*. *Pap*. 190, 385-399.

280. Rampino, M.R. and Volk, T. (1988) Mass extinctions, atmospheric sulphur and climatic warming at the K/T boundary. *Nature* 332, 63-65.

281. Barnett, T.P. and Baker, D.J., Jr. (1982) Possibilities of detecting CO_2-induced effects: Ocean physics. *In* Moses, H. and MacCracken, M. (eds) *Proceedings of the Workshop on First Detection of Carbon Dioxide Effects*. U.S. Dept. Energy, Washington, DC, pp. 301-342.

282. MacCracken, M. and Moses, H. (1982) The first detection of carbon dioxide effects: Workshop summary. *Bull*. *Amer*. *Meteorol*. *Soc*. 63, 1164-1178.

283. Kerr, R.A. (1988) Is the greenhouse here? *Science* 239, 559-561.

284. Solow, A.R. (1988) Detecting changes through time in the variance of a long-term hemispheric temperature record: An application of robust locally weighted regression. *J*. *Climate* 1, 290-296.

285. Solow, A.R. (1988) A Bayesian approach to statistical inference about climate change. *J*. *Climate* 1, 512-521.

286. Manabe, S. and Wetherald, R.T. (1980) On the distribution of climate change resulting from an increase in CO_2 content of the atmosphere. *J*. *Atmos*. *Sci*. 37, 99-118.

287. National Research Council, U.S. (1983) *Changing Climate*. National Academy Press, Washington, DC.

288. Bradley, R.S., Diaz, H.F., Eischeid, J.K., Jones, P.D., Kelly, P.M. and Goodess, C.M. (1987) Precipitation fluctuations over Northern Hemisphere land areas since the mid-19th century. *Science* 237, 171-175.

289. Diaz, H.F., Bradley, R.S. and Eischeid, J.K. (1989) Precipitation fluctuations over global land areas since the late 1800's. *J*. *Geophys*. *Res*. 94, 1195-1210.

290. Barnett, T.P. and Schlesinger, M.E. (1987) Detecting changes in global climate induced by greenhouse gases. *J*. *Geophys*. *Res*. 92(D12), 14,772-14,780.

291. Angell, J.K. (1988) Variations and trends in tropospheric and stratospheric global temperatures, 1958-87. *J*. *Climate* 1, 1296-1313.

292. Parker, D.E. (1985) On the detection of temperature changes induced by increasing atmospheric carbon dioxide. Quart. J. Roy. Meteorol. Soc. 111, 587-601.

293. Elling, W. and Schwentek, H. (1988) Is there a trend in stratospheric temperature from 1960 to 1986 detectable in Berlin data? Beitr. Phys. Atmos. 61, 50-55.

294. Paltridge, G. and Woodruff, S. (1981) Changes in global surface temperature from 1880 to 1977 derived from historical records of sea surface temperature. Mon. Wea. Rev. 109, 2427-2434.

295. Folland, C.K., Parker, D.E. and Kates, F.E. (1984) Worldwide marine temperature fluctuations, 1856-1981. Nature 310, 670-673.

296. Hense, A., Krahe, P. and Flohn, H. (1988) Recent fluctuations of tropospheric temperature and water vapour content in the tropics. Meteorol. Atmos. Phys. 38, 215-227.

297. Kiehl, J.T. (1983) Satellite detection of effects due to increased atmospheric carbon dioxide. Science 222, 504-506.

298. Charlock, T.P. (1984) CO_2 induced climatic change and spectral variations in the outgoing terrestrial infrared radiation. Tellus 36B, 139-148.

299. Barnett, T.P. (1985) Variations in near-global sea level pressure. J. Atmos. Sci. 41, 478-501.

300. Etkins, R. and Epstein, E.S. (1982) The rise of global sea level as an indication of climate change. Science 215, 287-289.

301. Gornitz, V., Lebedeff, S. and Hansen, J. (1982) Global sea level trend in the past century. Science 215, 1611-1614.

302. Cartwright, D.E., Barnett, T.P., Garrett, C.J.R., Carter, W.E., Peltier, R. and Thompson, K.R. (1985) Changes in relative mean sea level. Trans. Amer. Geophys. Union 66, 754-756.

303. Peltier, W.R. (1988) Global sea level and Earth rotation. Science 240, 895-901.

304. Street-Perrott, F.A. and Harrison, S.P. (1984) Temporal variations in lake levels since 30,000 yr BP -- An index of the global hydrologic cycle. In Hansen, J.E. and Takahashi, T. (eds) Geophys. Monogr. 29: Climate Processes and Climate Sensitivity. Amer. Geophys. Union, Washington, DC, pp. 118-129.

305. Eischeid, J.K., Bradley, R.S. and Shao, X. (1985) Secular climatic fluctuations in the Great Salt Lake Basin. In Kay, P.A. and Diaz, H.F. (eds) Problems and Prospects for Predicting Great Salt Lake Levels. Univ. Utah, Salt Lake City, UT, pp. 111-122.

306. Larsen, C.E. (1985) Geoarcheological interpretation of Great Lakes coastal environments. In Stein, J.K. and Farrand, W.R. (eds) Archeological Sediments in Context. Center for the Study of Early Man, Inst. Quat. Studies, Univ. Maine, Orono, ME, pp. 91-110.

307. Matson, M., Ropelewski, C.F. and Varnadore, M.S. (1986) An Atlas of Satellite-Derived Northern Hemisphere Snow Cover Frequency. National Weather Service, Washington, DC.

308. Yates, H., Strong, A., McGinnis, D. and Tarpley, D. (1986) Terrestrial observations from NOAA operational satellites. Science 231, 463-470.

309. Chang, A.T.C., Foster, J.L. and Hall, D.K. (1987) Nimbus-7 SMMR derived global snow cover parameters. Ann. Glaciol. 9, 39-44.

310. Kukla, G. and Gavin, J. (1981) Summer ice and carbon dioxide. Science 214, 497-503.

311. Ropelewski, C.F. (1983) Spatial and temporal variations in Antarctic sea ice (1973-1983). J. Clim. Appl. Meteorol. 22, 470-473.

312. Zwally, H.J., Parkinson, C.L. and Comiso, J.C. (1983) Variability of Antarctic sea ice and changes in carbon dioxide. Science 220, 1005-1012.

313. Parkinson, C.L. and Bindschadler, R.A. (1984) Response of Antarctic sea ice to uniform atmospheric temperature increases. In Hansen, J.E. and

REFERENCES   SECTION

      Takahashi, T. (eds) Geophys. Monogr. 29: Climate Processes and Climat
      Sensitivity. Amer. Geophys. Union, Washington, DC, pp. 254- 264.
314. Reynaud, L. (1984) European glaciological data and their relation wit
      climate. In Berger, A.L. (ed) New Perspectives in Climate Modeling
      Elsevier, New York, NY, pp. 47-60.
315. Hall, D.K. and Martinec, J. (1985) Remote Sensing of Ice and Snow
      Chapman and Hall, London, UK.
316. Williams, R.S., Jr. (1985) Monitoring the area and volume of ice cap
      and ice sheets: Present and future opportunities using satellit
      remote sensing technology. In Meier, M., Aubrey, D.G., Bentley, C.R.
      Broecker, W.S., Hansen, J.E., Peltier, W.R. and Somerville, R.C.J
      (eds) Glaciers, Ice Sheets, and Sea Level: Effect of a CO_2-Induce
      Climatic Change. U.S. Dept. Energy, Washington, DC, pp. 232-240.
317. Bradley, R.S. and Serreze, M.C. (1987) Mass balance of two high Arcti
      plateau ice caps. J. Glaciol. 33, 123-128.
318. Wood, F.B., Jr. (1988) Global alpine glacier trends, 1960s to 1980s
      Arctic Alpine Res. 20, 404-413.
319. Hall, D.K. (1988) Assessment of polar climate change using satellit
      technology. Rev. Geophys. 26, 26-39.
320. Weller, G. (1985) A monitoring strategy to detect carbon dioxide
      induced climatic changes in the polar regions. In McBeath, J.H. (ed
      The Potential Effects of Carbon-Dioxide-Induced Climate Changes i
      Alaska. School of Agriculture and Land Resources Management, Univ
      Alaska, Fairbanks, AK.
321. Palecki, M.A. and Barry, R.G. (1986) Freeze-up and break-up of lakes a
      an index of temperature changes during the transition seasons: A cas
      study for Finland. J. Clim. Appl. Meteorol. 25, 893-902.
322. Maslanik, J.A. and Barry, R.G. (1987) Lake ice formation and breakup as
      an indicator of climate change: Potential for monitoring using remote
      sensing techniques. IAHS Pub. 168, 153-161.
323. Ramesh, R., Bhattacharya, S.K. and Gopalan, K. (1986) Climatic
      correlations in the stable isotope records of silver fir (Abies
      pindrow) trees from Kashmir, India. Earth Planet. Sci. Lett. 79, 66-
      74.
324. Friedman, I., Carrara, P. and Gleason, J. (1988) Isotopic evidence of
      Holocene climatic change in the San Juan Mountains, Colorado. Quat.
      Res. 30, 350-353.
325. Ramesh, R., Bhattacharya, S.K. and Pant, G.B. (1989) Climatic
      significance of D variations in a tropical tree species from India.
      Nature 337, 149-150.
326. Barnett, T.P. (1986) Detection of changes in the global troposphere
      temperature field induced by greenhouse gases. J. Geophys. Res. 91,
      6659-6667.
327. Emery, K.O. (1980) Relative sea levels from tide gauge records. Proc.
      Nat. Acad. Sci. 77, 6968-6972.
328. Bryant, E.A. (1987) CO_2-warming, rising sea-level and retreating coasts:
      Review and critique. Austr. Geogr. 18, 101-113.
329. Aubrey, D.G. and Emery, K.O. (1983) Eigenanalysis of recent United
      States sea levels. Cont. Shelf Res. 2, 21-33.
330. Aubrey, D.G. and Emery, K.O. (1986) Relative sea levels of Japan from
      tide-gauge records. Bull. Geol. Soc. Amer. 97, 194-205.
331. Aubrey, D.G. and Emery, K.O. (1986) Australia -- An unstable platform
      for tide-gauge measurements of changing sea levels. J. Geol. 94, 699-
      712.
332. Emery, K.O. and Aubrey, D.G. (1986) Relative sea-level changes from tide
      gauge records of eastern Asia mainland. Mar. Geol. 72, 33-46.
333. IAPSO Advisory Committee on Tides and Mean Sea Level. (1985) Changes in
      relative mean sea level. Trans. Amer. Geophys. Union 66, 754-756.
334. Meade, R.H. and Emery, K.O. (1971) Sea level as affected by river

runoff, eastern United States. Science 173, 425-428.

35. Enfield, D.B. and Allen, J.S. (1980) On the structure and dynamics of monthly mean sea level anomalies along the Pacific coast of North and South America. J. Phys. Oceanogr. 10, 557-578.

36. Harrison, D.E. and Crane, M.A. (1984) Changes in the Pacific during the 1982-83 event. Oceanus 27(2), 21-28.

37. Komar, P.D. (1986) The 1982-83 El Nino and erosion on the coast of Oregon. Shore Beach 54, 3-12.

38. Bryant, E.A. (1989) Sea-surface temperature and high-tide beach change, Stanwell Park, Australia, 1943-1978. J. Coast. Res., in press.

39. Bryant, E. A. (1989) Storminess and high-tide beach change, Stanwell Park, Australia, 1943-1987. Mar. Geol., in press.

40. Bryant, E.A. (1985) Rainfall and beach erosion relationships, Stanwell Park, Australia, 1895-1900: Worldwide implications for coastal erosion. Zeit. Geomorphol. Suppl. 57, 51-66.

41. Mountain, G. and Watkins, J. (1989) Sea level/ODP workshop. Trans. Amer. Geophys. Union 70, 155-156.

42. Idso, S.B. (1986) Implications of sea level trends. CO_2/Clim. Dial. 1(1), 70-71.

43. Allison, I., Ed. (1981) Sea Level, Ice and Climatic Change. Internat. Assoc. Hydrol. Sci., Wallingford, UK.

44. Tanner, W.F. (1988) Two scales of climate change. In Smith, L.A. and Coleman, G. (eds) Symposium: Global Climate and the Future of the High Plains Aquifer. Institute for Tertiary-Quaternary Studies, Nebraska Acad. Sci., Lincoln, NE.

45. National Research Council, U.S. (1976) Halocarbons: Effects on Stratospheric Ozone. National Academy Press, Washington, DC.

46. National Research Council, U.S. (1979) Stratospheric Ozone Depletion by Halocarbons: Chemistry and Transport. National Academy Press, Washington, DC.

47. National Research Council, U.S. (1982) Causes and Effects of Stratospheric Ozone Reductions: An Update. National Academy Press, Washington, DC.

48. Beardsley, T. (1986) Ozone depletion: US gets tough on CFC emissions. Nature 324, 102.

49. Dickson, D. and Marshall, E. (1986) Europe recognizes the ozone threat. Science 243, 1279.

50. McGourty, C. (1989) London ozone meeting wins some hearts. Nature 338, 101.

51. Singer, S.F. (1988) Does the Antarctic ozone hole have a future? Trans. Amer. Geophys. Union 69, 1588.

52. Farman, J.C., Gardiner, B.G. and Shanklin, J.D. (1988) How deep is an "ozone hole"? Nature 336, 198.

53. Schoeberl, M.R. (1988) Ozone depletion: Dynamics weaken the polar hole. Nature 336, 420.

54. Jukes, M.N. and McIntyre, M.E. (1987) A high-resolution one-layer model of breaking planetary waves in the stratosphere. Nature 328, 590-596.

55. Barnett, J.J., Houghton, J.T. and Pyle, J.A. (1975) The temperature dependence of the ozone concentration near the stratopause. Quart. J. Roy. Meteorol. Soc. 101, 245-257.

56. Keating, G.M., Brasseur, G.P., Nicholson, J.Y.,III and De Rudder, A. (1985) Detection of the response of ozone in the middle atmosphere to short-term solar ultraviolet variations. Geophys. Res. Lett. 12, 449-452.

57. Gribbin, J. (1988) CFCs could alter world's heat balance. New Sci. 118(1613), 36.

58. Brasseur, G. and Hitchman, M.H. (1988) Stratospheric response to trace gas perturbations: Changes in ozone and temperature distributions. Science 240, 634-637.

359. Heath, D.F. (1988) Non-seasonal changes in total column ozone fro
     satellite observations, 1970-86. Nature 332, 219-227.
360. Poole, L.R. and McCormick, M.P. (1988) Airborne lidar observations o
     Arctic polar stratospheric clouds: Indications of two distinct growt
     stages. Geophys. Res. Lett. 15, 21-23.
361. Palca, J. and Lloyd, P. (1989) Ozone hole looms large. Nature 337, 492.
362. Maggs, W.W. (1989) Arctic ripe for ozone depletion. Trans. Amer
     Geophys. Union 70, 131.
363. Pearce, F. and Anderson, I. (1989) Is there an ozone hole over the nort
     pole? New Sci. 121(1653), 32-33.
364. MacKenzie, D. (1988) Coming soon: The next ozone hole. New Sci
     119(1628), 38-39.
365. Frederick, J.E. (1980) Seasonal variations in high-latitude ozone an
     metastable molecular oxygen emissions: A theoretical interpretation
     J. Geophys. Res. 85, 1611-1617.
366. Crutzen, P.J. and Schmailzl, U. (1983) Chemical budgets of th
     stratosphere. Planet. Space Sci. 31, 1009-1032.
367. Solomon, S., Rusch, D.W., Thomas, R.J. and Eckman, R.S. (1983
     Comparison of mesospheric ozone abundances measured by the Sola
     Mesosphere Explorer and model calculations. Geophys. Res. Lett. 10
     249-252.
368. Frederick, J.E., Serafino, G.N. and Douglass, A.R. (1984) An analysis o
     the annual cycle in upper stratospheric ozone. J. Geophys. Res. 89
     9547-9555.
369. Froidevaux, L., Allen, M. and Young, Y.L. (1985) A critical analysis o
     ClO and O_3 in the mid-latitude stratosphere. J. Geophys. Res. 90
     12,999-13,029.
370. Rusch, D.W. and Clancy, R.T. (1988) Trends in atmospheric ozone
     Conflicts between models and SBUV data. J. Geophys. Res. 93, 8431-
     8437.
371. McElroy, M.B. and Salawitch, R.J. (1989) Changing composition of the
     global stratosphere. Science 243, 763-770.
372. Kerr, R.A. (1989) Arctic ozone is poised for a fall. Science 243, 1007-
     1008.
373. Walker, H.M. (1985) Ten-year ozone trends in California and Texas. J.
     Air Poll. Contr. Assoc. 35, 903-912.
374. Walker, H.M. (1986) Author's reply. J. Air Poll. Contr. Assoc. 36, 600-
     603.
375. Bojkov, R.D. (1986) Surface ozone during the second half of the
     nineteenth century. J. Clim. Appl. Meteorol. 25, 343-352.
376. Schell, R.C. and Rosson, R.M., Eds. (1987) Geophysical Monitoring for
     Climatic Change No. 15: Summary Report 1986. Nat. Ocean. Atmos.
     Admin., Boulder, CO.
377. Penkett, S.A. (1988) Atmospheric chemistry: Increased tropospheric
     ozone. Nature 332, 204.
378. Grant, W.B. (1988) Global stratospheric ozone and UVB radiation. Science
     242, 1111.
379. Volz, A. and Kley, D. (1988) Evaluation of the Montsouris series of
     ozone measurements made in the nineteenth century. Nature 332, 240-
     242.
380. Bischof, W., Borchers, R., Fabian, P. and Kruger, B.C. (1985) Increased
     concentration and vertical distribution of carbon dioxide in the
     stratosphere. Nature 316, 708-710.
381. Rowland, F.S. (1989) Chlorofluorocarbons and the depletion of
     stratospheric ozone. Amer. Sci. 77, 36-45.
382. Maggs, W.W. (1988) Storms blamed for rural ozone. Trans. Amer.
     Geophys. Union 69, 833.
383. Kirchhoff, V.W.J.H., Browell, E.V. and Gregory, G.L. (1988) Ozone
     measurements in the troposphere of an Amazonian rain forest

environment. J. Geophys. Res. 93, 15,850-15,860.

384. Danielsen, E.F. (1968) Stratospheric tropospheric exchange based on radioactivity, ozone and potential vorticity. J. Atmos. Sci. 25, 502-518.

385. Danielsen, E.F., Bleck, R., Shedlovsky, J., Wartburg, A., Haagenson, P. and Pollock, W. (1970) Observed distribution of radioactivity, ozone and potential vorticity associated with tropopause folding. J. Geophys. Res. 75, 2353-2361.

386. Vaughan, G. and Tuck, A.F. (1985) Aircraft measurements near jet streams. In Zerefos, C.S. and Ghazi, A. (eds) Atmospheric Ozone. D. Riedel, Hingham, MA, pp. 572-579.

387. Scotto, J., Cotton, G., Urbach, F., Berger, D. and Fears, T. (1988) Biologically effective ultraviolet radiation: Surface measurements in the United States. Science 239, 762-764.

388. Hobbs, P.V. (1989) Research on clouds and precipitation: Past, present, and future, part I. Bull. Amer. Meteorol. Soc. 70, 282-285.

389. Barkstrom, B.R. (1984) The Earth Radiation Budget Experiment (ERBE). Bull. Amer. Meteorol. Soc. 65, 1170-1185.

390. ERBE Science Team. (1986) First data from the Earth Radiation Budget Experiment (ERBE). Bull. Amer. Meteorol. Soc. 67, 818-824.

391. Lee, R.B., III, Barkstrom, B.R. and Cess, R.D. (1987) Characteristics of the earth radiation budget experiment solar monitors. Appl. Opt. 26, 3090-3096.

392. Henderson-Sellers, A. (1986) Increasing cloud in a warming world. Clim. Change 9, 267-309.

393. Henderson-Sellers, A. (1986) Cloud changes in a warmer Europe. Clim. Change 8, 25-52.

394. McGuffie, K. and Henderson-Sellers, A. (1988) Is Canadian cloudiness increasing? Atmos.-Ocean 26, 608-633.

395. Barry, R.G., Henderson-Sellers, A. and Shine, K.P. (1984) Climate sensitivity and the marginal cryosphere. In Hansen, J.E. and Takahashi, T. (eds) Geophys. Monogr. 29: Climate Processes and Climate Sensitivity. Amer. Geophys. Union, Washington, DC, pp. 221-237.

REFERENCES   SECTION 6

1. Kasting, J.R., Toon, O.B. and Pollack, J.B. (1988) How climate evolved on the terrestrial planets. Sci. Amer. 258(2), 90-97.

2. Idso, S.B. (1982) The anthropic principle, gaia, carbon dioxide and the human condition. Spec. Sci. Tech. 5, 455-459.

3. Sundquist, E.T. (1986) Geologic analogues: Their value and limitations in carbon dioxide research. In Trabalka, J.R. and Reichle, D.E. (eds) The Changing Carbon Cycle: A Global Analysis. Springer-Verlag, New York, NY, pp. 371-402.

4. Keeling, C.D. and Bacastow, R.B. (1977) Impact of industrial gases on climate. In National Research Council, U.S. (eds) Energy and Climate. National Academy Press, Washington, DC, pp. 72-95.

5. Siegenthaler, U. and Oeschger, H. (1978) Predicting future atmospheric carbon dioxide levels. Science 199, 388-395.

6. Perry, A.M., Araj, K., Fulkerson, W., Rose, D.J., Miller, M.M. and Rotty, R.M. (1982) Energy supply and demand implications of CO_2. Energy 7: 991-1004.

7. Houghton, R.A. and Woodwell, G.M. (1989) Global climatic change. Sci. Amer. 260(4), 36-44.

8. Stevenson, R.E. (1988) AMSOC's albatross award to Joe Reid. Trans. Amer. Geophys. Union 69, 1628.

REFERENCES   SECTION 7

1. de Saussure, T. (1804) Recherches Chemiques sur la Vegetation, Paris,
    France.
2. Wieler, A. (1890) Chemische Untersuchungen uber die Vegetation.
    Engelmann, Leipzig, East Germany.
3. Brown, H.T. and Escombe, F. (1902) The influence of varying amounts of
    carbon dioxide in the air on the photosynthetic process of leaves and
    on the mode of growth of plants. Proc. Roy. Soc. London 70, 397-413.
4. Demoussy, E. (1903) Sur la vegetation dans des atmospheres riches en
    acide carbonique. Compt. Rend. Acad. Sci. Paris 136, 325-328.
5. Demoussy, E. (1904) Influence sur la vegetation de l'acide carbonique.
    Compt. Rend. Acad. Sci. Paris 138, 291-293.
6. Demoussy, E. (1904) Sur la vegetation dans des atmospheres riches en
    acide carbonique. Compt. Rend. Acad. Sci. Paris 139, 883-885.
7. Fischer, H. (1912) Pflanzenernahrung mittels Kohlensaure. Gartenflora
    61, 298-307, 336.
8. Berkowski, W. (1913) Beobachtungen uber das Wachstum der Pflanzen in
    kohlensaurereicher Luft. Gartenwelt 17, 707-709.
9. Winter, E. (1913) Kohlensaure zur Ernahrung der Pflanzen. Gartenflora
    62, 402-404.
10. Lobner, M. (1913) Nochmals uber Dungung der Pflanzen mit Kohlensaure.
    Mollers Dtsch. Gartner-Zeitung 28, 434-435.
11. Berkowski, W. (1914) Pflanzenwachstum in kohlensaurereicher Luft.
    Gartenwelt 18, 445-446.
12. Werth, A.J. (1914) Gemuseanbauversuche auf Schleswig-Holsteinischen
    Mooren. Gartenwelt 18, 447-449.
13. Cummings, M.B. and Jones, C.H. (1918) The Aerial Fertilization of Plants
    with Carbon Dioxide. Vermont Agric. Exp. Sta. Bull. No. 211.
14. Fischer, H. (1919) Die Kohlenstoffernahrung der Kulturpflanzen.
    Gartenflora 68, 165-168.
15. Gerlach, M. (1919) Kohlensauredungung. Mitt. Dtsch. Landwirtschafts-
    gesellschaft 34, 54-62.
16. Strain, B.R. and Cure, J.D. (1986) Direct Effects of Atmospheric CO_2
    Enrichment on Plants and Ecosystems: A Bibliography with Abstracts.
    Oak Ridge Nat. Lab., Oak Ridge, TN.
17. Wittwer, S.H. and Robb, W. (1964) Carbon dioxide enrichment of green-
    house atmospheres for food crop production. Econ. Bot. 18, 34-56.
18. Strain, B.R. and Cure, J.D., Eds. (1985) Direct Effects of Increasing
    Carbon Dioxide on Vegetation. U.S. Dept. Energy, Washington, DC.
19. Pallas, J.E., Jr. (1970) Theoretical aspects of CO_2 enrichment. Trans.
    Amer. Soc. Agric. Eng. 13, 240-245.
20. Wittwer, S.H. (1978) Carbon dioxide fertilization of crop plants. In
    Gupta, U.S. (ed) Crop Physiology. Oxford & IBH, New Delhi, India, pp.
    310-333.
21. Allen, L.H., Jr. (1979) Potentials for carbon dioxide enrichment. In
    Barfield, B.J. and Gerber, J.F. (eds) Modification of the Aerial
    Environment of Crops. Amer. Soc. Agric. Eng., St. Joseph, MI, pp. 500-
    519.
22. Rosenberg, N.J. (1981) The increasing CO_2 concentration in the
    atmosphere and its implication on agricultural productivity. I.
    Effects on photosynthesis, transpiration and water use efficiency.
    Clim. Change 3, 265-279.
23. Gates, D.M. (1983) Plants gas up. Nat. Hist. 92(7), 4-8.
24. Lemon, E.R., Ed. (1983) CO_2 and Plants: The Response of Plants to
    Rising Levels of Atmospheric Carbon Dioxide. Westview Press, Boulder,
    CO.
25. Enoch, H.Z. (1984) Primary plant production during increasing global CO_2
    concentration. In Sybesma, C. (ed) Advances in Photosynthesis
    Research, Vol. IV. Nijhoff/Junk, The Hague, The Netherlands, pp. 201-
    207.

REFERENCES

26. Waggoner, P.E. (1984) Agriculture and carbon dioxide. Amer. Sci. 72, 179-184.

27. Idso, S.B. (1985) The search for global CO_2 etc. "greenhouse effects." Environ. Conserv. 12, 29-35.

28. Dahlman, R.C., Strain, B.R. and Rogers, H.H. (1985) Research on the response of vegetation to elevated atmospheric carbon dioxide. J. Environ. Qual. 14, 1-8.

29. Enoch, H.Z. and Kimball, B.A., Eds. (1986) Carbon Dioxide Enrichment of Greenhouse Crops. Vol. I and II. CRC Press, Boca Raton, FL.

30. Wittwer, S.H. (1986) Worldwide status and history of CO_2 enrichment-- An overview. In Enoch, H.Z. and Kimball, B.A. (eds) Carbon Dioxide Enrichment of Greenhouse Crops. Vol. I: Status and CO_2 Sources. CRC Press, Boca Raton, FL, pp. 3-15.

31. Kretchman, D.W. and Howlett, F.S. (1970) CO_2 enrichment for vegetable production. Trans. Amer. Soc. Agric. Engin. 13, 252-256.

32. van Berkel, N. (1984) CO_2 enrichment in the Netherlands. Acta Hort. 162, 197-205.

33. Bauerle, W.L. and Kimball, B.A. (1984) CO_2 enrichment in the United States. Acta Hort. 162, 207-216.

34. Moe, R. (1984) CO_2 enrichment in Scandinavia. Acta Hort. 162, 217-225.

35. Garab, G., Rozsa, Z. and Govindjee (1988) Carbon dioxide affects charge accumulation in leaves. Naturwissenschaften 75, 517-519.

36. Trebst, A. and Avron, M., Eds. (1977) Photosynthesis I: Photosynthetic Electron Transport and Photophosphorylation. Springer-Verlag, New York, NY.

37. Gibbs, M. and Latzko, E., Eds. (1979) Photosynthesis II: Photosynthetic Carbon Metabolism and Related Processes. Springer-Verlag, New York, NY.

38. Gaastra, P. (1959) Photosynthesis of crop plants as influenced by light, carbon dioxide, temperature and stomatal diffusion resistance. Meded. Landbouwhogesch. Wageningen 59, 1-68.

39. Gaastra, P. (1962) Photosynthesis of leaves and field crops. Neth. J. Agric. Sci. 10, 311-324.

40. Gale, J., Kohl, H.C. and Hagen, R.M. (1966) Mesophyll and stomatal resistances affecting photosynthesis under varying conditions of soil water and evaporation demand. Israel J. Bot. 15, 64-71.

41. Gale, J. (1972) Availability of carbon dioxide for photosynthesis at high altitudes: Theoretical considerations. Ecology 53, 494-497.

42. Hofstra, G. and Hesketh, J.D. (1975) The effects of temperature and CO_2 enrichment on photosynthesis in soybean. In Marcelle, R. (ed) Environmental and Biological Control of Photosynthesis. W. Junk, The Hague, The Netherlands, pp. 71-80.

43. Mauney, J.R., Guinn, G., Fry, K.E. and Hesketh, J.D. (1979) Correlation of photosynthetic carbon dioxide uptake and carbohydrate accumulation in cotton, soybean, sunflower, and sorghum. Photosynthetica 13, 260-266.

44. Thomas, J.F., Ells, J.M. and Miksche, J.P. (1983) Microdensitometer determinations of starch in palisade cells of soybean grown under CO_2 enrichment. Plant Physiol. Suppl. 72, 165.

45. Allen, L.H., Jr., Vu, J.C.V., Valle, R.R., Boote, K.J. and Jones, P.H. (1988) Nonstructural carbohydrates and nitrogen of soybean grown under carbon dioxide enrichment. Crop Sci. 28, 84-94.

46. Campbell, W.J., Allen, L.H., Jr. and Bowes, G. (1988) Effects of CO_2 concentration on rubisco activity, amount, and photosynthesis in soybean leaves. Plant Physiol. 88, 1310-1316.

47. Nijs, I., Impens, I. and Behaeghe, T. (1988) Effects of elevated atmospheric carbon dioxide on gas exchange and growth of white clover. Photosyn. Res. 15, 163-176.

48. Ehleringer, J. and Bjorkman, O. (1977) Quantum yields for CO_2 uptake in

C_3 and C_4 plants. Dependence on temperature, CO_2, and O_2 concentration. Plant Physiol. 59, 86-90.

49. Pearcy, R.W. and Bjorkman, O. (1983) Physiological effects. In Lemon, E.R. (ed) CO_2 and Plants: The Response of Plants to Rising Levels of Atmospheric Carbon Dioxide. Westview Press, Boulder, CO, pp. 65-105.

50. Sharkey, T.D. (1985) Photosynthesis in intact leaves of C_3 plants: Physics, physiology, and rate limitations. Bot. Rev. 51, 54-105.

51. McHale, N.A., Zelitch, I. and Peterson, R.B. (1987) Effects of CO_2 and O_2 on photosynthesis and growth of autotrophic tobacco callus. Plant Physiol. 84, 1055-1058.

52. Idso, S.B. (1982) Carbon Dioxide: Friend or Foe? IBR Press, Tempe, AZ.

53. Ford, M.A. and Thorne, G.N. (1967) Effect of CO_2 concentrations on growth of sugar-beet, barley, kale and maize. Ann. Bot. 31, 629-644.

54. Sionit, N., Strain, B.R. and Hellmers, H. (1981) Effects of atmospheric CO_2 concentration on growth and yield components of wheat. J. Agric. Sci. Camb. 79, 335-339.

55. Rogers, H.H., Bingham, G.E., Cure, J.D., Smith, J.M. and Surano, K.A. (1983) Responses of selected plant species to elevated carbon dioxide in the field. J. Environ. Qual. 12, 569-574.

56. Idso, S.B., Kimball, B.A. and Anderson, M.G. (1985) Atmospheric CO_2 enrichment of water hyacinths: Effects on transpiration and water use efficiency. Water Resources Res. 21, 1787-1790.

57. Imai, K. and Murata, Y. (1978) Effect of carbon dioxide concentration on growth and dry matter production of crop plants. IV. After-effects of carbon dioxide treatments on the apparent photosynthesis, dark respiration and dry matter production. Jap. J. Crop. Sci. 47, 330-335.

58. Paez, A., Hellmers, H. and Strain, B.R. (1980) CO_2 effects on apical dominance in Pisum-sativum. Physiol. Plant. 50, 43-46.

59. Sionit, N., Hellmers, H. and Strain, B.R. (1982) Interaction of atmospheric CO_2 enrichment and irradiance on plant growth. Agron. J. 74, 721-725.

60. Sionit, N. (1983) Response of soybean to two levels of mineral nutrition in CO_2-enriched atmosphere. Crop Sci. 23, 329-333.

61. Imai, K. and Murata, Y. (1976) Effect of carbon dioxide concentration on growth and dry matter production of crop plants. Proc. Crop Sci. Soc. Jap. 45, 598-606.

62. Morison, J.I.L. and Gifford, R.M. (1984) Plant growth and water use with limited water supply in high CO_2 concentrations. I. Leaf area, water use and transpiration. Austr. J. Plant Physiol. 11, 361-374.

63. Rogers, H.H., Cure, J.D., Thomas, J.F. and Smith, J.M. (1984) Influence of elevated CO_2 on growth of soybean plants. Crop Sci. 24, 361-366.

64. Rogers, H.H., Heck, W.W. and Heagle, A.S. (1983) A field technique for the study of plant responses to elevated carbon dioxide concentrations. J. Air Poll. Contr. Assoc. 33, 42-44.

65. Thomas, J.F. (1983) Soybean leaf development under CO_2 enrichment. J. Elisha Mitchell Sci. Soc. 99(4).

66. Thomas, J.F. and Harvey, C.N. (1983) Leaf anatomy of four species grown under continuous CO_2 enrichment. Bot. Gaz. 144, 303-309.

67. Leadley, P.W., Reynolds, J.A., Thomas, J.F. and Reynolds, J.F. (1987) Effects of CO_2 enrichment on internal leaf surface area in soybeans. Bot. Gaz. 148, 137-140.

68. Khavari-Nejad, R.A. (1986) Carbon dioxide enrichment preconditioning effects on chlorophylls contents and photosynthetic efficiency in tomato plants. Photosynthetica 20, 315-317.

69. Enoch, H.Z., Rylski, J. and Spigelman, M. (1976) CO_2 enrichment of strawberry and cucumber plants grown in unheated greenhouses in Israel. Sci. Hort. 5, 33-41.

70. Spencer, W. and Bowes, G. (1986) Photosynthesis and growth of water hyacinth under CO_2 enrichment. Plant Physiol. 82, 528-533.

REFERENCES

71. Goldsberry, K.L. (1986) CO_2 fertilization of carnations and some other flower crops. In Enoch, H.Z. and Kimball, B.A. (eds) Carbon Dioxide Enrichment of Greenhouse Crops. Vol. II. Physiology, Yield, and Economics. CRC Press, Boca Raton, FL, pp. 117-140.

72. Mortensen, L.M. (1986) Effect of intermittent as compared to continuous CO_2 enrichment on growth and flowering of Chrysanthemum x morifolium Ramat. and Saintpaulia ionantha H. Wentl. Scien. Hort. 29, 283-289.

73. Hanan, J.J. (1986) CO_2 enrichment for greenhouse rose production. In Enoch, H.Z. and Kimball, B.A. (eds) Carbon Dioxide Enrichment of Greenhouse Crops. Vol. II. Physiology, Yield, and Economics. CRC Press, Boca Raton, FL, pp. 141-149.

74. Hardman, L.L. and Brun, W.A. (1971) Effect of atmospheric carbon dioxide enrichment at different developmental stages on growth and yield components of soybeans. Agron J. 11, 886-888.

75. Cock, J.H. and Yoshida, S. (1973) Changing sink and source relations in rice (Oryza sativa L.) using carbon dioxide enrichment in the field. Soil Sci. Plant Nutr. 19, 229-234.

76. Krenzer, E.G., Jr. and Moss, D.N. (1975) Carbon dioxide enrichment effects upon yield and yield components in wheat. Crop Sci. 15, 71-74.

77. Fischer, R.A. and Aguilar, M.I. (1976) Yield potential in a dwarf spring wheat and the effect of carbon dioxide fertilization. Agron. J. 68, 749-752.

78. Leonard, O.A. and Pinckard, J.A. (1946) Effect of various oxygen and carbon dioxide concentrations on cotton root development. Plant Physiol. 21, 18-36.

79. Geisler, G. (1963) Morphogenetic influence of $(CO_2 + HCO_3^-)$ on roots. Plant Physiol. 38, 77-80.

80. Grable, A.R. and Danielson, R.E. (1965) Influence of CO_2 on growth of corn and soybean seedlings. Soil Sci. Soc. Amer. Proc. 29, 233-238.

81. Rogers, H.H., Peterson, C.M., McCrimmon, J.N. and Cure, J.D. (1987) Response of soybean roots to elevated atmospheric carbon dioxide. Agron. Abstr. 79, 100.

82. Kimball, B.A. (1983) Carbon dioxide and agricultural yield: An assemblage and analysis of 430 prior observations. Agron. J. 75, 779-788.

83. Kimball, B.A. (1983) Carbon Dioxide and Agricultural Yield: An Assemblage and Analysis of 770 Prior Observations. U.S. Water Conservation Laboratory, Phoenix, AZ.

84. Allen, L.H., Jr., Boote, K.J., Jones, J.W., Jones, P.H., Valle, R.R., Acock, B., Rogers, H.H. and Dahlman, R.C. (1987) Response of vegetation to rising carbon dioxide: Photosynthesis, biomass, and seed yield of soybean. Global Biogeochem. Cycles 1, 1-14.

85. Prioul, J.L. and Chartier, P. (1977) Partitioning of transfer and carboxylation components of intracellular resistance to photosynthetic CO_2 fixation: A critical analysis of the methods used. Ann. Bot. 41, 789-800.

86. Muller, J. (1986) Ecophysiological characterization of CO_2 exchange in leaves of winter wheat (Triticum aestivum L.). I. Model. Photosynthetica 20, 454-465.

87. Breen, P.J., Hesketh, J.D. and Peters, D.B. (1986) Field measurements of leaf photosynthesis of C_3 and C_4 species under high irradiance and enriched CO_2. Photosynthetica 20, 281-285.

88. Bergmann, L. (1967) Growth of green suspension cultures of Nicotiana tabacum var. "Samsun" with CO_2 as carbon source. Planta 74, 243-249.

89. Horn, M.E. and Widholm, J.M. (1984) Aspects of photosynthetic plant tissue cultures. In Collins, G.B. and Petolino, J. (eds) Applications of Genetic Engineering to Crop Improvements. Martinus Nijhoff Dr. W. Junk, Boston, MA, pp. 113-161.

90. Neumann, K.H. and Bender, L. (1987) Photosynthesis in cell and tissue

culture systems. In Green, C.E., Somers, D.A., Hackett, W.P. and Biesboer, D.D. (eds) Plant Tissue and Cell Culture. Alan R. Liss, New York, NY, pp. 151-165.

91. Xu, C., Blair, L.C., Rogers, S.M.D., Govindjee and Widholm, J.M. (1988) Characteristics of five new photoautrophic suspension cultures including two Amaranthus species and a cotton strain growing on ambient CO_2 levels. Plant Physiol. 88, 1297-1302.

92. Keeling, C.D. (1982) The oceans and terrestrial biosphere as future sinks for fossil fuel CO_2. In Reck, R.A. and Hummel, J.R. (eds) AIP Conf. Proc. No. 82: Interpretation of Climate and Photochemical Models, Ozone and Temperature Measurements. Amer. Instit. Physics., New York, NY, pp. 47-82.

93. Perry, A.M. (1986) Possible changes in future use of fossil fuels to limit environmental effects. In Trabalka, J.R. and Reichle, D.E. (eds) The Changing Carbon Cycle: A Global Analysis. Springer-Verlag, New York, NY, pp. 561-574.

94. Trabalka, J.R., Edmonds, J.A., Reilly, J.M., Gardner, R.H. and Reichle, D.E. (1986) Atmospheric CO_2 projections with globally averaged carbon cycle models. In Trabalka, J.R. and Reichle, D.E. (eds) The Changing Carbon Cycle: A Global Analysis. Springer-Verlag, New York, NY, pp. 534-560.

95. Bugbee, B.G. and Salisbury, F.B. (1988) Exploring the limits of crop productivity. I. Photosynthetic efficiency of wheat in high irradiance environments. Plant Physiol. 88, 869-878.

96. Garrels, R.M., Lerman, A. and Mackenzie, F.T. (1976) Controls of atmospheric O_2 and CO_2: Past, present, and future. Amer. Sci. 64, 306-315.

97. Budyko, M.E. and Ronov, A.B. (1979) Chemical evolution of the atmosphere in the Phanerozoic. Geochem. Int. 16, 1-9.

98. Berner, R.A., Lasaga, A.C. and Garrels, R.M. (1983) The carbonate-silicate geochemical cycle and its effect on atmospheric carbon dioxide over the past 100 million years. Amer. J. Sci. 283, 641-683.

99. Volk, T. (1989) Sensitivity of climate and atmospheric CO_2 to deep-ocean and shallow-ocean carbonate burial. Nature 337, 637-640.

100. Jefferson, T.H. (1982) Fossil forests from the lower Cretaceous of Alexander Island, Antarctica. Palaeontology 25, 681-708.

101. Axelrod, D.I. (1984) An interpretation of Cretaceous and Tertiary biota in polar regions. Palaeogeog. Palaeoclimatol. Palaeoecol. 45, 105-147.

102. Morris, S.C. (1985) Polar forests of the past. Nature 313, 739.

103. Seyfert, C.K. and Sirkin, L.A. (1973) Earth History and Plate Tectonics: An Introduction to Historical Geology. Harper & Row, New York, NY.

104. Berner, R.A. and Lasaga, A.C. (1989) Modeling the geochemical carbon cycle. Sci. Amer. 260(3), 74-81.

105. Idso, S.B. (1986) Industrial age leading to the greening of the Earth? Nature 320, 22.

106. Cure, J.D. (1985) Carbon dioxide doubling responses: A crop survey. In Strain, B.R. and Cure, J.D. (eds) Direct Effects of Increasing Carbon Dioxide on Vegetation. U.S. Dept. Energy, Washington, DC, pp. 99-116.

107. Cure, J.D. and Acock, B. (1986) Crop responses to carbon dioxide doubling: A literature survey. Agric. For. Meteorol. 38, 127-145.

108. Pallas, J.E. (1965) Transpiration and stomatal opening with changes in carbon dioxide content of the air. Science 147, 171-173.

109. Akita, S. and Moss, D.N. (1972) Differential stomatal response between C_3 and C_4 species to atmospheric CO_2 concentration and light. Crop Sci. 12, 789-793.

110. Louwerse, W. (1980) Effect of CO_2 concentration and irradiance on stomatal behavior of maize, barley and sunflower plants in the field. Plant Cell Environ. 3, 391-398.

REFERENCES   SECTION 7

111. Sionit, N., Rogers, H.H., Bingham, G.E. and Strain, B.R. (1984)
     Photosynthesis and stomatal conductance with CO_2-enrichment of
     container- and field-grown soybeans. Agron. J. 76, 447-451.
112. Darwin, F. (1898) Observations on stomata. Phil. Trans. Roy. Soc. Lond.
     B Biol. Sci. 190, 531-621.
113. Linsbauer, K. (1916) Beitrage zur Kenntnis der Spaltoffnungsbewegung.
     Flora 9, 100-143.
114. Freudenberger, H. (1940) Die Reaktion der Schliesszellen auf Kohlensaure
     und Sauerstoffentzug. Protoplasma 35, 15-54.
115. Heath, O.V.S. (1949) Studies in stomatal behavior. V. The role of carbon
     dioxide in the light response of stomata. J. Exp. Bot. 1, 29-62.
116. Jones, R.J. and Mansfield, T.A. (1970) Increases in the diffusion
     resistances of leaves in a carbon dioxide-enriched atmosphere. J. Exp.
     Bot. 21, 951-958.
117. Farquhar, G.D., Dubbe, D.R. and Raschke, K. (1978) Gain of the feedback
     loop involving carbon dioxide and stomata, theory and measurement.
     Plant Physiol. 62, 406-412.
118. Bell, C.J. (1982) A model of stomatal control. Photosynthetica 16, 486-
     495.
119. Havelka, U.D., Ackerson, R.C., Boyle, M.G. and Wittenbach, V.A. (1984)
     CO_2-enrichment effects on soybean physiology. I. Effects of long-term
     CO_2 exposure. Crop Sci. 24, 1146-1150.
120. Morison, J.I.L. (1985) Sensitivity of stomata and water use efficiency
     to high CO_2. Plant Cell Environ. 8, 467-474.
121. Raschke, K. (1986) The influence of the CO_2 content of the ambient air
     on stomatal conductance and the CO_2 concentration in leaves. In Enoch,
     H.Z. and Kimball, B.A. (eds) Carbon Dioxide Enrichment of Greenhouse
     Crops. Vol. II. Physiology, Yield, and Economics. CRC Press, Boca
     Raton, FL, pp. 87-102.
122. Morison, J.I.L. (1987) Intercellular CO_2 concentration and stomatal
     response to CO_2. In Zeiger, E., Farquhar, G.D. and Cowan, I.R. (eds)
     Stomatal Function. Stanford Univ. Press, Stanford, CA, pp. 229-251.
123. Mott, K.A. (1988) Do stomata respond to CO_2 concentrations other than
     intercellular? Plant Physiol. 86, 200-203.
124. Kimball, B.A. and Idso, S.B. (1983) Increasing atmospheric CO_2: Effects
     on crop yield, water use and climate. Agric. Water Manag. 7, 55-72.
125. Van Bavel, C.H.M. (1974) Antitranspirant action of carbon dioxide on
     intact sorghum plants. Crop Sci. 14, 208-212.
126. Idso, S.B., Kimball, B.A. and Clawson, K.L. (1984) Quantifying effects
     of atmospheric CO_2 enrichment on stomatal conductance and
     evapotranspiration of water hyacinth via infrared thermometry. Agric.
     For. Meteorol. 33, 15-22.
127. Osmond, C.B., Bjorkman, O. and Anderson, O.J. (1980) Physiological
     processes in plant ecology: Toward a synthesis with Atriplex. Ecol.
     Studies: Analysis Synthesis 36, 1-468.
128. Wong, S.C. (1980) Effects of elevated partial pressure of CO_2 on rate of
     CO_2 assimilation and water use efficiency in plants. In Pearman, G.I.
     (ed) Carbon Dioxide and Climate: Australian Research. Austr. Acad.
     Sci., Canberra, Australia, pp. 159-166.
129. Jurik, T.W., Weber, J.A. and Gates, D.M. (1984) Short-term effects of
     CO_2 on gas exchange of leaves of bigtooth aspen (Populus
     grandidentata) in the field. Plant Physiol. 75, 1022-1026.
130. Valle, R., Mishoe, J.W., Jones, J.W. and Allen, L.H., Jr. (1985)
     Transpiration rate and water use efficiency of soybean leaves adapted
     to different CO_2 environments. Crop Sci. 25, 477-482.
131. Wong, S.C. (1979) Elevated atmospheric partial pressure of CO_2 and plant
     growth. I. Interactions of nitrogen nutrition and photosynthetic
     capacity in C_3 and C_4 plants. Oecologia 44, 68-74.
132. Acock, B. and Allen, L.H., Jr. (1985) Crop responses to elevated carbon

dioxide concentrations. In Strain, B.R. and Cure, J.D. (eds) Direct
Effects of Increasing Carbon Dioxide on Vegetation. U.S. Dept. Energy,
Washington, DC, pp. 53-97.

133. Rogers, H.H., Thomas, J.R. and Bingham, G.E. (1983) Response of
agronomic and forest species to elevated atmospheric carbon dioxide.
Science 220, 428-429.

134. Wright, R.D. (1974) Rising atmospheric CO_2 and photosynthesis of San
Bernadino Mountain plants. Amer. Midl. Nat. 91, 360-370.

135. Downton, W.J.S., Bjorkman, O. and Pike, C.S. (1980) Consequences of
increased atmospheric concentrations of carbon dioxide for growth and
photosynthesis of higher plants. In Pearman, G.I. (ed) Carbon Dioxide
and Climate: Australian Research. Austr. Acad. Sci., Canberra,
Australia, pp. 143-151.

136. Idso, S.B. (1984) What if increases in atmospheric CO_2 have an inverse
greenhouse effect? I. Energy balance considerations related to surface
albedo. J. Climatol. 4, 399-409.

137. Idso, S.B. (1984) Atmospheric CO_2 variability: A cause for concern in
the reconstruction of past climates. Spec. Sci. Tech. 7, 37-40.

138. Sveinbjornsson, B. (1984) Alaskan plants and atmospheric carbon dioxide.
In McBeath, J.H. (ed) The Potential Effects of Carbon Dioxide-Induced
Climatic Changes in Alaska. Sch. Agric. Land Res. Manag., Univ.
Alaska, Fairbanks, AK, pp. 149-154.

139. Idso, S.B. and Quinn, J.A. (1983) Vegetational Redistribution in
Arizona and New Mexico in Response to a Doubling of the Atmospheric
CO_2 Concentration. Climatol. Pub. Sci. Pap. No. 17. Lab. Climatol.,
Arizona State Univ., Tempe, AZ.

140. Willmott, C.J. (1977) WATBUG: A Fortran IV Algorithm for Calculating
the Climatic Water Budget. Water Res. Ctr., Univ. Delaware, Newark,
DE.

141. Thornthwaite, C.W. (1948) An approach toward a rational classification
of climate. Geogr. Rev. 38, 55-94.

142. Sedjo, R.A. (1988) The global carbon cycle. Science 241, 1737-1738.

143. Pain, S. (1988) Winners and losers -- carbon dioxide as a fertilizer.
New Sci. 120(1638), 39.

144. Botkin, D.B., Janek, J.R. and Wallis, J.R. (1973) Estimating the effects
of carbon fertilization on forest composition by ecosystems
simulation. In Woodwell, G.M. and Pecan, E.V. (eds) Carbon and the
Biosphere. U.S. Atomic Energy Comm., Washington, DC, pp. 328-344.

145. Strain, B.R. (1982) Ecological aspects of plant responses to carbon
dioxide enrichment. In Brown, S. (ed) Global Dynamics of Biospheric
Carbon. U.S. Dept. Energy, Washington, DC, pp. 46-55.

146. Gates, D.M., Strain, B.R. and Weber, J.A. (1983) Ecophysiological
effects of changing atmospheric carbon dioxide concentrations. In
Lange, O.L., Nobel, P.S., Osmond, C.B. and Ziegler, H. (eds)
Physiological Plant Ecology IV. Springer-Verlag, New York, NY, pp.
503-526.

147. Overdieck, D., Bossemeyer, D. and Lieth, H. (1984) Long-term effects of
an increased CO_2 concentration level on terrestrial plants in model-
ecosystems. I. Phytomass production and competition of Trifolium
repens L. and Lolium perenne L. Prog. Biometeorol. 3, 344-352.

148. Shugart, H.H. and Emanuel, W.R. (1985) Carbon dioxide increase: The
implications at the ecosystem level. Plant Cell Environ. 8, 381-386.

149. Strain, B.R. (1985) Physiological and ecological controls on carbon
sequestering in terrestrial ecosystems. Biogeochem. 1, 219-232.

150. Carlson, R.W. and Bazzaz (1980) The effects of elevated carbon dioxide
concentrations on growth, photosynthesis, transpiration, and water use
efficiency of plants. In Singh, J.J. and Deepack, A. (eds)
Environmental and Climatic Impact of Coal Utilization. Academic Press,
New York, NY, pp. 609-622.

REFERENCES

151. Bazzaz, F.A. and Carlson, R.W. (1984) The response of plants to elevated CO_2. I. Competition among an assemblage of annuals at different levels of soil moisture. Oecologia 62, 196-198.

152. Tolley, L.C. and Strain, B.R. (1984) Effects of CO_2 enrichment and water stress on growth of Liquidambar styraciflua and Pinus taeda seedlings. Can. J. Bot. 62, 2135-2139.

153. Sionit, N., Strain, B.R., Helmers, H.H., Riechers, G.H. and Jaeger, C.H. (1985) Long-term atmospheric CO_2 enrichment affects the growth and development of Liquidambar styraciflua and Pinus taeda seedlings. Can. J. For. Res. 15, 468-471.

154. Tolley, L.C. and Strain, B.R. (1985) Effects of CO_2 enrichment and water stress on gas exchange of Liquidambar styraciflua and Pinus taeda seedlings grown under different irradiance levels. Oecologia 65, 166-172.

155. Overdieck, D. and Reining, F. (1986) Effect of atmospheric CO_2 enrichment on perennial ryegrass (Lolium perenne L.) and white clover (Trifolium repens L.) competing in managed model-ecosystems. Acta Oecol./Oecol. Plant. 7, 357-366.

156. Williams, W.E., Garbutt, K., Bazzsaz, F.A. and Vitousek, P.M. (1986) The response of plants to elevated CO_2. IV. Two deciduous-forest tree communities. Oecologia 69, 454-459.

157. Smith, S.D., Strain, B.R. and Sharkey, D. (1987) Effects of CO_2 enrichment on four Great Basin grasses. Funct. Ecol. 1, 139-143.

158. Bazzaz, F.A. and Garbutt, K. (1988) The response of annuals in competitive neighborhoods: Effects of elevated CO_2. Ecology 69, 937-946.

159. Roberts, S. (1988) Is there life after climate change? Science 242, 1010-1012.

160. Goldsmith, E. (1988) Gaia: Some implications for theoretical ecology. The Ecologist 18(2/3), 64-74.

161. Carter, D.R. and Peterson, K.M. (1983) Effects of a CO_2-enriched atmosphere on the growth and competitive interaction of a C_3 and a C_4 grass. Oecologia 58, 188-193.

162. Zangerl, A.R. and Bazzaz, F.A. (1984) The response of plants to elevated CO_2. II. Competitive interactions among annual plants under varying light and nutrients. Oecologia 62, 412-417.

163. Wray, S.M. and Strain, B.R. (1986) Response of two old field perennials to interactions of CO_2 enrichment and drought stress. Amer. J. Bot. 73, 1486-1491.

164. Wray, S.M. and Strain, B.R. (1987) Competition in old-field perennials under CO_2 enrichment. Ecology 68, 1116-1120.

165. Riechers, G.H. and Strain, B.R. (1988) Growth of blue grama (Bouteloua gracilis) in response to atmospheric CO_2 enrichment. Can. J. Bot. 66, 1570-1573.

166. Patterson, D.T., Flint, E.P. and Beyers, J.L. (1984) Effects of CO_2 enrichment on competition between a C_4 weed and C_3 crop. Weed Sci. 32, 101-105.

167. Patterson, D.T. (1986) Responses of soybean (Glycine max) and three C_4 grass weeds to CO_2 enrichment during drought. Weed Sci. 34, 203-210.

168. Patterson, D.T. and Flint, E.P. (1980) Potential effects of global atmospheric carbon dioxide enrichment on the growth and competitiveness of carbon 3 and carbon 4 weed and crop plants. Weed Sci. 28, 71-75.

169. Black, C.C., Chen, T.M. and Brown, R.H. (1969) Biochemical basis for plant competition. Weed Sci. 17, 338-344.

170. Patterson, D.T. (1985) Comparative ecophysiology of weeds and crops. In Duke, S.O. (ed) Weed Physiology. Vol. I. CRC Press, Boca Raton, FL, pp. 101-129.

171. Pain, S. (1988) No escape from the global greenhouse. New Sci. 120(1638),

38-43.

172. May, R.M. (1988) How many species are there on Earth? Science 241, 1441-1442.

173. May, R.M. (1988) This week's citation classic. Current Contents 19(50), 22.

174. Boucher, D.H., James, S. and Keeler, K.H. (1982) The ecology of mutualism. Ann. Rev. Ecol. Syst. 13, 315-347.

175. Axelrod, D.I. (1944) The Oakdale flora (California). Carnegie Institute of Washington Pub. 553, 147-166.

176. Axelrod, D.I. (1944) The Sonoma flora (California). Carnegie Institute of Washington Pub. 553, 167-200.

177. Axelrod, D.I. (1956) Mio-Pliocene floras from west-central Nevada. Univ. Calif. Pub. Geol. Sci. 33, 1-316.

178. Axelrod, D.I. (1976) Evolution of the Santa Lucia fir (Abies bracteata) ecosystem. Ann. Missouri Bot. Garden 63, 24-41.

179. Axelrod, D.I. (1987) The Late Oligocene Creede flora, Colorado. Univ. Calif. Pub. Geol. Sci. 130, 1-235.

180. Axelrod, D.I. (1988) An interpretation of high montane conifers in western Tertiary floras. Paleobiol. 14, 301-306.

181. Volk, T. (1987) Feedbacks between weathering and atmospheric CO_2 over the last 100 million years. Amer. J. Sci. 287, 763-779.

182. Brown, A.F.H. and Harrison, A.F. (1983) Effects of tree mixtures on earthworm populations and nitrogen and phosphorus status in Norway Spruce (Picea abies) stands. In Lebrum, P.H., Andre, H.M., De Medts, A., Gregoire-Wibo, C. and Wauthy, G. (eds) New Trends in Soil Biology. Proc. VIII Internat. Colloq. Soil Zool. Louvain-la-Neure, Belgium, pp. 101-108.

183. Carlyle, J.C. and Malcolm, D.C. (1986) Nitrogen availability beneath pure spruce and mixed larch + spruce stands growing on a deep peat I. Net mineralization measured by field and laboratory incubations. Plant Soil 93, 95-113.

184. Carlyle, J.C. and Malcolm, D.C. (1986) Nitrogen availability beneath pure spruce and mixed larch + spruce stands growing on a deep peat II. A comparison of N availability as measured by plant uptake and long-term laboratory incubations. Plant Soil 93, 115-122.

185. Chapman, K., Whittaker, J.B. and Heal, O.W. (1988) Metabolic and faunal activity in litters of tree mixtures compared with pure stands. Agric. Ecosys. Environ. 24, 33-40.

186. Gerdemann, J.W. (1968) Vesicular-arbuscular mycorrhizae and plant growth. Ann. Rev. Phytopath. 6, 397-418.

187. Read, D.J., Koucheki, H.K. and Hodgson, T. (1976) Vesicular-arbuscular mycorrhizae in natural ecosystems. I. The occurrence of infection. New Phytol. 77, 641-653.

188. Gerdemann, J.W. and Nicolson, T.H. (1963) Spores of mycorrhizal Endogone species extracted from soil by wet sieving and decanting. Trans. Brit. Mycol. Soc. 46, 235-244.

189. Ingham, R.E. (1988) Interactions between nematodes and vesicular-arbuscular mycorrhizae. Agric. Ecosys. Environ. 24, 169-182.

190. Johnson, N.C. and McGraw, A.-C. (1988) Vesicular-arbuscular mycorrhizae in taconite tailings. II. Effects of reclamation practices. Agric. Ecosys. Environ. 21, 143-152.

191. Edwards, C.A. and Stinner, B.R. (1988) Interactions between soil-inhabiting invertebrates and microorganisms in relation to plant growth and ecosystem processes: An introduction. Agric. Ecosys. Environ. 24, 1-3.

192. Chiariello, N., Hickman, J.C. and Mooney, H.A. (1982) Endomycorrhizal role for interspecific transfer of phosphorus in a community of annual plants. Science 217, 941-943.

193. Reid, C.P.P., and Woods, F.W. (1969) Translocation of ^{14}C-labelled

compounds in mycorrhizae and its implications in interplant nutrient cycling. Ecology 50, 179-187.

194. Read, D.J., Francis, R. and Finlay, R.D. (1985) Mycorrhizal mycelia and nutrient cycling in plant communities. In Fitter, A.H., Atkinson, D., Read, D.J. and Usher, M.B. (eds) Brit. Ecol. Soc. Spec. Pub. 4: Ecological Interactions in Soil, pp. 193-217.

195. Warcup, J.H. (1988) Mycorrhizal associations and seedling development in Australian Lobelioideae (Campanulaceae). Aust. J. Bot. 36, 461-472.

196. Clements, M.A. and Ellyyard, R.K. (1979) The symbiotic germination of Australian terrestrial orchids. Amer. Orchid Soc. Bull. 48, 810-816.

197. Masuhara, G. and Katsuya, K. (1989) Effects of mycorrhizal fungi on seed germination and early growth of three Japanese terrestrial orchids. Sci. Hort. 37, 331-337.

198. Moore, J.C. (1988) The influence of microarthropods on symbiotic and non-symbiotic mutualism in detrital-based below-ground food webs. Agric. Ecosys. Environ. 24, 147-159.

199. Arachevaleta, M., Bacon, C.W., Hoveland, C.S. and Radcliffe, D.E. (1989) Effect of the tall fescue endophyte on plant response to environmental stress. Agron. J. 81, 83-90.

200. Siegel, M.R., Latch, G.C.M. and Johnson, M.C. (1987) Fungal endophytes of grasses. Ann. Rev. Phytopathol. 25, 293-315.

201. Bacon, C.W. and Siegel, M.R. (1988) Endophyte parasitism of tall fescue. J. Prod. Agric. 1, 45-55.

202. Bacon, C.W., Lyons, P.C., Porter, J.K. and Robbins, J.D. (1986) Ergot toxicity from endophyte-infected grasses: A review. Agron. J. 78, 106-116.

203. Latch, G.C.M., Hunt, W.F. and Musgrave, D.R. (1985) Endophytic fungi affect growth of perennial ryegrass. New Zeal. J. Agric. Res. 28, 165-168.

204. Porter, J.K., Bacon, C.W., Cutler, H.G., Arrendale, R.F. and Robbins, J.D. (1985) In vitro auxin production by Balansia epichloe. Phytochem. 24, 1429-1431.

205. Read, J.C. and Camp, B.J. (1986) The effect of the fungal endophyte Acremonium coenophialum in tall fescue on animal performance, toxicity, and stand maintenance. Agron. J. 78, 848-850.

206. Atkinson, D., Naylor, D. and Coldrick, G.A. (1976) The effect of tree spacing on the apple root system. Hort. Res. 16, 89-105.

207. Odum, E. (1969) The strategy of ecosystem development. Science 164, 262-269.

208. Brown, D., Hallman, R.G., Lee, C.R., Skogerboe, J.G., Eskew, K., Price, R.A., Page, N.R., Clar, M., Kort, R. and Hopkins, H. (1986) Reclamation and Vegetative Restoration of Problem Soils and Disturbed Lands. Noyes Data Corp., Park Ridge, NJ.

209. Dymond, J.R. and Hawley, J.G. (1988) How much do trees reduce landsliding? J. Soil Water Conserv. 43, 495-498.

210. Kebin, Z. and Kaiguo, Z. (1989) Afforestation for sand fixation in China J. Arid Environ. 16, 3-10.

211. Brown, L.R. and Wolf, E.C. (1984) Soil Erosion: Quiet Crisis in the World Economy. Worldwatch Institute, Washington, DC.

212. Harlin, J.M. and Berardi, G.M. (1987) Agricultural soil loss: An introduction. In Harlin, J.M. and Berardi, G.M. (eds) Agricultural Soil Loss: Processes, Policies, and Prospects. Westview Press, Boulder, CO, pp. 1-5.

213. Idso, S.B. (1976) Dust storms. Sci. Amer. 235(4), 108-114.

214. Greeley, R. and Iverson, J.D. (1985) Wind as a Geological Process on Earth, Mars, Venus, and Titan. Cambridge Univ. Press, Cambridge, UK.

215. Kalma, J.D., Speight, J.D. and Wasson, R.J. (1988) Potential wind erosion in Australia: A continental perspective. J. Climatol. 8, 411-428.

192

REFERENCES

216. Food Agric. Org. (1965) Soil Erosion by Water: Some Measures for its
     Control on Cultivated Lands. Food and Agriculture Organization of the
     United Nations, Rome, Italy.
217. Foster, G.R., Ed. (1977) Soil Erosion: Prediction and Control. Soil
     Conservation Soc. Amer., Ankeny, IA.
218. Troeh, F.R., Hobbs, J.A. and Donahue, R.L. (1980) Soil and Water
     Conservation for Productivity and Environmental Protection. Prentice-
     Hall, Englewood Cliffs, NJ.
219. IECA (1987) Erosion Control: Procedings of Conference XVIII. Internat.
     Erosion Contr. Assoc., Pinole, CA.
220. Tate, R.L., III. (1987) Soil Organic Matter: Biological and Ecological
     Effects. John Wiley & Sons, New York, NY.
221. Coleman, D.C., Andrews, R., Ellis, J.E. and Singh, J.S. (1976) Energy
     flow and partitioning in selected man-managed and natural ecosystems.
     Agro-Ecosys. 3, 45-54.
222. Woodwell, G.M. (1988) CO_2 reduction and reforestation. Science 242,
     1493.
223. Lowrance, R. and Williams, R.G. (1988) Carbon movement in runoff and
     erosion under simulated rainfall conditions. Soil Sci. Soc. Amer. J.
     52, 1445-1448.
224. Langdale, G.W. and Lowrance, R. (1984) Effects of soil erosion on
     agroecosystems of the humid United States. In Lowrance, R. (ed)
     Agricultural Ecosystems. John Wiley & Sons, New York, NY, pp. 133-144.
225. Schreiber, J.D. and McGregor, K.C. (1979) The transport and oxygen
     demand of organic carbon released to runoff from crop residues. Prog.
     Water Tech. 11, 253-261.
226. Cook, R.J., Boosalis, M.G. and Doupnik, B. (1978) Influence of crop
     residues on plant diseases. In Oschwald, W.R. (ed) Crop Residue
     Management Systems. Amer. Soc. Agron., Madison, WI, pp. 147-163.
227. Lumsden, R.D., Lewis, J.A. and Papavizas, G.C. (1983) Effect of organic
     amendments on soilborne plant diseases and pathogen antagonists. In
     Lockeretz, W. (ed) Environmentally Sound Agriculture. Praeger Press,
     New York, NY, pp. 51-70.
228. Hoitink, H.A.J. and Fahy, P.C. (1986) Basis for the control of soilborne
     plant pathogens with composts. Ann. Rev. Phytopathol. 24, 93-114.
229. Stolwijk, J.A.J. and Thimann, K.V. (1957) On the uptake of carbon
     dioxide and bicarbonate by roots and its influence on growth. Plant
     Physiol. 32, 513-520.
230. Voznesenskii, V.L. (1958) The absorption of carbon dioxide by plant
     roots. Fiziol. Rast. 5, 325-332.
231. Skok, J., Chorney, W. and Broecker, W.S. (1962) Uptake of CO_2 by roots
     of Xanthium plants Bot. Gaz. 124, 118-120.
232. Splittstoesser, W.E. (1966) Dark CO_2 fixation and its role in the growth
     of plant tissue. Plant Physiol. 41, 755-759.
233. Erickson, L.C. (1946) Growth of tomato roots as influenced by oxygen in
     the nutrient solution. Amer. J. Bot. 33, 551-556.
234. Leonard, O.A. and Pinckard, J.A. (1946) Effect of various oxygen and
     carbon dioxide concentrations on cotton root development. Plant
     Physiol. 21, 18-36.
235. Geisler, G. (1963) Morophogentic influence of (CO_2 + HCO_3^-) on roots.
     Plant Physiol. 38, 77-80.
236. Yorgalevitch, C.M. and Janes, W.H. (1988) Carbon dioxide enrichment of
     the root zone of tomato seedlings. J. Hort. Sci. 63, 265-270.
237. Kursanov, A.L., Kuzin, A.M. and Mamul, Y.V. (1951) On the possibility
     for assimilation by plants of carbonates taken in with the soil
     solution. Dokl. Akad. Nauk SSSR 79, 685-687.
238. Grinfeld, E.G. (1954) On the nutrition of plants with carbon dioxide
     through the roots. Dokl. Akad. Nauk SSSR 94, 919-922.
239. Berquist, N.O. (1964) Absorption of carbon dioxide by plant roots. Bot.

REFERENCES

<u>Not</u>. 117, 249-261.

240. Nakayama, F.S. and Bucks, D.A. (1980) Using subsurface trickle system for carbon dioxide enrichment. <u>In</u> Jensen, M.H. and Oebker, N.F. (eds) <u>Proc</u>. <u>15th</u> <u>Nat</u>. <u>Agric</u>. <u>Plastic Congr</u>. Nat. Agric. Plastics Assoc., Manchester, MO, pp. 13-18.

241. Baron, J.J. and Gorski, S.F. (1986) Response of eggplant to a root environment enriched with CO_2. <u>HortScience</u> 21, 495-498.

242. Misra, R.K. (1951) Further studies on the carbon dioxide factor in the air and soil layers near the ground. <u>Indian</u> <u>J</u>. <u>Meteorol</u>. <u>Geophys</u>. 2, 284-292.

243. Arteca, R.N., Pooviah, B.W. and Smith, O.E. (1979) Changes in carbon fixation, tuberization, and growth induced by CO_2 applications to the root zones of potato plants. <u>Science</u> 205, 1279-1280.

244. Mauney, J.R. and Hendrix, D.L. (1988) Responses of glasshouse grown cotton to irrigation with carbon dioxide-saturated water. <u>Crop Sci</u>. 28, 835-838.

245. Jackson, W.A. and Coleman, N.T. (1959) Fixation of carbon dioxide by plant roots through phosphoenolpyruvate carboxylase. <u>Plant</u> <u>Soil</u> 11, 1-16.

246. Mitz, M.A. (1979) CO_2 biodynamics: A new concept of cellular control. <u>J</u>. <u>Theoret</u>. <u>Biol</u>. 80, 537-551.

247. Gilmore, A.E. (1971) The influence of endotrophic mycorrhizae on the growth of peach seedlings. <u>J</u>. <u>Amer</u>. <u>Soc</u>. <u>Hort</u>. <u>Sci</u>. 96, 35-38.

248. LaRue, J.H., McClellan, W.D. and Peacock, W.L. (1975) Mycorrhizal fungi and peach nursery nutrition. <u>Calif</u>. <u>Agirc</u>. 29, 5-7.

249. Cooper, K.M. and Tinker, P.B. (1978) Translocation and transfer of nutrients in vesicular-arbuscular mycorrhizas. II. Uptake and translocation of phosphorus, zinc, and sulphur. <u>New</u> <u>Phytol</u>. 73, 901-912.

250. Swaminathan, K. and Verma, B.C. (1979) Responses of three crop species to vesicular-arbuscular mycorrhizal infection on zinc-deficient Indian soils. <u>New</u> <u>Phytol</u>. 82, 481-487.

251. Rogers, R.D. and Williams, S.E. (1986) Vesicular-arbuscular mycorrhiza. Influence on plant uptake of cesium and cobolt. <u>Soil</u> <u>Biol</u>. <u>Biochem</u>. 18, 371-376.

252. Brown, M.T. and Wilkins, D.A. (1985) Zinc tolerance of mycorrhizal <u>Betula</u>. <u>New</u> <u>Phytol</u>. 99, 101-106.

253. Dueck, T.A., Visser, P., Ernest, W.H.O. and Schat, H. (1986) Vesicular-arbuscular mycorrhizae decrease zinc toxicity to grasses in zinc polluted soil. <u>Soil</u> <u>Biol</u>. <u>Biochem</u>. 18, 331-333.

254. Jones, M.D. and Hutchinson, T.C. (1986) The effect of mycorrhizal infection on the responses of <u>Betula</u> <u>papyrifera</u> to nickel and copper. <u>New</u> <u>Phytol</u>. 102, 429-442.

255. Heggo, A., Angle, J.S. and Chaney, R.L. (1989) Vesicular-arbuscular mycorrhiza effects on heavy metal uptake by soybeans. <u>Soil</u> <u>Biol</u>. <u>Biochem</u>., submitted.

256. Kucey, R.M. and Paul, E.A. (1982) Carbon flow, photosynthesis and N_2-fixation in mycorrhizal and nodulated faba beans (<u>Vicia</u> <u>faba</u> L.) <u>Soil</u> <u>Biol</u>. <u>Biochem</u>. 14, 407-412.

257. Darwin, C. (1881) <u>The</u> <u>Formation</u> <u>of</u> <u>Vegetable</u> <u>Mould</u> <u>through</u> <u>the</u> <u>Action</u> <u>of</u> <u>Worms</u>, <u>with</u> <u>Observations</u> <u>on</u> <u>their</u> <u>Habits</u>. John Murray, London, UK.

258. Edwards, C.A. (1988) Earthworms and agriculture. <u>Agron</u>. <u>Abstr</u>. 80, 274.

259. Sharpley, A.N., Syers, J.K. and Springett, J. (1988) Earthworm effects on the cycling of organic matter and nutrients. <u>Agron</u>. <u>Abstr</u>. 80, 285.

260. Bertsch, P.M., Peters, R.A., Luce, H.D. and Claude, D. (1988) Comparison of earthworm activity in long-term no-tillage and conventionally tilled corn systems. <u>Agron</u>. <u>Abstr</u>. 80, 271.

261. McCabe, D., Protz, R. and Tomlin, A.D. (1988) Earthworm influence on soil quality in native sites of southern Ontario. <u>Agron</u>. <u>Abstr</u>. 80,

281.
262. Zachmann, J.E. and Molina, J.A. (1988) Earthworm-microbe interactions in
     soil. Agron. Abstr. 80, 289.
263. Graham, R.C., Wood, H.B. and Lueking, M.A. (1988) Soil morphologic
     development in a 40 year old chaparral biosequence. Agron. Abstr. 80,
     258.
264. Johnson, D.L. (1988) Biomantle evolution and the redistribution of earth
     materials and artifacts. Agron. Abstr. 80, 259.
265. Kemper, W.D. (1988) Earthworm burrowing and effects on soil structure
     and transmissivity. Agron. Abstr. 80, 278.
266. Hall, R.B. and Dudas, M.J. (1988) Effects of chromium loading on
     earthworms in an amended soil. Agron. Abstr. 80, 275.
267. Logsdon, S.D. and Linden, D.L. (1988) Earthworm effects on root growth
     and function, and on crop growth. Agron. Abstr. 80, 280.
268. Hendrix, P.F., Mueller, B.R., Van Vliet, P., Bruce, R.R. and Langdale,
     G.W. (1988) Earthworm abundance and distribution in agricultural
     landscapes of the Georgia piedmont. Agron. Abstr. 80, 276.
269. Kladivko, E.J. (1988) Soil management effects on earthworm populations
     and activity. Agron. Abstr. 80, 278.
270. Lal, R. (1988) Effects of macrofauna on soil properties in tropical
     ecosystems. Agric. Ecosys. Environ. 24, 101-116.
271. Peterson, H. and Luxton, M. (1982) A comparataive analysis of soil fauna
     populations and their role in decomposition processes. Oikos 39, 287-
     388.
272. Swift, M.J., Heal, O.W. and Anderson, J.M. (1979) Studies in Ecology:
     Decomposition in Terrestrial Ecosystems, Vol. 5. Univ. Calif. Press,
     Berkeley, CA.
273. Seastedt, T.R. (1984) The role of microarthropods in decomposition and
     mineralization process. Ann. Rev. Entomol. 29, 25-46.
274. Witkamp, M. and Ausmus, B.S. (1976) Processes in decomposition and
     nutrient transfer in forest systems. In Anderson, J.M. and Macfadyen,
     A. (eds) The Role of Terrestrial and Aquatic Organisms in
     Decomposition Processes. Blackwell, Oxford, UK, pp. 375-396.
275. Vossbrinck, C.R., Coleman, D.C. and Woolley, T.A. (1979) Abiotic and
     biotic factors in litter decomposition in a semi-arid grassland.
     Ecology 60, 265-271.
276. Persson, T., Baath, E., Clarholom, M., Lundkvist, H., Soderstrom, B.E.
     and Sohlenius, B. (1980) Trophic structure, biomass dymanics and
     carbon metabolism of soil organisms in a Scots pine forest. Ecol.
     Bull. 32, 419-459.
277. Reichle, D.E. (1977) The role of invertebrates in nutrient cycling.
     Ecol. Bull. 25, 145-154.
278. Cole, C.V., Elliot, E.T., Hunt, H.W. and Coleman, D.C. (1978) Trophic
     interactions in soils as they affect energy and nutrient dynamics. V.
     Phosphorus transformations. Microbiol. Ecol. 4, 381-387.
279. Anderson, J.M., Ineson, P. and Huish, J.A. (1983) Nitrogen and cation
     mobilization by soil fauna feeding on leaf litter and soil organic
     matter from deciduous woodlands. Soil Biol. Biochem. 15, 463-467.
280. Patrick, R., Ford, E. and Quarles, J. (1987) Groundwater Contamination
     in the United States. Univ. Pennsyl. Press, Philadelphia, PA.
281. Macalady, D.L., Ed. (1986) Transport and transformations of organic
     contaminants. J. Contam. Hydrol. 1, 1-261.
282. Bouwer, E., Mercer, J., Kavanaugh, M. and DiGiano, F. (1988) Coping
     with groundwater contamination. J. Water Poll. Contr. Fed. 60, 1415-
     1427.
283. Wood, E.F., Ferrara, R.A., Gray, W.G. and Pinder, G.F. (1984) Groundwater
     Contamination from Hazardous Wastes. Prentice-Hall, Englewood Cliffs,
     NJ.
284. Switzenbaum, M.S. and Alleman, J.E., Eds. (1988) Annual literature

review. J. Water Poll. Contr. Fed. 60, 868-924.

85. Young, C.P., Oakes, D.B. and Wilkinson, W.B. (1983) The impact of
    agricultural practices on the nitrate content of groundwater in the
    principal United Kingdom aquifers. In Goluber, G. (ed) Environmental
    Management of Agricultural Watersheds. Internat. Inst. Appl. Syst.
    Analysis, Laxenburg, Austria, pp. 165-197.

86. Spalding, R.F. and Kitcher, L.A. (1988) Nitrate in the intermediate
    vadose zone beneath irrigated cropland. Ground Water Monitor. Rev.
    8(2), 89-95.

87. Croll, B.T. (1986) The effects of the agricultural use of herbicides on
    fresh waters. In Solbe, J.R.deL.G. (ed) Effects of Land Use on Fresh
    Waters. Ellis Horwood Ltd., Chichester, UK, pp. 201-209.

88. Burkart, M.R., Ragone, S.E., Thurman, E.M. and Perry, C.A. (1988)
    Herbicides in Ground and Surface Water: A Mid-Continent Initiative.
    U.S. Geological Survey, Reston, VA.

89. Holden, P.W. (1986) Pesticides and Ground Water Quality. National
    Academy Press, Washington, DC.

90. Younos, T.M. and Weigmann, D.L. (1988) Pesticides: A continuing dilemma.
    J. Water Poll. Contr. Fed. 60, 1199-1205.

91. Bouwer, H. (1989) Agricultural chemicals and groundwater quality--
    problems and solutions. Proc. Second Pan-American Regional Conference
    on Irrigation and Drainage: Toxic Substances in Agricultural Water
    Supply and Drainage -- An International Environmental Perspective.
    U.S. Committee on Irrigation and Drainage, Denver, CO, in press.

92. Sierra Club Legal Defense Fund (1989) The Poisoned Well: New Strategies
    for Groundwater Protection. Island Press, Covelo, CA.

93. Bouwer, H. (1987) Effect of irrigated agriculture on groundwater. J.
    Irrig. Drain. Engin. 113, 4-15.

94. Sheets, T.J., Crafts, A.S. and Drever, H.R. (1962) Influence of soil
    properties on the phytotoxicities of the s-triazine herbicides. J.
    Agric. Food Chem. 10, 458-462.

95. LeBaron, H.M., McFarland, J.E. and Simoneaux, B.J. (1988) Metolachlor.
    In Kearney, P.C. and Kaufman, D.D. (eds) Herbicides: Chemistry,
    Degradation, and Mode of Action. Marcel Dekker, New York, NY, pp. 335-
    382.

96. Kearney, P.C. and Kaufman, D.C. (1988) Herbicides: Chemistry,
    Degradation, and Mode of Action. Vol. 3. Marcel Dekker, New York, NY.

97. Alexander, M. (1981) Biodegradation of chemicals of environmental
    concern. Science 211, 132-138.

98. Alexander, M. (1985) Biodegradation of organic chemicals. Environ. Sci.
    Technol. 18, 106-111.

99. Hutchins, S.R., Tomson, M.B. and Ward, C.H. (1985) Microbial involvement
    in trace organic removal during rapid infiltration recharge of ground
    water. In Caldwell, D.E., Brierley, J.A. and Brierley, C.L. (eds)
    Planetary Ecology. Van Nostrand Reinhold, New York, NY, pp. 370-382.

300. Kunishi, H.M. (1988) Sources of nitrogen and phosphorus in an estuary of
     the Chesapeake Bay. J. Environ. Qual. 17, 185-188.

301. Smith, A.E., Waite, D., Grover, R., Kerr, L.A., Milward, L.J. and
     Sommerstad, H. (1988) Persistance and movement of picloram in a
     northern Saskatchewan watershed. J. Environ. Qual. 17, 262-268.

302. Wauchope, R.D. (1978) The pesticide content of surface water draining
     from agricultural fields -- a review. J. Environ. Qual. 7, 459-472.

303. Leonard, R.A. (1988) Herbicides in surface waters. In Grover, R. (ed)
     Environmental Chemistry of Herbicides. Vol. I. CRC Press, Boca Raton,
     FL, pp. 45-87.

304. Patrick, W.H., Jr., Delaune, R.D., Engler, R.M. and Gotoh, S. (1976)
     Nitrate Removal from Water at the Water-Mud Interface in Wetlands.
     U.S. Environmental Protection Agency, Washington, DC.

305. Fenchel, T.M. and Jorgensen, B.B. (1977) Detritus food chains of aquatic

REFERENCES   SECTION '

ecosystems: The role of bacteria. In Alexander, M. (ed) Advances in
      Microbial Ecology, Vol. 1. Plenum Press, New York, NY, pp. 1-58.
306. Reddy, K.R. and Smith, W.H., Eds. (1987) Aquatic Plants for Water
      Treatment and Resource Recovery. Magnolia Publ., Orlando, FL.
307. Moorhead, K.K., Reddy, K.R. and Graetz, D.A. (1988) Nitrogen
      transformations in a waterhyacinth-based water treatment system. J.
      Environ. Qual. 17, 71-76.
308. Dacey, J.W.H. (1980) Internal winds in water lilies: An adaptation for
      life in anerobic sediments. Science 210, 1017-1019.
309. Moorhead, K.K. and Reddy, K.R. (1988) Oxygen transport through selected
      aquatic macrophytes. J. Environ. Qual. 17, 138-142.
310. Allen, S.G., Idso, S.B., Kimball, B.A. and Anderson, M.G. (1988)
      Interactive effects of CO_2 and environment on photosynthesis of
      Azolla. Agric. For. Meteorol. 42, 209-217.
311. Allen, S.G., Idso, S.B. and Kimball, B.A. (1989) Interactive effects of
      CO_2 and environment on net photosynthesis of water lily. Agric.
      Ecosys. Environ., in press.
312. Idso, S.B., Kimball, B.A., Anderson, M.G. and Mauney, J.R. (1987)
      Effects of atmospheric CO_2 enrichment on plant growth: The interactive
      role of air temperature. Agric. Ecosys. Environ. 20, 1-10.
313. Farquhar, G.D. and Sharkey, T.D. (1982) Stomatal conductance and
      photosynthesis. Ann. Rev. Plant Physiol. 33, 317-345.
314. Wong, S.C., Cowan, I.R. and Farquhar, G.D. (1979) Stomatal conductance
      correlates with photosynthetic capacity. Nature 282, 424-426.
315. Field, C.B. (1987) Leaf-age effects on stomatal conductance. In Zeiger,
      E., Farquhar, G.D. and Cowan, I.R. (eds) Stomatal Function. Stanford
      Univ. Press, Stanford, CA, pp. 367-384.
316. Jordan, W.R. and Ritchie, J.T. (1971) Influence of soil water stress on
      evaporation, root absorption, and internal water status of cotton.
      Plant Physiol. 48, 783-788.
317. Ackerson, R.C. and Krieg, D.R. (1977) Stomatal and nonstomatal
      regulation of water use in cotton, corn, and sorghum. Plant Physiol.
      60, 850-853.
318. Radin, J.W., Kimball, B.A., Hendrix, D.L. and Mauney, J.R. (1987)
      Photosynthesis of cotton plants exposed to elevated levels of carbon
      dioxide in the field. Photosyn. Res. 12, 191-203.
319. Hutmacher, R.B. and Krieg, D.R. (1983) Photosynthetic rate control in
      cotton. Stomatal and nonstomatal factors. Plant Physiol. 73, 658-661.
320. Radin, J.W., Hartung, W., Kimball, B.A. and Mauney, J.R. (1988)
      Correlation of stomatal conductance with photosynthetic capacity of
      cotton only in a CO_2-enriched atmosphere: Mediation by abscisic acid?
      Plant Physiol. 88, 1058-1062.
321. Leadley, P.W. and Reynolds, J.F. (1988) Effects of elevated carbon
      dioxide on estimation of leaf area and leaf dry weight of soybean.
      Amer. J. Bot. 75, 1771-1774.
322. Lemon, E.R. (1977) The land's response to more carbon dioxide. In
      Anderson, N.R. and Malahoff, A. (eds) The Fate of Fossil Fuel CO_2 in
      the Ocean. Plenum, New York, NY, pp. 97-130.
323. Lemon, E.R. (1983) Interpretive summary. In Lemon, E.R. (ed) CO_2 and
      Plants: The Response of Plants to Rising Levels of Atmospheric Carbon
      Dioxide. Westview Press, Boulder, CO, pp. 1-5.
324. Idso, S.B., Kimball, B.A. and Mauney, J.R. (1988) Atmospheric CO_2
      enrichment and plant dry matter content. Agric. For. Meteorol. 43,
      171-181.
325. Acock, B. and Pasternak, D. (1986) Effects of CO_2 concentrations on
      composition, anatomy, and morphology of plants. In Enoch, H.Z. and
      Kimball, B.A. (eds) Carbon Dioxide Enrichment of Greenhouse Crops.
      Vol. II. Physiology, Yield, and Economics. CRC Press, Boca Raton, FL,
      pp. 41-52.

EFERENCES

26. Idso, S.B., Kimball, B.A. and Mauney, J.R. (1988) Effects of atmospheric CO_2 enrichment on root:shoot ratios of carrot, radish, cotton and soybean. Agric. Ecosys. Environ. 21, 293-299.

27. Knecht, G.S. (1975) Response of radish to high CO_2. HortSci. 10, 274-275.

28. Bhattacharya, N.C., Biswas, P.K., Bhattacharya, S., Sionit, N. and Strain, B.R. (1985) Growth and yield response of sweet potato (Ipomoea batatas) to atmospheric CO_2 enrichment. Crop Sci. 25, 975-981.

29. Daley, P.F., Surano, K.A. and Shinn, J.H. (1989) Long-term exposure of alfalfa (Medicago sativa L.) to elevated atmospheric carbon dioxide. I. Photosynthesis, yield, and growth analysis. Plant Cell Environ., submitted.

30. Chapin, F.S., III. (1980) The mineral nutrition of wild plants. Ann. Rev. Ecol. Syst. 11, 233-260.

31. Larigauderie, A., Hilbert, D.W. and Oechel, W.C. (1988) Effect of CO_2 enrichment and nitrogen availability on resource aquisition and resource allocation in a grass, Bromus mollis. Oecologia 77, 544-549.

32. Grant Lipp, A.E. and Ballard, L.A.T. (1959) The breaking of seed dormancy of some legumes by carbon dioxide. Austr. J. Agric. Res. 10, 495-499.

33. Ballard, L.A.T. (1961) Studies of dormancy in the seeds of subterranean clover (Trifolium subterraneum L.). II. The interaction of time, temperature, and carbon dioxide during passage out of dormancy. Austr. J. Biol. Sci. 14, 173-186.

34. James, A.L. (1968) Some influences of soil atmosphere on germination of annual weeds. Dissert. Abstr. 29, 4479-B.

35. Rijven, A.H.G. and Parkash, V. (1970) Cytokinin-induced growth responses by fenugreek cotyledons. Plant Physiol. 46, 638-640.

36. Heichel, G.H. and Jaynes, R.A. (1974) Stimulating emergence and growth of Kalmia genotypes with CO_2. HortScience 9, 60-62.

37. Wulff, R.D. and Alexander, H.M. (1985) Intraspecific variation in the response to CO_2 enrichment in seeds and seedlings of Plantago ianceolata L. Oecologia 66, 458-460.

38. Schonbeck, M.W. and Egley, G.H. (1980) Redroot pigweed (Amaranthus retroflexus) seed germination responses to afterripening, temperature, ethylene, and some other environmental factors. Weed Sci. 28, 543-548.

39. St. Omer, L. and Horvath, S.M. (1983) Potential effects of elevated carbon dioxide levels on seed germination of three native plant species. Bot. Gaz. 144, 477-480.

340. Edwards, M.E. (1977) Carbon dioxide and ethylene control of spore germination in Onoclea sensibilis L. Plant Physiol. 59, 756-758.

341. Esashi, Y. and Katoh, H. (1975) Dormancy and impotency of cocklebur seeds. III. CO_2- and C_2H_4-dependent growth of the embryonic axes and cotyledon segments. Plant Cell Physiol. 16, 707-718.

342. Esashi, Y., Watanabe, K., Ohhara, Y. and Katoh, H. (1976) Dormancy and impotency of cocklebur seeds. V. Growth and ethylene production of axial segments in respoonse to O_2. Austr. J. Plant Physiol. 3, 701-710.

343. Esashi, Y., Kotaki, K. and Ishidoya, O. (1976) Dormancy and impotency of cocklebur seeds. VI. Growth and ethlylene production of cotyledonary segments in response to O_2. Austr. J. Plant Physiol. 3, 711-719.

344. Negm, F.B., Smith, O.E. and Kumamoto, J. (1972) Interaction of carbon dioxide and ethylene in overcoming thermodormancy of lettuce seeds. Plant Physiol. 49, 869-872.

345. Katoh, H. and Esashi, Y. (1975) Dormancy and impotency of cocklebur seeds. I. CO_2, C_2H_4, O_2 and high temperature. Plant Cell Physiol. 16, 687-696.

346. Esashi, Y., Ooshima, Y., Abe, M., Kurota, A. and Satoh, S. (1986) CO_2-enhanced C_2H_4 production in tissues of imbibed cocklebur seeds. Austr.

   J. Plant Physiol. 13, 417-429.
347. Esashi, Y., Saijoh, Y., Ishida, S., Oota, H. and Isizawa, K. (1986) Reversal of ethylene action on cocklebur seed germination in relation to duration of pre-treatment soaking and temperature. Plant Cell Environ. 9, 121-126.
348. Saini, H.S., Bassi, P.K., Consolacion, E.D. and Spencer, M.S. (1986) Interactions among plant hormones, carbon dioxide, and light in the relief of thermoinhibition of lettuce seed germination: Studies in a flow-through gaseous system. Can. J. Bot. 64, 2322-2326.
349. Saini, H.S., Consolacion, E.D., Bassi, P.K. and Spencer, M.S. (1986) Requirement for ethylene synthesis and action during relief of thermoinhibition of lettuce seed germination by combinations of gibberellic acid, kinetin, and carbon dioxide. Plant Physiol. 81, 950-953.
350. Tittle, F.L. and Spencer, M.S. (1986) Interactions between ethylene, CO_2, and ABA on GA_3-induced amylase synthesis in barley aleurone tissue. Plant Physiol. 80, 1034-1037.
351. Esashi, Y., Fuwa, N., Kurotz, A., Oota, H. and Abe, M. (1987) Interrelationship between ethylene and carbon dioxide in relation to respiration and adenylate content in the pre-germination period of cocklebur seeds. Plant Cell Physiol. 28, 141-150.
352. Esashi, Y., Hase, S., Kojima, K. and Fuwa, N. (1987) Differential responsiveness in zonal growth of cocklebur seed axes to CO_2, C_2H_4, O_2 and respiration inhibitors. Plant Cell Physiol. 28, 163-169.
353. Ishizawa, K., Hoshina, M., Kawabe, K. and Esashi, Y. (1988) Effects of 2,5-Norbornadiene on cocklebur seed germination and rice coleoptile elongation in response to CO_2 and C_2H_4. J. Plant Growth Regul. 7, 45-58.
354. Sisler, E.C. and Wood, C. (1988) Interaction of ethylene and CO_2. Physiol. Plant. 73, 440-444.
355. Krizek, D.T., Zimmerman, R.H., Klueter, H.H. and Bailey, W.A. (1971) Growth of crabapple seedlings in controlled environments: Effects of carbon dioxide level, and time and duration of carbon dioxide treatment. J. Amer. Soc. Hort. Sci. 96, 285-288.
356. Krizek, D.T., Bailey, W.A., Klueter, H. and Liu, R.C. (1974) Maximizing growth of vegetable seedlings in controlled environments at elevated temperature, light and CO_2. Acta Hort. 39, 89-102.
357. Tolley, L.C. and Strain, B.R. (1984) Effects of atmospheric CO_2 enrichment on growth of Liquidambar styraciflua and Pinus taeda seedlings under different irradiance levels. Can. J. For. Res. 14, 343-350.
358. Ito, T. (1973) Plant growth and physiology of vegetable plants as influenced by carbon dioxide environment. Chiba Daigaku Engeigakubu Gakujutsu Hokoko 7, 1-134.
359. Neales, T.F. and Nicholls, A.O. (1978) Growth responses of young wheat plants to a range of ambient CO_2 levels. Aust. J. Plant Physiol. 5, 45-59.
360. Hicklenton, P.R. and Jolliffe, P.A. (1980) Alterations in the physiology of CO_2 exchange in tomato plants grown in CO_2-enriched atmospheres. Can. J. Bot. 58, 2181-2189.
361. Idso, S.B. (1989) Three stages of plant response to atmospheric CO_2 enrichment. Plant Physiol. Biochem. 27, 131-134.
362. Hardh, J.E. (1968) Trials with carbon dioxide, light and growth substances on forest tree plants. Acta For. Fen. 81, 1-10.
363. Molnar, J.M. and Cumming, W.A. (1968) Effect of carbon dioxide on propagation of softwood, conifer, and herbaceous cuttings. Can. J. Plant Sci. 48, 595-599.
364. Lin, W.C. and Molnar, J.M. (1980) Carbonated mist and high intensity supplementary lighting for propagation of selected woody ornamentals.

Proc. Int. Plant Prop. Soc. 30, 104-109.

365. Davis, T.D. and Porter, J.R. (1983) High CO_2 applied to cuttings. Effects on rooting and subsequent growth of ornamental species. HortScience 18, 194-196.

366. Harrison, A. (1965) Auxanometer experiments on extension growth of Avena coleoptiles in different carbon dioxide concentrations. Physiol. Plant. 18, 321-328.

367. McIntyre, G.I. (1987) Inhibition by carbon dioxide of the phototropic response of the Avena coleoptile to blue light. Can. J. Bot. 65, 488-490.

368. St. Omer and Horvath, S.M. (1983) Elevated carbon dioxide concentrations and whole plant senescence. Ecology 64, 1311-1314.

369. Bhattacharya, S., Bhattacharya, N.C., Biswas, P.K. and Strain, B.R. (1985) Response of cowpea (Vigna unguiculata L.) to CO_2-enriched environment on growth, dry matter production and yield components at different stages of vegetative and reproductive growth. J. Agric. Sci. Camb. 105, 527-534.

370. Hughes, A.P. and Cockshull, K.E. (1969) Effects of carbon dioxide concentration on the growth of Callistephus chinensis cultivar Johannistag. Ann. Bot. 33, 351-365.

371. Hand, D.W. and Postlethwaite, J.D. (1971) The response to CO_2 enrichment of capillary-watered single-truss tomatoes at different plant densities and seasons. J. Hort. Sci. 46, 461-470.

372. Backhaus, R., Jones, D. and Kimball, B. (1979) The influence of CO_2 fertilization on seedling growth of guayule. HortScience 14, 463-464.

373. Garbutt, K. and Bazzaz, F.A. (1984) The effects of elevated CO_2 on plants. III. Flower, fruit and seed production and abortion. New Phytol. 98, 433-446.

374. Marc, J. and Gifford, R.M. (1984) Floral initiation in wheat, sunflower, and sorghum under carbon dioxide enrichment. Can. J. Bot. 62, 9-14.

375. Slack, G. (1986) CO_2 enrichment of tomato crops. In Enoch, H.Z. and Kimball, B.A. (eds) Carbon Dioxide Enrichment of Greenhouse Crops. Vol. II. Physiology, Yield, and Economics. CRC Press, Boca Raton, FL, pp. 151-163.

376. Hesketh, J.D. and Hellmers, H. (1973) Floral initiation in four plant species growing in CO_2 enriched air. Environ. Contr. Biol. 11, 51-53.

377. Hartman, H.D. (1966) The effect of CO_2 concentration on head lettuce with special regard to the response of varieties. Proc. 13th Internat. Hort. Congr. 1, 467.

378. Daunicht, H.-J. (1971) The CO_2 requirements of greenhouse crops. Acta Hort. 17, 62-71.

379. Calvert, A. (1972) Effects of day and night temperatures and carbon dioxide enrichment on yield of glasshouse tomatoes. J. Hort. Sci. 47, 231-247.

380. Raper, C.D., Jr., Weeks, W.W., Downs, R.J. and Johnson, W.H. (1973) Chemical properties of tobacco leaves as affected by carbon dioxide depletion and light intensity. Agron. J. 65, 988-992.

381. Chang, C.W. (1975) Carbon dioxide and senescence in cotton plants. Plant Physiol. 55, 515-519.

382. Drake, B.G. (1988) A field study of the effects of elevated ambient CO_2 on ecosystem processes in Chesapeake Bay wetlands. In Koomanoff, F.A. (ed) Carbon Dioxide and Climate: Summaries of Research in FY 1988. U.S. Dept. Energy, Washington, DC, pp. 48-49.

383. Curtis, P.S., Drake, B.G., Leadley, P.W., Arp, W.J. and Whigham, D.F. (1989) Growth and senescence in plant communities exposed to elevated CO_2 concentrations on an estuarine marsh. Oecologia 78, 20-26.

384. Mulder, E.G. and Van Veen, W.L. (1960) The influence of carbon dioxide on symbiotic nitrogen fixation. Plant Soil 13, 265-278.

385. Marx, J.L. (1974) Nitrogen fixation: Research efforts intensify. Science

185, 132-136.

386. Masterson, C.L. and Sherwood, M.T. (1978) Some effects of increased atmospheric carbon dioxide on white clover (Trifolium repens) and pea (Pisum sativum). Plant Soil 49, 421-426.

387. Shivashankar, K. and Vlassak, K. (1978) Influence of straw and CO_2 on N_2-fixation and yield of field-grown soybeans. Plant Soil 49, 259-256.

388. Acock, B., Wallace, S.U. and Lambert, J.R. (1987) The effects of increased carbon dioxide concentrations on N-fixation rates in soybeans. Agron. Abstr. 79, 99.

389. Bothe, H., Debruijn, F.J. and Newton, W.E., Eds. (1988) Nitrogen Fixation: Hundred Years After. VCH Pub., Deerfield Beach, FL.

390. Dobereiner, J. and Pedrosa, F.O. (1988) Nitrogen-fixing Bacteria in Nonleguminous Crop Plants. Sci. Tech. Pub., Madison, WI.

391. Daubenmire, R.F. (1964) Plants and Environment. John Wiley & Sons, New York, NY.

392. Torrey, J.G. (1978) Nitrogen fixation by actinomycete-nodulated angiosperms. BioScience 28, 586-592.

393. Mitsui, A. (1988) Marine aerobic nitrogen-fixing unicellular cyanobacteria. Trans. Amer. Geophys. Union 69, 1089.

394. Hardy, R.W.F. and Havelka, U.D. (1975) Nitrogen fixation research: A key to world food? Science 188, 633-643.

395. Vance, C.P. (1983) Rhizobium infection and nodulation: A beneficial plant disease? Ann. Rev. Microbiol. 37, 399-424.

396. Vance, C.P., Stade, S. and Maxwell, C.A. (1983) Alfalfa root nodule carbon dioxide fixation. I. Association with nitrogen fixation and incorporation into amino acids. Plant Physiol. 72, 469-473.

397. Lowe, R.H. and Evans, H.J. (1962) Carbon dioxide requirement for growth of legume nodule bacteria. Soil Sci. 94, 351-356.

398. Burk, D. (1961) On the use of carbonic anhydrase in carbonate and amine buffers for CO_2 exchange in manometric vessels, atomic submarines, and industrial CO_2 scrubbers. Ann. New York Acad. Sci. 92, 372-400.

399. Sinclair, T.R. and deWit, C.T. (1975) Photosynthate and nitrogen requirements for seed productivity by various crops. Science 189, 565-567.

400. Hardy, R.W.F., Havelka, U.D. and Quebedeaux, B. (1976) Opportunities for improved seed yield and protein production: N_2 fixation, CO_2 fixation, and O_2 control of reproductive growth. In National Research Council, U.S. (eds) Genetic Improvement of Seed Protein. National Academy Press, Washington, DC, pp. 196-228.

401. Finn, G.A. and Brun, W.A. (1982) Effect of atmospheric CO_2 enrichment on growth, nonstructural carbohydrate content, and root nodule activity in soybean. Plant Physiol. 69, 327-331.

402. Quebedeaux, B., Havelka, U.D., Livak, K.L. and Hardy, R.W.F. (1975) Effect of altered pO_2 on the aerial part of soybean on symbiotic N_2 fixation. Plant Physiol. 56, 761-764.

403. Murphy, P.M. (1986) Effect of light and atmospheric carbon dioxide concentration on nitrogen fixation by herbage legumes. Plant Soil 95, 399-409.

404. Hardy, R.W.F. and Havelka, U.D. (1973) Symbiotic N_2 fixation: Multifold enhancement by CO_2-enrichment of field-grown soybeans. Plant Physiol. Suppl. 48, 35.

405. Hardy, R.W.F. and Havelka, U.D. (1975) Photosynthate as a major factor limiting nitrogen fixation by field-grown legumes with emphasis on soybeans. In Nutman, P.S. (ed) Symbiotic Nitrogen Fixation in Plants. Cambridge Univ. Press, Cambridge, UK, pp. 421-439.

406. Havelka, U.D. and Hardy, R.W.F. (1976) Legume N_2 fixation as a problem in carbon nutrition. In Newton, W.E. and Nyman, C.J. (eds) Proc. 1st Internat. Symp. Nitrogen Fixation, Vol. 2. Washington State Univ. Press, Pullman, WA, pp. 456-475.

REFERENCES

407. MacDowall, F.D.H. (1983) Effects of light intensity and CO_2 concentration on the kinetics of 1st month growth and nitrogen fixation of alfalfa. Can. J. Bot. 61, 731-740.

408. Wilson, P.W., Fred, E.B. and Salmon, M.R. (1933) Relation between carbon dioxide and elemental nitrogen assimilation in leguminous plants. Soil Sci. 35, 145-165.

409. Phillips, D.A., Newell, K.D., Hassel, S.A. and Felling, C.E. (1976) The effect of CO_2 enrichment on root nodule development and symbiotic N_2 reduction in Pisum sativum L. Amer. J. Bot. 63, 356-362.

410. Hardy, R.W.F., Havelka, U.D. and Quebedeaux, B. (1978) The opportunity for and significance of alteration of ribulose 1,5-bisphosphate carboxylase activities in crop production. In Siegelman, H.W. and Hind, G. (eds) Photosynthetic Carbon Assimilation. Plenum Pub. Co., New York, NY, pp. 165-178.

411. Shearer, G., Kohl, D.H., Virginia, R.A., Bryan, B.A., Skeeters, J.L., Nilsen, E.T., Sharifi, M.R. and Rundel, P.W. (1983) Estimates of N_2-fixation from variation in the natural abundance of ^{15}N in Sonoran Desert ecosystems. Oecologia 56, 365-373.

412. Hoegberg, P. (1986) Nitrogen-fixation and nutrient relations in savanna woodland trees (Tanzania). J. Appl. Ecol. 23, 675-688.

413. Magid, H.M.A., Singleton, P.W. and Tavares, J.W. (1988) Sesbania-Rhizobium specificity and nitrogen fixation. Desert Plants 9, 45-48.

414. Crosswhite, F.S. and Crosswhite, C.D. (1988) Nitrogen fixation by desert legumes. Desert Plants 9, 64.

415. Jenkins, M.B., Virginia, R.A. and Jarrell, W.M. (1988) Depth distribution and seasonal populations of mesquite-nodulating rhizobia in warm desert ecosystems. Soil Sci. Soc. Amer. J. 52, 1644-1650.

416. National Academy of Sciences, U.S. (1979) Tropical Legumes: Resources for the Future. National Academy Press, Washington, DC.

417. Schulman, H.M., Lewis, M.C., Tipping, E.M. and Bordeleau, L.M. (1988) Nitrogen fixation by three species of leguminosae in the Canadian High Arctic Tundra. Plant Cell Environ. 11, 721-728.

418. Kimball, B.A. (1986) CO_2 stimulation of growth and yield under environmental constraints. In Enoch, H.Z. and Kimball, B.A. (eds) Carbon Dioxide Enrichment of Greenhouse Crops. Vol. II. Physiology, Yield, and Economics. CRC Press, Boca Raton, FL, pp. 53-67.

419. Campbell, D.E. and Young, R. (1986) Short-term CO_2 exchange response to temperature, irradiance, and CO_2 concentration in strawberry. Photosyn. Res. 8, 31-40.

420. Sionit, N., Strain, B.R. and Flint, E.P. (1987) Interaction of temperature and CO_2 enrichment on soybean: Growth and dry matter partitioning. Can. J. Plant Sci. 67, 59-67.

421. Sionit, N., Strain, B.R. and Flint, E.P. (1987) Interaction of temperature and CO_2 enrichment on soybean: Photosynthesis and seed yield. Can. J. Plant Sci. 67, 629-636.

422. Drake, B.G., Curtis, P.S., Arp, W.J., Leadley, P.W., Johnson, J. and Whigham, D. (1988) Effects of elevated CO_2 on Chesapeake Bay wetlands. III. Ecosystem and whole plant responses in the first year of exposure, April-November 1987. Response of Vegetation to Carbon Dioxide No. 44. U.S. Dept. Energy, Washington, DC.

423. Idso, S.B., Kimball, B.A., Anderson, M.G. and Mauney, J.R. (1989) Greenhouse warming could magnify positive effects of CO_2 enrichment on plant growth. CDIAC Communications Winter, 8-9.

424. Baker, J.T., Allen, L.H., Jr., Boote, K.J., Jones, P. and Jones, J.W. (1989) Response of soybean to air temperature and carbon dioxide concentration. Crop Sci. 29, 98-105.

425. Forrester, M.L., Krotkow, E. and Nelson, C.D. (1966) Effect of oxygen on photosynthesis, photorespiration and respiration in detached leaves. I. Soybean. Plant Physiol. 41, 422-427.

REFERENCES SECTION 7

426. Hesketh, J. (1967) Enhancement of photosynthetic CO_2 assimilation in the absence of oxygen, as dependent upon species and temperature. Planta 76, 371-374.

427. Ku, S. and Edwards, G.E. (1977) Oxygen inhibition of photosynthesis. Plant Physiol. 59, 986-999.

428. Ehleringer, J.R. (1979) Photosynthesis and photorespiration. Biochemistry, physiology, and ecological implications. HortScience 14, 217-222.

429. Maleszewski, S., Kaminska, Z., Kondracka, A. and Mikulska, M. (1988) Response of net photosynthesis in bean (Phaseolus vulgaris) leaves to the elevation of the partial pressures of oxygen and carbon dioxide. Physiol. Plant. 74, 221-224.

430. Joliffe, P.A. and Tregunna, E.B. (1968) Effect of temperature, CO_2 concentration, and light intensity on oxygen inhibition of photosynthesis in wheat leaves. Plant Physiol. 43, 902-906.

431. Laing, W.A., Ogren, W.L. and Hageman, R.H. (1974) Regulation of soybean net photosynthetic CO_2 fixation by the interaction of CO_2, O_2, and ribulose 1,5-diphosphate carboxylase. Plant Physiol. 54, 678-685.

432. Monson, R.K., Stidham, M.A., Williams, G.J., III, Edwards, G.E. and Uribe, E.G. (1982) Temperature dependence of photosynthesis in Agropyron smithii Rydb. Plant Physiol. 69, 921-928.

433. Jensen, R.G. (1977) Ribulose-1,5-bisphosphate carboxylase oxygenase. Ann. Rev. Plant Physiol. 28, 379-400.

434. Mortensen, L.M. (1983) Growth responses of some greenhouse plants to environment. VIII. Effect of CO_2 on photosynthesis and growth of Norway spruce. Meld. Nor. Landbrukshoegsk. 62, 1-13.

435. Mortensen, L.M. and Moe, R. (1983) Growth responses of some greenhouse plants to environment. V. Effect of CO_2, O_2 and light on net photosynthetic rate in Chrysanthemum morifolium Ramat. Scien. Hort. 19, 133-140.

436. Mortensen, L.M. and Ulsaker, R. (1985) Effect of CO_2 concentration and light levels on growth, flowering and photosynthesis of Begonia x hiemalis Fotsch. Scien. Hort. 27, 133-141.

437. Huner, N.P.A., Migus, W. and Tollenaar, M. (1986) Leaf CO_2 exchange rates in winter rye grown at cold-hardening and nonhardening temperatures. Can. J. Plant Sci. 66, 443-452.

438. Mortensen, L.M. (1987) Review: CO_2 enrichment in greenhouses. Crop responses. Scien. Hort. 33, 1-25.

439. Sage, R.F. and Sharkey, T.D. (1987) The effect of temperature on the occurrence of O_2 and CO_2 insensitive photosynthesis in field grown plants. Plant Physiol. 84, 658-664.

440. Idso, S.B. and Kimball, B.A. (1989) Growth response of carrot and radish to atmospheric CO_2 enrichment. Environ. Exp. Bot. 28, 135-139.

441. Hardh, J.E. (1966) CO_2 enrichment in raising young vegetable plants. Acta Hort. 4, 126-128.

442. Hanan, J.J. (1973) Effect of elevated temperatures and CO_2 levels on rose production. HortSci. 8, 266-267.

443. Enoch, H.Z. and Hurd, R.G. (1977) Effect of light intensity, carbon dioxide concentration, and leaf temperature on gas exchange of spray carnation plants. J. Exp. Bot. 28, 84-95.

444. Jurik, T.W., Webber, J.A. and Gates, D.M. (1984) Short term effects of CO_2 in gas exchange of leaves of Bigtooth Aspen (Populus grandidentata) in the field. Plant Physiol. 75, 1022-1026.

445. Imai, K., Coleman, D.F. and Yanagisawa, T. (1984) Elevated atmospheric partial pressure of carbon dioxide and dry matter production of cassava (Manihot esculenta Crantz). Jap. J. Crop Sci. 53, 479-485.

446. Patterson, D.T., Highsmith, M.T. and Flint, E.P. (1988) Effects of temperature and CO_2 concentration on the growth of cotton (Gossypium hirsutum), spurred anoda (Anoda cristata), and velvetleaf (Abutilon

theophrasti). Weed Sci. 36, 751-757.

447. Sinclair, T.R. and Rand, R.H. (1979) Mathematical analysis of cell CO_2 exchange under high CO_2 concentrations. Photosynthetica 13, 239-244.

448. Kidd, F. (1915) The controlling influence of CO_2. Part III. The retarding effect of CO_2 on respiration. Proc. Roy. Soc. Lond. Ser. B 89, 136-150.

449. Young, R.E., Romain, R.J. and Biale, J.B. (1962) Carbon-dioxide effects on fruit respiration. II. Response of avocados, bananas and lemons. Plant Physiol. 37, 416-422.

450. Nilovskaya, N.T. (1968) Photosynthesis and respiration in vegetable plants exposed to various CO_2 concentrations. Soviet Plant Physiol. 15, 853-858.

451. Cornic, J. and Jarvis, P.G. (1972) Effect of O_2 on CO_2 exchange and stomatal resistance in Sitka spruce and maize at low irradiances. Photosynthetica 6, 225-239.

452. Kaplan, A., Gale, J. and Poljakoff-Mayber, A. (1977) Effect of O_2 and CO_2 concentration on gross dark CO_2 fixation and dark respiration in Bryphylum diagremontianum. Aust. J. Plant Physiol. 4, 745-752.

453. Gale, J. (1982) Evidence for essential maintenance respiration of leaves of Xanthium strumarium at high temperature. J. Exp. Bot. 33, 471-476.

454. Spencer, W. and Bowes, G. (1986) Photosynthesis and growth of water hyacinth under CO_2 enrichment. Plant Physiol. 82, 528-533.

455. Gifford, R.M., Lambers, H. and Morison, J.I.L. (1985) Respiration of crop species under CO_2 enrichment. Physiol. Plant. 63, 351-356.

456. Reuveni, J. and Gale, J. (1985) The effect of high levels of carbon dioxide on dark respiration and growth of plants. Plant Cell Environ. 8, 623-628.

457. Poorter, H., Pot, S. and Lambers, H. (1988) The effect of an elevated atmospheric CO_2 concentration on growth, photosynthesis and respiration of Plantago major. Physiol. Plant. 73, 553-559.

458. Ting, I.P. and Hanscom, Z. (1977) Induction of acid metabolism in Portulacaria afra. Plant Physiol. 59, 511-514.

459. Hanscom, Z. and Ting, I.P. (1978) Responses of succulents to plant water stress. Plant Physiol. 61, 327-330.

460. Osmond, C.B. (1978) Crassulacean acid metabolism: A curiosity in context. Ann. Rev. Plant Physiol. 29, 379-414.

461. Winter, K. and von Willert, D.J. (1972) NaCl-induzierter Crassulaceensaurestoffwechsel bei Mesembryanthemum crystallinum. Zeit. Pflanzenphysiol. 267, 166-170.

462. Winter, K. (1979) Effect of different CO_2 regimes on the induction of crassulacean acid metabolism in Mesembryanthemum crystallinum L. Austr. J. Plant Physiol. 6, 589-594.

463. Huerta, A.J. and Ting, I.P. (1988) Effects of various levels of CO_2 on the induction of Crassulacean acid metabolism in Portulacaria afra (L.) Jacq. Plant Physiol. 88, 183-188.

464. Bjorkman, O., Hiesey, W.M., Nobs, M.A., Nicholson, F. and Hart, R.W. (1968) Effect of oxygen concentration in higher plants. Carnegie Inst. Wash. Year Book 66, 228-232.

465. Bjorkman, O., Gauhl, E., Hiesey, W.M., Nicholson, F. and Nobs, M.A. (1969) Growth of Mimulus, Marchantia and Zea under different oxygen and carbon dioxide levels. Carnegie Inst. Wash. Year Book 67, 477-478.

466. Quebedeaux, B. and Hardy, R.W.F. (1975) Reproductive growth and dry matter production of Glycine max (L.) Merr. in response to oxygen concentration. Plant Physiol. 55, 102-107.

467. Quebedeaux, B. and Hardy, R.W.F. (1976) Oxygen concentration: Regulation of crop growth and productivity. In Burris, R.H. and Black, C.C. (eds) CO_2 Metabolism and Plant Productivity. Univ. Park Press, Baltimore, MD, pp. 185-204.

468. Musgrave, M.E. and Strain, B.R. (1988) Response of two wheat cultivars to CO_2 enrichment under subambient oxygen conditions. Plant Physiol. 87, 346-350.

469. Kilpatrick, R.A. (1966) Induced sporulation of fungi on filter paper. Phytopathology 56, 789.

470. Rotem, J. and Bashi, E. (1969) Induction of sporulation of Alternaria porri f. sp. solani by inhibition of its vegetative development. Trans. Brit. Mycol. Soc. 53, 433-439.

471. Griffin, D.H. (1981) Fungal Physiology. John Wiley & Sons, New York, NY.

472. Dahlberg, K.R. and Van Etten, J.L. (1982) Physiology and biochemistry of fungal sporulation. Ann. Rev. Phytopathol. 20, 281-301.

473. Neiderpruem, D.J. (1963) Role of carbon dioxide in the control of fruiting of Schizophyllum commune. J. Bacteriol. 85, 1300-1308.

474. Misaghi, I.J., Grogan, R.G., Duniway, J.M. and Kimble, K.A. (1978) Influence of environment and culture media on spore morphology of Alternaria alternata. Phytopathology 68, 29-34.

475. Adams, G.C., Jr. and Butler, E.E. (1983) Environmental factors influencing the formation of basidia and basidiospores in Thanatephorus cucumeris. Phytopathology 73, 152-155.

476. Cotty, P.J. (1985) Carbon dioxide modulates the sporulation of Alternaria species. Phytopathology 75, 1297.

477. Cotty, P.J. (1987) Modulation of sporulation of Alternaria tagetica by carbon dioxide. Mycologia 79, 508-513.

478. Ascher, P.D. and Peloquin, S.J. (1966) Effect of floral aging on the growth of compatible and incompatible pollen tubes in Lilium longiflorum. Amer. J. Bot. 53, 99-102.

479. Linskens, H.F. (1975) Incompatibility in Petunia. Proc. Roy. Soc. Lond. Ser. B 188, 299-311.

480. De Nettancourt, D. (1977) Incompatibility in Angiosperms. Springer-Verlag, Berlin, West Germany.

481. Mau, S.-L., Williams, E.G., Atkinson, A., Anderson, M.A., Cornish, E.C., Grego, B., Simpson, R.J., Kheyr-Pour, A. and Clark, A.E. (1986) Style proteins of a wild tomato (Lycopersicon peruvianum) associated with expression of self-incompatibility. Planta 169, 184-191.

482. Nakanishi, T., Esashi, Y. and Hinata, K. (1969) Control of self incompatibility by CO_2 gas in Brassica. Plant Cell Physiol. 10, 925-927.

483. Sfakiotakis, E.M., Simons, D.H. and Dilley, D.R. (1972) Pollen germination and tube growth: Dependent on carbon dioxide and independent of ethylene. Plant Physiol. 49, 963-967.

484. Nakanishi, T. and Hinata, K. (1973) An effective time for CO_2 gas treatment in overcoming self-incompatibility in Brassica. Plant Cell Physiol. 14, 873-879.

485. Nakanishi, T. and Hinata, K. (1975) Self seed production by CO_2 gas treatment in self-incompatible cabbage. Euphytica 24, 117-120.

486. Taylor, J.P. (1982) Carbon dioxide treatment as an effective aid to the production of selfed seed in kale and brussel sprouts. Euphytica 31, 957-964.

487. O'Neill, P., Singh, M.B., Neales, T.F., Knox, R.B. and Williams, E.G. (1984) Carbon dioxide blocks the stigma callose response following incompatible pollinations in Brassica. Plant Cell Environ. 7, 285-288.

488. Ducos, J.P. and Pareilleux, A. (1986) Effect of aeration rate and influence of pCO_2 in large-scale cultures of Catharanthus roseus cells. Appl. Microbiol. Biotechnol. 25, 101-105.

489. Ducos, J.P., Feron, G. and Pareilleux, A. (1988) Growth and activities of enzymes of primary metabolism in batch cultures of Catharanthus roseus cell suspension under different pCO_2 conditions. Plant Cell Tiss. Organ Cult. 13, 167-177.

490. Nesius, K.K. and Fletcher, J.S. (1973) Carbon dioxide and pH requirements

REFERENCES

of non-photosynthetic tissue cell cultures. Physiol. Plant. 28, 259-263.

491. Constabel, F., Kurz, W.G.W., Chatson, K.B. and Kirkpatrick, J.W. (1977) Partial synchrony in soybean cell suspension culture induced by ethylene. Exp. Cell Res. 105, 263-268.

492. Maurel, B. and Pareilleux, A. (1985) Effect of carbon dioxide on the growth of cell suspensions of Catharanthus roseus. Biotechnol. Lett. 7, 313-318.

493. Gathercole, R.W.E., Mansfield, K.J. and Street, H.E. (1976) Carbon dioxide as an essential requirement for cultured sycamore cells. Physiol. Plant. 37, 213-217.

494. Akita, S. and Moss, D.N. (1973) Differential stomatal response between C_3 and C_4 species to atmospheric CO_2 concentration and light. Crop Sci. 12, 789-793.

495. Couchat, P. (1977) Effet de l'oxygene sur la transpiration. Compt. Rend. Acad. Sci. Paris Ser. D 285, 1303-1306.

496. Vavasseur, A., Lasceve, G. and Couchat, P. (1988) Oxygen-dependent stomatal opening in Zea mays leaves. Effect of light and carbon dioxide. Physiol. Plant. 73, 547-552.

497. Brainerd, K.E. and Fuchigami, L.H. (1981) Acclimatization of aseptically cultured apple plants to low relative humidity. J. Amer. Soc. Hort. Sci. 106, 515-518.

498. Sutter, E. and Langhans, R.W. (1982) Formation of epicuticular wax and its effect on water loss in cabbage plants regenerated from shoot tip culture. Can. J. Bot. 60, 2896-2902.

499. Wardle, K., Dobbs, E.B. and Short, K.C. (1983) In vivo acclimatization of aseptically cultured plantlets to humidity. J. Amer. Soc. Hort. Sci. 108, 386-389.

500. Dunstan, D.E. and Turner, K.E. (1985) The acclimatization of micropropagated plants. In Vasil, I.K. (ed) Cell Culture and Somatic Cell Genetics of Plants, Vol. 1. Academic Press, New York, NY, pp. 123-129.

501. Grout, B.W.W. and Aston, M.J. (1978) Transplanting of cauliflower plants regenerated from meristem culture: II. Carbon dioxide fixation and the development of photosynthetic ability. Hort. Res. 17, 65-71.

502. Ziv, M., Meir, G. and Halevy, A.H. (1983) Factors influencing the production of hardened glaucous carnation plantlets. Plant Cell Tiss. Organ Cult. 2, 55-65.

503. Hu, C.Y. and Wang, P.J. (1983) Meristem, shoot tip and bud culture. In Evans, D.A., Sharp, W.R., Ammirato, P.V. and Yamada, Y. (eds) Handbook of Plant Cell Culture, Vol. 1. Macmillan, New York, NY, pp. 177-227.

504. Lakso, A.M., Reisch, B.I., Mortensen, J. and Roberts, M.H. (1986) Carbon dioxide enrichment for stimulation of growth of in vitro-propagated grapevines after transfer from culture. J. Amer. Soc. Hort. Sci. 111, 634-638.

505. Eng, R.Y.N., Tsujita, M.J., Grodzinski, B. and Dutton, R.G. (1983) Production of chrysanthemum cuttings under supplemental lighting and carbon dioxide enrichment. HortScience 18, 878-879.

506. Tsujita, M.J. (1983) Tips on carbon dioxide and high intensity lighting for good winter production. Minn. State Flor. Bull. 32(6), 1-6.

507. Moe, R. (1977) Effect of light, temperature and CO_2 on the growth of Campanula isophylla stock plants and the subsequent growth and development of their cuttings. Sci. Hort. 6, 129-141.

508. Molitor, H.-D. and von Hentig, W.-D. (1987) Effect of carbon dioxide enrichment during stock plant cultivation. HortScience 22, 741-746.

509. Marcus, Y., Schwarz, R., Friedberg, D. and Kaplan, A. (1986) High CO_2 requiring mutant of Anacystis nidulans R_2. Plant Physiol. 82, 610-612.

510. Ogawa, T., Kaneda, T. and Omata, T. (1987) A mutant of Synechococcus PCC7942 incapable of adapting to low CO_2 concentration. Plant Physiol.

84, 711-715.

511. Omata, T., Ogawa, T., Marcus, Y., Friedberg, D. and Kaplan, A. (1987) Adaptation to low CO_2 level in a mutant of Anacystis nidulans R_2 which requires high CO_2 for growth. Plant Physiol. 83, 892-894.

512. Haughn, G.W. and Somerville, C.R. (1986) Sulfonylurea-resistant mutants of Arabidopsis thaliana. Mol. Gen. Genet. 20, 430-434.

513. Somerville, C.R. (1986) Analysis of photosynthesis with mutants of higher plants and algae. Ann. Rev. Plant Physiol. 37, 467-507.

514. Artus, N.N. and Somerville, C.R. (1987) A high CO_2-requiring mutant that displays photooxidation in air. In Biggins, J. (ed) Advances in Photosynthesis Research Vol. IV. Martinus Nijhoff, Dordrecht, Holland, pp. 67-70.

515. Somerville, C.R., Portis, A.R. and Ogren, W.L. (1982) A mutant of Arabidopsis thaliana which lacks activation of RuBP carboxylase oxygenase in vivo. Plant Physiol. 70, 381-387.

516. Salvucci, M.E., Portis, A.R. and Ogren, W.L. (1985) A soluble chloroplast protein catalyzes activation of ribulosebisphosphate carboxylase/oxygenase in vivo. Photosyn. Res. 7, 193-201.

517. Artus, N.N. and Somerville, C. (1988) A mutant of Arabidopsis thaliana that exhibits chlorosis in air but not in atmospheres enriched in CO_2. Plant Physiol. 87, 83-88.

518. Price, G.D. and Badger, M.R. (1989) Ethoxyzolamide inhibition of CO_2 uptake in the cyanobacterium Synechococcus PCC7942 without apparent inhibition of internal carbonic anhydrase activity. Plant Physiol. 89, 37-43.

519. Price, G.D. and Badger, M.R. (1989) Ethoxyzolamide inhibition of CO_2-dependent photosynthesis in the cyanobacterium Synechococcus PCC7942. Plant Physiol. 89, 44-50.

520. Badger, M.R. and Price, G.D. (1989) Carbonic anhydrase activity associated with the cyanobacterium Synechococcus PCC7942. Plant Physiol. 89, 51-60.

521. Spalding, M.H. and Jeffrey, M. (1989) Membrane-associated polypeptides induced in Chlamydomonas by limiting CO_2 concentrations. Plant Physiol. 89, 133-137.

522. Idso, S.B., Kimball, B.A. and Clawson, K.L. (1984) Quantifying effects of atmospheric CO_2 enrichment on stomatal conductance and evapotranspiration of water hyacinth via infrared thermometry. Agric. For. Meteorol. 33, 15-22.

523. Smith, W.K. (1978) Temperatures of desert plants: Another perspective on the adaptability of leaf size. Science 201, 614-616.

524. Idso, S.B., Kimball, B.A. and Anderson, M.G. (1986) Foliage temperature increases in water hyacinth caused by atmospheric CO_2 enrichment. Archiv. Meteorol. Geophys. Bioclimatol. Ser. B 36, 365-370.

525. Idso, S.B., Clawson, K.L. and Anderson, M.G. (1986) Foliage temperature: Effects of environmental factors with implications for plant water stress assessment and the CO_2/climate connection. Water Resources Res. 22, 1702-1716.

526. Idso, S.B., Kimball, B.A. and Mauney, J.R. (1987) Atmospheric carbon dioxide enrichment effects on cotton midday foliage temperature: Implications for plant water use and crop yield. Agron. J. 79, 667-672.

1. Botkin, D.B. (1977) Forests, lakes and the anthropogenic production of carbon dioxide. BioScience 27, 325-331.

2. Goudriaan, J. and Atjay, G.L. (1979) The possible effects of increased

CO_2 on photosynthesis. In Bolin, B., Degens, E.T., Kempe, S. and Ketner, P. (eds) The Global Carbon Cycle. John Wiley & Sons, London, UK.

3. Kramer, P.J. (1981) Carbon dioxide concentration, photosynthesis and dry matter production. BioScience 31, 29-33.

4. Lugo, A. (1983) Influence of green plants on the world carbon budget. In Veziroglu, T.N. (ed) Alternative Energy Sources V. Part E: Nuclear/Conservation/Environmnent. Elsevier, Amsterdam, The Netherlands, pp. 391-398.

5. Hobbie, J., Cole, J., Dugan, J., Houghton, R.A. and Peterson, B. (1984) Role of biota in global CO_2 balance: The controversy. BioScience 34, 492-498.

6. Kimball, B.A. (1986) CO_2 stimulation of growth and yield under environmental constraints. In Enoch, H.Z. and Kimball, B.A. (eds) Carbon Dioxide Enrichment of Greenhouse Crops. Vol. II. Physiology, Yield, and Economics. CRC Press, Boca Raton, FL, pp. 53-67.

7. Gifford, R.M. (1980) Carbon storage by the biosphere. In Pearman, G.I. (ed) Carbon Dioxide and Climate: Australian Research. Austr. Acad. Sci., Canberra, Australia, pp. 167-181.

8. Sionit, N., Hellmers, H. and Strain, B.R. (1980) Growth and yield of wheat under carbon dioxide enrichment and water stress conditions. Crop Sci. 20, 687-690.

9. Sionit, N., Strain, B.R., Hellmers, H. and Kramer, P.J. (1981) Effects of atmospheric carbon dioxide concentration and water stress on water relations of wheat. Bot. Gaz. 142, 191-196.

10. Paez, A., Hellmers, H. and Strain, B.R. (1983) CO_2 enrichment, drought stress and growth of Alaska pea plants (Pisum sativum). Plant Physiol. 58, 161-165.

11. Paez, A., Hellmers, H. and Strain, B.R. (1984) Carbon dioxide enrichment and water stress interaction on growth of two tomato cultivars. J. Agric. Sci. Camb. 102, 687-693.

12. Sionit, N. and Patterson, D.T. (1985) Responses of C_4 grasses to atmospheric CO_2 enrichment. II. Effect of water stress. Crop Sci. 25, 533-537.

13. Goudriaan, J. and Bijlsma, R.J. (1987) Effect of CO_2 enrichment on growth of faba beans at two levels of water supply. Neth. J. Agric. Sci. 35, 189-191.

14. Cure, J.D. (1985) Carbon dioxide doubling responses: A crop survey. In Strain, B.R. and Cure, J.D. (eds) Direct Effects of Increasing Carbon Dioxide on Vegetation. U.S. Dept. Energy, Washington, DC, pp. 99-116.

15. Hurt, P. and Wright, R.D. (1976) CO_2 compensation point for photosynthesis: Effect of variable CO_2 and soil moisture levels. Amer. Midl. Nat. 95, 450-455.

16. Gifford, R.M. (1979) Carbon dioxide and plant growth under water and light stress: Implications for balancing the global carbon budget. Search 10, 316-318.

17. Gifford, R.M. (1979) Growth and yield of carbon dioxide-enriched wheat under water-limited conditions. Austr. J. Plant Physiol. 6, 367-378.

18. Huber, S.C., Rogers, H.H. and Mowry, F.L. (1984) Effects of water stress on photosynthesis and carbon partitioning in soybean (Glycine max [L.] Merr.) plants grown in the field at different CO_2 levels. Plant Physiol. 76, 244-249.

19. Conway, J.P., Virgona, J.M., Smillie, R.M. and Barlow, E.W. (1988) Influence of drought acclimation and CO_2 enrichment on osmotic adjustment and chlorophyll a fluorescence of sunflower during drought. Plant Physiol. 86, 1108-1115.

20. Chaudhuri, U.N., Burnett, R.B., Kanemasu, E.T. and Kirkham, M.B. (1986) Effect of Elevated Levels of CO_2 on Winter Wheat under Two Moisture Regimes. U.S. Dept. Energy, Washington, DC.

REFERENCES

21. Conroy, J.P., Barlow, E.W.R. and Bevege, D.I. (1986) Response of Pinus radiata seedlings to carbon dioxide enrichment at different levels of water and phosphorus: Growth, morphology and anatomy. Ann. Bot. 57, 165-177.

22. Conroy, J.P., Smillie, R.M., Kuppers, M., Bevege, D.I. and Barlow, E.W. (1986) Chlorophyll a fluorescence and photosynthetic and growth responses of Pinus radiata to phosphorus deficiency, drought stress, and high CO_2. Plant Physiol. 81, 423-429.

23. Chaudhuri, U.N., Burnett, R.B., Kanemasu, E.T. and Kirkham, M.B. (1987) Effect of Elevated Levels of CO_2 on Winter Wheat under Two Moisture Regimes. U.S. Dept. Energy, Washington, DC.

24. Doyle, T.W. (1987) Seedling response to CO_2 enrichment under stressed and non-stressed conditions. In Jacoby, G.C., Jr. and Hornbeck, J.W. (1987) Proc. Internat. Symp. Ecological Aspects of Tree-Ring Analysis. U.S. Dept. Energy, Washington, DC, pp. 501-510.

25. Prior, S.A., Rogers, H.H. and Sionit, N. (1987) Water relations and growth responses of soybean in carbon dioxide-enriched atmospheres. Agron. Abstr. 79, 16.

26. Schonfeld, M.A., Johnson, R.C. and Ferris, D.M. (1987) Growth and photosynthesis of winter wheat seedlings under elevated CO_2 and dehydration stress. Agron. Abstr. 79, 100.

27. Conroy, J.P., Kuppers, M., Kuppers, B., Virgona, J. and Barlow, E.W.R. (1988) The influence of CO_2 enrichment, phosphorus deficiency and water stress on the growth, conductance and water use of Pinus radiata D.Don. Plant Cell Environ. 11, 91-98.

28. Gifford, R.M. and Morison, J.I.L. (1985) Photosynthesis, water use and growth of a C_4 grass stand at high CO_2 concentration. Photosyn. Res. 7, 69-76.

29. Idso, S.B., Kimball, B.A., Anderson, M.G. and Szarek, S.R. (1986) Growth response of a succulent plant, Agave vilmoriniana, to elevated CO_2. Plant Physiol. 80, 796-797.

30. Idso, S.B. (1988) Three phases of plant response to atmospheric CO_2 enrichment. Plant Physiol. 87, 5-7.

31. Idso, S.B., Kimball, B.A. and Anderson, M.G. (1985) Atmospheric CO_2 enrichment of water hyacinths: Effects on transpiration and water use efficiency. Water Resources Res. 21, 1787-1790.

32. Allen, S.G., Idso, S.B., Kimball, B.A. and Anderson, M.G. (1988) Interactive effects of CO_2 and environment on photosynthesis of Azolla. Agric. For. Meteorol. 42, 209-217.

33. Idso, S.B., Kimball, B.A., Anderson, M.G. and Mauney, J.R. (1987) Effects of atmospheric CO_2 enrichment on plant growth: The interactive role of air temperature. Agric. Ecosys. Environ. 20, 1-10.

34. Kimball, B.A., Mauney, J.R., Guinn, G., Nakayama, F.S., Pinter, P.J., Clawson, K.L., Reginato, R.J. and Idso, S.B. (1983) Effects of Increasing Atmospheric CO_2 on the Yield and Water Use of Crops. U.S. Dept. Energy, Washington, DC.

35. Kimball, B.A., Mauney, J.R., Guinn, G., Nakayama, F.S., Pinter, P.J., Clawson, K.L., Idso, S.B., Butler, G.D. and Radin, J.W. (1984) Effects of Increasing Atmospheric CO_2 on the Yield and Water Use of Crops. U.S. Dept. Energy, Washington, DC.

36. Kimball, B.A., Mauney, J.R., Guinn, G., Nakayama, F.S., Idso, S.B., Radin, J.W., Hendrix, D.L., Butler, G.D., Zarembinski, T.I. and Nixon, P.E. (1985) Effects of Increasing Atmospheric CO_2 on the Yield and Water Use of Crops. U.S. Dept. Energy, Washington, DC.

37. Kimball, B.A., Mauney, J.R., Radin, J.W., Nakayama, F.S., Idso, S.B., Hendrix, D.H., Akey, D.H., Allen, S.G. and Anderson, M.G. (1986) Effects of Increasing Atmospheric CO_2 on the Growth, Water Relations, and Physiology of Plants Grown under Optimal and Limiting Levels of Water and Nitrogen. U.S. Dept. Energy, Washington, DC.

REFERENCES

38. Idso, S.B., Kimball, B.A. and Anderson, M.G. (1986) Foliage temperature increases in water hyacinth caused by atmospheric CO_2 enrichment. Arch. Meteorol. Geophys. Bioclimatol. Ser. B 36, 365-370.

39. Idso, S.B., Kimball, B.A. and Mauney, J.R. (1983) Atmospheric carbon dioxide enrichment effects on cotton midday foliage temperature: Implications for plant water use and crop yield. Agron. J. 79, 667-672.

40. Allen, S.G., Idso, S.B., Kimball, B.A. and Anderson, M.G. (1988) Relationship between growth rate and net photosynthesis of Azolla in ambient and elevated CO_2 concentrations. Agric. Ecosys. Environ. 20, 137-141.

41. Longenecker, D.E. and Lyerly, P.J. (1969) Moisture content of cotton leaves and petioles as related to environmental moisture stress. Agron. J. 61, 687-690.

42. Reginato, R.J. and Howe, J. (1985) Irrigation scheduling using crop indicators. J. Irrig. Drain. Engin. 111, 125-133.

43. Kriedemann, P.E., Sward, R.J. and Downton, W.J.S. (1976) Vine response to carbon dioxide enrichment during heat therapy. Austr. J. Plant Physiol. 3, 605-618.

44. Converse, R.H. and George, R.A. (1987) Elimination of mycoplasmalike organisms in Cabot highbush blueberry with high-carbon dioxide thermotherapy. Plant Disease 71, 36-38.

45. Sherman, H. (1987) Carbon dioxide makes heat therapy work. Agric. Res. 35, 5.

46. Idso, S.B., Allen, S.G., Anderson, M.G. and Kimball, B.A. (1989) Atmospheric CO_2 enrichment enhances survival of Azolla at high temperatures Environ. Exp. Bot., in press.

47. Rowley, J.A. and Taylor, A.O. (1972) Plants under climatic stress. IV. Effects of CO_2 and O_2 on photosynthesis under high-light, low-temperature stress. New Phytol. 71, 477-481.

48. Potvin, C., Goeschl, J.D. and Strain, B.R. (1984) Effects of temperature and CO_2 enrichment on carbon translocation of the C_4 grass species Echinochola crus-galli (L.) Beauv. from cool and warm environments. Plant Physiol. 75, 1054-1057.

49. Potvin, C. (1985) Amelioration of chilling effects by atmospheric CO_2 enrichment. Physiol. Veg. 23, 345-352.

50. Potvin, C., Strain, B.R. and Goeschl, J.D. (1985) Low night temperature effect on photosynthate translocation of two C_4 grasses. Oecologia 67, 305-309.

51. Loach, K. and Whalley, D.N. (1975) Use of light, carbon dioxide enrichment and growth regulators in the overwintering of hardy ornamental nursery stock cuttings. Acta Hort. 54, 105-116.

52. Sionit, N., Strain, B.R. and Beckford, H.A. (1981) Environmental controls on the growth and yield of okra. I. Effects of temperature and of carbon dioxide enrichment at cool temperature. Crop Sci. 21, 885-888.

53. Maas, E.V. and Nieman, R.H. (1978) Physiology of plant tolerance to salinity. In Jung, G.A. (ed) Crop Tolerance to Suboptimal Land Conditions. Amer. Soc. Agron., Madison, WI, pp. 277-299.

54. Hoffman, G.L. and Van Genuchten, M.Th. (1983) Soil properties and efficient water use: Water management for salinity control. In Taylor, H.M., Jordan, W.R. and Sinclair, T.R. (eds) Limitations to Efficient Water Use in Crop Production. Amer. Soc. Agron., Madison, WI, pp. 73-85.

55. Maas, E.V. (1986) Salt tolerance of plants. Appl. Agric. Res. 1, 12-26.

56. Schwarz, M. and Gale, J. (1984) Growth response to salinity at high levels of carbon dioxide. J. Exp. Bot. 35, 193-196.

57. Acock, B. and Allen, L.H., Jr. (1985) Crop responses to elevated carbon dioxide concentrations. In Strain, B.R. and Cure, J.D. (eds) Direct

Effects _of_ _Increasing_ _Carbon_ _Dioxide_ _on_ _Vegetation_. U.S. Dept. Energy, Washington, DC, pp. 53-97.

58. Enoch, H.Z., Zieslin, N., Biran, Y., Halevy, A.H., Schwarz, M., Kesler, B. and Shimshi, D. (1972) Principles of CO_2 nutrition research. _Acta_ _Hort_. 32, 97-118.

59. Gale, J. (1982) Use of brackish and solar desalinated water in closed system agriculture. _In_ San Pietro, A. (ed) _Biosaline_ _Research_: _A_ _Look_ _to_ _the_ _Future_. Plenum, New York, pp. 315-324.

60. Gale, J. and Zeroni, M. (1985) Cultivation of plants in brackish water in controlled environment agriculture. _In_ Staples, R.C. and Toenniessen, G. (eds) _Salinity_ _Tolerance_ _in_ _Plants_ -- _Strategies_ _for_ _Crop_ _Improvement_. Wiley Interscience, New York, NY, pp. 363-380.

61. Guy, R.D. and Reid, D.M. (1986) Photosynthesis and the influence of CO_2-enrichment on δ ^{13}C values in a C_3 halophyte. _Plant_ _Cell_ _Environ_. 9, 65-72.

62. Hill, C.A. (1971) Vegetation: A sink for atmospheric pollutants. _J_. _Air_ _Poll_. _Contr_. _Assoc_. 21, 341-346.

63. Heath, R.L. (1980) Initial events in injury to plants by air pollutants. _Ann_. _Rev_. _Plant_ _Physiol_. 31, 395-431.

64. Unsworth, M.H. and Ormrod, D.P., Eds. (1982) _Effects_ _of_ _Gaseous_ _Air_ _Pollution_ _in_ _Agriculture_ _and_ _Horticulture_. Butterworth Scientific, London, UK.

65. Reinert, R.A. (1984) Plant responses to air pollutant mixtures. _Ann_. _Rev_. _Phytopathol_. 22, 421-442.

66. Heck, W.W., Taylor, O.C. and Tingey, D.T., Eds. (1988) Response of crops to air pollutants. _Environ_. _Poll_. 53(1-4), 1-478.

67. Darrall, N.M. (1989) The effect of air pollutants on physiological processes in plants. _Plant_ _Cell_ _Environ_. 12, 1-30.

68. Dugger, W.M. and Ting, I.P. (1970) Air pollution oxidants -- Their effects on metabolic processes in plants. _Ann_. _Rev_. _Plant_ _Physiol_. 21, 215-234.

69. Evans, L.S. and Ting, I.P. (1973) Ozone induced permeability changes. _Amer_. _J_. _Bot_. 60, 155-162.

70. Mudd, J.B., Banesjee, S.K., Dooley, M.M. and Knight, K.L. (1984) Pollutants and plant cells: Effects on membranes. _In_ Koziol, M.J. and Whatley, F.R. (eds) _Gaseous_ _Air_ _Pollutants_ _and_ _Plant_ _Metabolism_. Buttworth Scientific, London, UK, pp. 106-116.

71. Hill, A.C., Heggestad, H.E. and Linzon, S.N. (1970) Ozone. _In_ Jacobson, J.S. and Hill, A.C. (eds) _Recognition_ _of_ _Air_ _Pollution_ _Injury_ _to_ _Vegetation_: _A_ _Pictorial_ _Atlas_. Air Poll. Contr. Assoc., Pittsburgh, PA, pp. B1-B22.

72. Coyne, P.I. and Bingham, G.E. (1982) Variation in photosynthesis and stomatal conductance in an ozone-stressed ponderosa pine stand: Light response. _For_. _Sci_. 28, 257-273.

73. Amthor, J.S. (1988) Growth and maintenance respiration in leaves of bean (_Phaseolus_ _vulgaris_ L.) exposed to ozone in open-top chambers in the field. _New_ _Phytol_. 110, 319-325.

74. Heck, W.W., Taylor, O.C., Adams, R., Bingham, G., Miller, J., Preston, E. and Weinstein, L. (1982) Assessment of crop loss from ozone. _J_. _Air_ _Poll_. _Contr_. _Assoc_. 32, 353-361.

75. Haegle, A.S., Heck, W.W., Rawlings, J.O. and Philbeck, R.B. (1983) Effects of chronic doses of ozone and sulfur dioxide on injury and yield of soybeans in open-top field chambers. _Crop_ _Sci_. 23, 1184-1191.

76. Hill, A.C. and Bennett, J.H. (1970) Inhibition of apparent photosynthesis by nitrogen oxides. _Atmos_ _Environ_. 4, 341-348.

77. Heck, W.W. (1984) Defining gaseous pollution problems in North America. _In_ Koziol, M.J. and Whatley, F.R. (eds) _Gaseous_ _Air_ _Pollutants_ _and_ _Plant_ _Metabolism_. Butterworth Scientific, London, UK, pp. 35-48.

78. Pitelka, L.F. (1988) Evolutionary responses of plants to anthropogenic pollutants. Trends Ecol. Evol. 3, 233-236.

79. Lendzian, K.J. (1984) Permeability of plant cuticles to gaseous air pollutants. In Koziol, M.J. and Whatley, F.R. (eds) Gaseous Air Pollutants and Plant Metabolism. Butterworth Scientific, London, UK, pp. 35-48.

80. Kluczewski, S.M., Bell, J.N.B., Brown, K.A. and Minski, M.J. (1983) The uptake of (^{35}S)-carbonyl sulphide by plants and soils in ecological aspects of radionuclide release. Spec. Pub. Brit. Ecol. Soc. III, 91-104.

81. Taylor, G.E., Jr., McLaughlin, S.B., Jr., Shriner, D.S. and Selvidge, W.J. (1983) The flux of sulfur-containing gases to vegetation. Atmos. Environ. 17, 789-796.

82. Kluczewski, S.M., Brown, K.A. and Bell, J.N.B. (1985) Deposition of [^{35}S]-carbonyl sulphide to vegetable crops. Radiat. Prot. Dosim. 11, 173-177.

83. Goldan, P.D., Fall, R., Kuster, W.C. and Fehsenfeld, F.C. (1988) Uptake of COS by growing vegetation: A major tropospheric sink. J. Geophys. Res. 93, 14,186-14,192.

84. Heck, W.W. and Dunning, J.A. (1967) The effects of ozone on tobacco and pinto bean as conditioned by several ecological factors. J. Air Poll. Contr. Assoc. 17, 112-114.

85. Majernik, O. and Mansfield, T.A. (1972) Stomatal responses to raised atmospheric CO_2 concentrations during exposure of plants to SO_2 pollution. Environ. Poll. 3, 1-7.

86. Coyne, P.I. and Bingham, G.E. (1977) Carbon dioxide correlation with oxident air pollution in the San Bernadino Mountains of California. J. Air Poll. Contr. Assoc. 27, 782-783.

87. Green, K. and Wright, R. (1977) Field response of photosynthesis to CO_2 enhancement in Ponderosa pine. Ecology 58, 687-692.

88. Hou, L., Hill, A.C. and Soleimani, A. (1977) Influence of CO_2 on the effects of SO_2 and NO_2 on alfalfa. Environ. Poll. 12, 7-16.

89. Carlson, R.W. and Bazzaz, F.A. (1982) Photosynthetic and growth response to fumigation with SO_2 at elevated CO_2 for C_3 and C_4 plants. Oecologia 54, 50-54.

90. Carlson, R.W. (1983) The effect of SO_2 on photosynthesis and leaf resistance at varying concentrations of CO_2. Environ. Poll. Ser. A 30, 309-321.

91. Carlson, R.W. and Bazzaz, F.A. (1985) Plant response to SO_2 and CO_2. In Winner, W.E., Mooney, H.A. and Goldstein, R.A. (eds) Sulfur Dioxide and Vegetation. Stanford Univ. Press, Standord, CA, pp. 313-331.

92. Bruggink, G.T., Wolting, H.G., Dassen, J.H.A. and Bus, V.G.M. (1988) The effect of nitric oxide fumgation at two CO_2 concentrations on net photosynthesis and stomatal resistance of tomato (Lycopersicum lycopersicum L. cv. Abunda). New Phytol. 110, 185-191.

93. Temple, P.J., Taylor, O.C. and Benoit, L.F. (1985) Cotton yield responses to ozone as mediated by soil moisture and evapotranspiration. J. Environ. Qual. 14, 55-60.

94. Miller, J.E., Heagle, A.S., Patterson, R.P., Pursley, W.A. and Corda, S.L. (1987) The influence of ozone on cotton yield under well-watered and water-stressed conditions. In Heck, W.W. (ed) National Crop Loss Assessment Network (NCLAN) 1985 Annual Report. U.S. Environ. Protection Agency, Corvallis, OR, pp. 26-48.

95. King, D.A. (1987) A model for predicting the influence of moisture stress on crop losses caused by ozone. Ecol. Model. 35, 39-44.

96. Temple, P.J., Kupper, R.S., Lennox, R.W. and Rohr, K. (1988) Injury and yield responses of differentially irrigated cotton to ozone. Agron. J. 80, 751-755.

97. Heagle, A.S., Miller, J.E., Heck, W.W. and Patterson, R.P. (1988) Injury

REFERENCES   SECTION

and yield response of cotton to chronic doses of ozone and soi
moisture deficit. J. Environ. Qual. 17, 627-635.

98. Mohnen, V.A. (1988) The challenge of acid rain. Sci. Amer. 259(2), 30-
38.

99. Kennedy, I.R. (1988) Acid Soil and Acid Rain: The Impact on the
Environment of Nitrogen and Sulphur Cycling. Research Studies Press,
Letchworth, Hertfordshire, UK.

100. Likens, G.E. and Bormann, F.H. (1974) Acid rain: A serious regiona
environmental problem. Science 184, 1176-1179.

101. Wright, R.F. and Gjessing, E.T. (1976) Acid precipitation: Changes i
the chemical composition of lakes. Ambio 5, 219-233.

102. Reuss, J.O. and Johnson, D.W. (1986) Acid Deposition and the
Acidification of Soils and Waters. Springer-Verlag, New York, NY.

103. White, J.C., Ed. (1987) Acid Rain: The Relationship Between Sources an
Receptors. Elsevier, New York, NY.

104. Johnson, R.W. and Gordon, G.E. (1987) The Chemistry of Acid Rain. Amer.
Chem. Soc., Washington, DC.

105. Schwartz, S.E. (1989) Acid deposition: Unraveling a regional phenomenon. Science 243, 753-763.

106. Baker, E.A. and Hunt, G.M. (1986) Erosion of waxes from leaf surfaces by
simulated rain. New Phytol. 102, 161-173.

107. Tukey, H.B., Jr. (1971) Leaching of substances from plants. In Preece,
T.F. and Dickinson, C.H. (eds) Ecology of Leaf Surface Micro-
Organisms. Academic Press, New York, NY, pp. 67-80.

108. Adams, C.M. and Hutchinson, T.C. (1984) A comparison of the ability of
leaf surfaces of three species to neutralize acidic rain drops. New
Phytol. 97, 463-478.

109. Cracker, L.E. and Bernstein, D. (1984) Buffering of acid rain by leaf
tissue of selected crop plants. Environ. Poll. Ser. A 36, 375-381.

110. Larsen, B.R. (1986) In vivo buffering and concentration of simulated
acidic rain drops on leaves of selected crops. Water Air Soil Poll.
31, 401-407.

111. Hutchinson, T.C., Adams, C.M. and Gaber, B.A. (1986) Neutralization of
acidic raindrops on leaves of agricultural crop and boreal forest
species. Water Air Soil Poll. 31, 475-484.

112. Elleman, C.J. and Entwistle, P.F. (1982) A study of glands on cotton
responsible for the high pH and cation concentration of the leaf
surface. Ann. Appl. Biol. 100, 553-558.

113. Gaber, B.A. and Hutchinson, T.C. (1988) The neutralization of acid rain
by the leaves of four boreal forest species. Can. J. Bot. 66, 1877-
1882.

114. Mecklenburg, R.A., Tukey, H.B., Jr. and Morgan, J.V. (1966) A mechanism
for the leaching of calcium from foliage. Plant Physiol. 41, 610-613.

115. Strapp, J.W., Leaitch, W.R., Anlauf, K.G., Bottenheim, J.W., Joe, P.,
Schemenauer, R.S., Wiebe, H.A., Isaac, G.A., Kelly, T.J. and Daum,
P.H. (1988) Winter cloudwater and air composition in central Ontario.
Proc.: Symposium on the Role of Clouds in Atmospheric Chemistry and
Global Climate. Amer. Meteorol. Soc., Boston, MA, pp. 231-235.

116. Mueller, S.F. and Weatherford, F.P. (1988) Chemical deposition to a high
elevation red spruce forest. Water Air Soil Poll. 38, 345-363.

117. Banwart, W.L. (1988) Field evaluation of an acid rain-drought stress
interaction. Environ. Poll. 53, 123-133.

118. Abeles, F.B. (1973) Stress ethylene. In Abeles, F.B. Ethylene in Plant
Biology. Academic Press, New York, NY, pp. 87-102.

119. Mehlhorn, H. and Wellburn, A.R. (1987) Stress ethylene formation
determines plant sensitivity to ozone. Nature 327, 417-418.

120. Hellpointner, E. and Gab, S. (1989) Detection of methyl, hydroxymethyl
and hydroxyethyl hydroperoxides in air and precipitation. Nature 337,
631-634.

121. Smith, E.F. (1911) Bacteria in Relation to Plant Disease. Vol. 2. Carnegie Institute of Washington, WA.

122. McLean, F.T. (1921) A study of the structures of the stomata of two species of citrus in relation to citrus canker. Bull. Torrey Bot. Club 48, 101-106.

123. McLean, F.T. and Lee, H.A. (1922) Pressures required to cause stomatal infections with the citrus-canker organism. Phillipine J. Sci. 20, 309-320.

124. Pool, V.W. and McKay, M.B. (1916) Relation of stomatal movement to infection by Cercospora beticola. J. Agric. Res. 5, 1011-1038.

125. Hart, H. (1929) Relation of stomatal behavior to stem-rust resistance in wheat. J. Agric. Res. 39, 929-948.

126. Hart, H. (1931) Tech. Bull. No. 266: Morphologic and Physiologic Studies on Stem-Rust Resistance in Cereals. U.S. Dept. Agric., Washington, DC.

127. Ullstrup, A.J. (1936) Leaf blight of china aster caused by Rhizoctonia solani. Phytopathol. 26, 981-990.

128. Duggar, B.M. and Johnson, B. (1933) Stomatal infection with the virus of typical tobacco mosaic. Phytopathol. 23, 934-948.

129. Goodey, T. (1933) Plant Parasitic Nematodes. Methuen, London, UK.

130. Winslow, R.D. (1960) Some aspects of the ecology of free-living and plant parasitic nematodes. In Sasser, J.N. and Jenkins, W.R. (eds) Nematology. N. Carolina Press, Chapel Hill, NC, pp. 341-415.

131. Rich, S. (1963) The role of stomata in plant disease. In Zelitch, I. (ed) Bull. No. 664: Stomata and Water Relations in Plants. Conn. Agric. Exp. Sta., New Haven, CT, pp. 102-114.

132. Yirgou, D. and Caldwell, R.M. (1963) Stomatal penetration of wheat seedlings by stem and leaf rust: Effect of light and carbon dioxide. Science 141, 272-273.

133. Lutman, B.R. (1922) The relation of the water pores and stomata of the potato leaf to the early stages and advance of tipburn. Phytopathol. 12, 305-333.

134. Curtis, L.C. (1943) Deleterious effects of guttated fluids on foliage. Amer. J. Bot. 30, 778-781.

135. Munnecke, D.A. and Chandler, P.A. (1957) A leaf spot of Philodendron related to stomatal exudation and to temperature. Phytopathol. 47, 299-303.

136. Ivanoff, S.S. (1963) Guttation injuries of plants. Bot. Rev. 29, 202-229.

137. Stefferud, A., Ed. (1952) Insects: The Yearbook of Agriculture. U.S. Dept. Agric., Washington, DC.

138. Hoffman, C.H. and Henderson, L.S. (1966) The fight against insects. In Hayes, J. (ed) Protecting Our Food: The Yearbook of Agriculture. U.S. Dept. Agric., Washington, DC.

139. Johnson, W.T. and Lyon, H.H. (1976) Insects that Feed on Trees and Shrubs. Cornell Univ. Press, Ithaca, NY.

140. Ahmad, S., Ed. (1983) Herbivorous Insects: Host-Seeking Behavior and Mechanisms. Academic Press, New York, NY.

141. Burkett, B.N. and Schneiderman, H.A. (1974) Roles of oxygen and carbon dioxide in the control of spiracular function in Cecropia pupae. Biol. Bull. 147, 274-293.

142. Franklin, R.T. (1970) Insect influences on the forest canopy. In Riechle, D.E. (ed) Analysis of Temperate Forest Ecosystems. Springer-Verlag, New York, NY, pp. 86-99.

143. Strain, B.R. and Bazzaz, F.A. (1983) Terrestrial plant communities. In Lemon, E.R. (ed) CO_2 and Plants: The Response of Plants to Rising Levels of Atmospheric Carbon Dioxide. Westview Press, Boulder, CO, pp. 177-222.

144. Rogers, H.H., Bingham, G.E., Cure, J.D., Smith, J.M. and Surano, K.A. (1983) Responses of selected plant species to elevated carbon dioxide

in the field. J. Environ. Qual. 12, 569-574.

145. Sionit, N. (1983) Response of soybean to two levels of mineral nutrition in CO_2-enriched atmosphere. Crop Sci. 23, 329-333.

146. Oechel, W.C. and Strain, B.R. (1985) Native species response to increased atmospheric carbon dioxide concentration. In Strain, B.R. and Cure, J.D. (eds) Direct Effects of Increasing Carbon Dioxide on Vegetation. U.S. Dept. Energy, Washington, DC, pp. 117-154.

147. Scriber, J.M. and Slansky, F. (1981) The nutritional ecology of immature insects. Ann. Rev. Entomol. 26, 183-211.

148. Lincoln, D.E., Sionit, N. and Strain, B.R. (1984) Growth and feeding response of Pseudoplusia includens (Lepidoptera: Noctuidae) to host plants grown in controlled carbon dioxide atmospheres. Environ. Entomol. 13, 1527-1530.

149. Butler, G.D., Jr., Kimball, B.A. and Mauney, J.R. (1986) Populations of Bemisia tabaci (Homoptera: Aleyrodidae) on cotton grown in open-top field chambers enriched with CO_2. Environ. Entomol. 15, 61-63.

150. Butler, G.D., Jr. (1985) Populations of several insects on cotton in open-top carbon dioxide enrichment chambers. Southwest. Entomol. 10, 264-267.

151. Akey, D.H. and Kimball, B.A. (1989) Growth and development of the beet armyworm on cotton grown in an enriched carbon dioxide atmosphere. Southwest Entomol., in press.

152. Lincoln, D.E., Couvet, D. and Sionit, N. (1986) Response of an insect herbivore to host plants grown on carbon dioxide enriched atmospheres. Oecologia 69, 556-560.

153. Osbrink, W.L.A., Trumble, J.T. and Wagner, R.E. (1987) Host suitability of Phaseolus lunata for Trichoplusia ni (Lepidoptera: Noctuidae) in controlled carbon dioxide atmospheres. Environ. Entomol. 16, 210-215.

154. Akey, D.H., Kimball, B.A. and Mauney, J.R. (1988) Growth and development of the pink bollworm, Pectinophora gossypiella (Lepidoptera: Gelechiidae) on bolls of cotton grown in enriched carbon dioxide atmospheres. Environ. Entomol., in press.

155. Fajer, E.D., Bowers, M.D. and Bazzaz, F.A. (1989) The effects of enriched carbon dioxide atmospheres on plant-insect herbivore interactions. Science 243, 1198-1200.

156. Price, P.W., Bouton, C.E., Gross, P., McPheron, B.A., Thompson, J.N. and Weis, A.E. (1980) Interactions among three trophic levels: Influence of plants on interactions between insect herbivores and natural enemies. Ann. Rev. Ecol. Syst. 11, 41-65.

157. Chew, F. (1975) Coevolution of Pierid butterflies and their Cruciferous foodplants. I. The relative quality of available resources. Oecologia 20, 117-127.

158. Dobkin, D.S., Olivieri, I. and Ehrlich, P.R. (1987) Rainfall and the interaction of microclimate with larval resources in the population dynamics of checkerspot butterflies (Euphydryas editha) inhabiting serpentine grassland. Oecologia 71, 161-166.

159. Bloom, A.J., Chapin, F.S., III and Mooney, H.A. (1985) Resource limitation in plants -- An economic analogy. Ann. Rev. Ecol. Syst. 16, 363-392.

160. Bazzaz, F.A., Chiariello, N.R., Coley, P.D. and Pitelka, L.F. (1987) Allocating resources to reproduction and defense. BioScience 37, 58-67.

161. Gershenzon, J. (1984) Changes in the level of plant secondary metabolites under water and nutrient stress. In Timmermann, B.N., Steelink, C. and Loewus, F.A. (eds) Phytochemical Adaptation to Stress. Plenum Press, New York, NY, pp. 273-320.

162. Coley, P.D., Bryant, J.P. and Chapin, F.S., III. (1985) Resource availability and plant antiherbivore defense. Science 230, 895-899.

163. Waterman, P.G., Ross, J.A.M. and McKey, D.B. (1984) Factors affecting

REFERENCES

levels of some phenolic compounds, digestibility, and nitrogen content of the mature leaves of Barteria fistulosa (Passifloraceae). J. Chem. Ecol. 10, 387-401.

164. Bryant, J.P., Chapin, F.S., III, Reichardt, P.B. and Clausen, T.P. (1987) Response of winter chemical defense in Alaska paper birch and green alder to manipulation of plant carbon/nutrient balance. Oecologia 72, 510-514.

165. Bryant, J.P., Clausen, T.P., Reichardt, P.B., McCarthy, M.C. and Werner, R.A. (1987) Effect of nitrogen fertilization upon the secondary chemistry and nutritional value of quaking aspen (Populus tremuloides Michx.) leaves for the larger aspen tortrix (Choristoneura conflictana (Walker)). Oecologia 73, 513-517.

166. McCrea, K.D. and Abrahamson, W.G. (1987) Variation in ball gall infestation in goldenrod: Historical vs genetic factors. Ecology 68, 822-827.

167. Abrahamson, W.G., Anderson, S.S. and McCrea, K.D. (1988) Effects of manipulation of plant carbon nutrient balance on tall goldenrod resistance to a gallmaking herbivore. Oecologia 77, 302-306.

168. Lincoln, D.E. and Couvet, D. (1989) The effect of carbon supply on allocation to allelochemicals and caterpillar consumption of peppermint. Oecologia 78, 112-114.

169. Aide, T.M. (1988) Herbivory as a selective agent on the timing of leaf production in a tropical understory community. Nature 336, 574-575.

170. Nagy, M. (1981) The effect of Lepidoptera larvae consumption on the leaf production of Quercus petraea (Matt) Liebl. Acta Bot. Hung. 27, 141-150.

171. Szabo, L, Varga, Z. and Lakatos, G. (1983) Die Rolle der Laubfressenden Lepidopterenlarven im Zerreichen-Traubeneichen Waldokosystem. Allattani Kozl. 70, 73-81.

172. McNaughton, S.J. (1979) Grassland-herbivore dynamics. In Sinclair, A.R.E. and Norton-Griffiths, M. (eds) Serengeti -- Dynamics of an Ecosystem. Univ. Chicago Press, Chicago, IL, pp. 46-81.

173. Rhoades, D.F. (1979) Evolution of plant chemical defense against herbivores. In Rosenthal, G.A. and Janzen, D.H. (eds) Herbivores: Their Interaction with Secondary Plant Metabolites. Academic Press, New York, NY, pp. 3-54.

174. Rhoades, D.F. (1983) Herbivores, population dynamics and plant chemistry. In Denno, R.F. and McClure, M.S. (eds) Variable Plants and Herbivores in Natural and Managed Systems. Academic Press, New York, NY, pp. 155-220.

175. Scriber, J.M. and Slansky, F. (1981) The nutritional ecology of immature insects. Ann. Rev. Entomol. 26, 183-211.

176. Lincoln, D.E. (1988) Herbivore responses to plants grown in enriched CO_2 atmosphere. In Koomanoff, F.A. (ed) Carbon Dioxide and Climate: Summaries of Research in FY 1988. U.S. Dept. Energy, Washington, DC, pp. 51-52.

177. Brady, N.C. (1980) The evaluation and removal of constraints to crop production. In Staples, R.C. and Kuhr, R.J. (eds) Linking Research to Crop Production. Plenum Press, New York, NY, pp. 11-34.

178. Lemon, E.R. (1977) The land's response to more carbon dioxide. In Anderson, N.R. and Malahoff, A. (eds) The Fate of Fossil Fuel CO_2 in the Ocean. Plenum, New York, NY, pp. 97-130.

179. Lemon, E.R. (1983) Interpretive summary. In Lemon, E.R. (ed) CO_2 and Plants: The Response of Plants to Rising Levels of Atmospheric Carbon Dioxide. Westview Press, Boulder, CO, pp. 1-5.

180. Bjorkman, O., Hiesey, W.M., Nobs, M.A., Nicholson, F. and Hart, R.W. (1968) Effect of oxygen concentration in higher plants. Carnegie Inst. Wash. Year Book 66, 228-232.

181. Bjorkman, O., Gauhl, E., Hiesey, W.M., Nicholson, F. and Nobs, M.A.

(1969) Growth of <u>Mimulus</u>, <u>Marchantia</u> and <u>Zea</u> under different oxygen and carbon dioxide levels. <u>Carnegie</u> <u>Inst</u>. <u>Wash</u>. <u>Year</u> <u>Book</u> 67, 477-478.

182. Quebedeaux, B. and Hardy, R.W.F. (1975) Reproductive growth and dry matter production of <u>Glycine</u> <u>max</u> (L.) Merr. in response to oxygen concentration. <u>Plant</u> <u>Physiol</u>. 55, 102-107.

183. Pearcy, R.W. and Bjorkman, O. (1983) Physiological effects. <u>In</u> Lemon, E.R. (ed) CO_2 <u>and</u> <u>Plants</u>: <u>The</u> <u>Response</u> <u>of</u> <u>Plants</u> <u>to</u> <u>Rising</u> <u>Levels</u> <u>of</u> <u>Atmospheric</u> <u>Carbon</u> <u>Dioxide</u>. Westview Press, Boulder, CO, pp. 65-106.

184. Madsen, E. (1968) Effect of CO_2-concentration on the accumulation of starch and sugar in tomato leaves. <u>Physiol</u>. <u>Plant</u>. 21, 168-175.

185. Epstein, E. (1972) <u>Mineral</u> <u>Nutrition</u> <u>of</u> <u>Plants</u>: <u>Principles</u> <u>and</u> <u>Perspectives</u>. John Wiley & Sons, New York, NY.

186. Baker, D.N. and Enoch, H.Z. (1983) Plant growth and development. <u>In</u> Lemon, E.R. (ed) CO_2 <u>and</u> <u>Plants</u>: <u>The</u> <u>Response</u> <u>of</u> <u>Plants</u> <u>to</u> <u>Rising</u> <u>Levels</u> <u>of</u> <u>Atmospheric</u> <u>Carbon</u> <u>Dioxide</u>. Westview Press, Boulder, CO, pp. 107-130.

187. Guinn, G. and Mauney, J. (1980) Analysis of CO_2 exchange assumptions: Feedback control. <u>In</u> Hesketh, J.D. and Jones, J.W. (eds) <u>Predicting</u> <u>Photosynthesis</u> <u>for</u> <u>Ecosystem</u> <u>Models</u>. <u>Vol</u>. <u>2</u>. CRC Press, Boca Raton, FL, pp. 1-16.

188. Herold, A. (1980) Regulation of photosynthesis by sink activity -- The missing link. <u>New</u> <u>Phytol</u>. 86, 131-144.

189. Von Caemmerer, S. and Farquhar, G.D. (1981) Some relationships between the biochemistry of photosynthesis and the gas exchange of leaves. <u>Planta</u> 153, 376-387.

190. Azcon-Bieto, J. (1983) Inhibition of photosynthesis by carbohydrates in wheat leaves. <u>Plant</u> <u>Physiol</u>. 73, 681-686.

191. Cave, G., Tolley, L.C. and Strain, B.R. (1981) Effect of carbon dioxide enrichment on chlorophyll content, starch content and starch grain structure in <u>Trifolium</u> <u>subterraneum</u> leaves. <u>Physiol</u>. <u>Plant</u>. 51, 171-174.

192. Wulff, R. and Strain, B.R. (1982) Effects of carbon dioxide enrichment on growth and photosynthesis of <u>Desmodium</u> <u>paniculatum</u>. <u>Can</u>. <u>J</u>. <u>Bot</u>. 60, 1084-1091.

193. Delucia, E.H., Sasek, T.W. and Strain, B.R. (1985) Photosynthetic inhibition after long-term exposure to elevated levels of atmospheric carbon dioxide. <u>Photosyn</u>. <u>Res</u>. 7, 175-184.

194. Lieth, H. (1975) Primary production of the major vegetation units of the world. <u>In</u> Lieth, H. and Whittaker, R.H. (eds) <u>Primary</u> <u>Productivity</u> <u>of</u> <u>the</u> <u>Biosphere</u>. Springer-Verlag, New York, NY, pp. 203-215.

195. Wilde, S.A. (1958) <u>Forest</u> <u>Soils</u>: <u>Their</u> <u>Properties</u> <u>and</u> <u>Relation</u> <u>to</u> <u>Silviculture</u>. Ronald Press, New York, NY.

196. Woodwell, G.M., Whittaker, R.H. and Houghton, R.A. (1975) Nutrient concentrations in plants in the Brookhaven oak-pine forest. <u>Ecology</u> 56, 318-332.

197. Kramer, P.J. and Kozlowski, T.T. (1979) <u>Physiology</u> <u>of</u> <u>Woody</u> <u>Plants</u>. Academic Press, New York, NY.

198. France, R.C. and Reid, C.P.P. (1983) Interactions of nitrogen and carbon in the physiology of ectomycorrhizae. <u>Can</u>. <u>J</u>. <u>Bot</u>. 61, 964-984.

199. Lavender, D.P. and Walker, R.B. (1979) Nitrogen and related elements in nutrition of forest trees. <u>In</u> Gessel, S.P., <u>et</u> <u>al</u>. (eds) <u>Contr</u>. <u>No</u>. <u>40</u>: <u>Proc</u>. <u>Forest</u> <u>Fertilization</u> <u>Conf</u>., <u>Union</u>. Univ. Washington, Seattle, WA, pp. 15-22.

200. Cole, D.W. and Rapp, M.R. (1980) Elemental cycling in forest ecosystems. <u>In</u> Reichle, D.E. (ed) <u>Dynamic</u> <u>Properties</u> <u>of</u> <u>Forest</u> <u>Ecosystems</u>. Cambridge Univ. Press, New York, NY, pp. 341-409.

201. Imai, K. and Murata, Y. (1978) Effect of carbon dioxide concentration on growth and dry matter production of crop plants. III. Relationship

REFERENCES

between CO_2 concentration and nitrogen nutrition in some C_3 and C_4 species. Jap. J. Crop Sci. 47, 118-123.

202. Patterson, D.T. and Flint, E.P. (1982) Interacting effects of CO_2 and nutrient concentration. Weed Sci. 30, 389-394.

203. Peet, M.M. and Willits, D.H. (1982) The effect of density and postplanting fertilization on response of lettuce to CO_2 enrichment. HortScience 17, 948-949.

204. Goudriaan, J. and de Ruiter, H.E. (1983) Plant growth in response to CO_2 enrichment at two levels of nitrogen and phosphorus supply. 1. Dry matter, leaf area and development. Neth. J. Agric. Sci. 31, 157-169.

205. Cure, J.D., Israel, D.W. and Rufty, T.W., Jr. (1988) Nitrogen stress effects on growth and seed yield of nonnodulated soybean exposed to elevated carbon dioxide. Crop Sci. 28, 671-677.

206. Skoye, D.A. and Toop, E.W. (1973) Relationship of temperature and mineral nutrition to carbon dioxide enrichment in the forcing of pot chrysanthemums. Can. J. Plant Sci. 53, 609-614.

207. Sionit, N., Mortensen, D.A., Strain, B.R. and Hellmers, H. (1981) Growth responses of wheat to carbon dioxide enrichment with different levels of mineral nutrition. Agron. J. 73, 1023-1027.

208. Sionit, N. (1983) Response of soybean to two levels of mineral nutrition in CO_2-enriched atmosphere. Crop Sci. 23, 329-333.

209. Zangerl, A.R. and Bazzaz, R.A. (1984) The response of plants to elevated CO_2. II. Competitive interactions among annual plants under varying light and nutrients. Oecologia 62, 412-417.

210. Conroy, J.P., Smillie, R.M., Kuppers, M., Bevege, D.I. and Barlow, E.W. (1986) Chlorophyll A fluorescence and photosynthetic and growth responses of Pinus radiata to phosphorus deficiency, drought stress, and high CO_2. Plant Physiol. 81, 423-429.

211. Hocking, P.J. and Meyer, C.P. (1985) Responses of Noogoora Burr (Xanthium occidentale Bertol) to nitrogen supply and carbon dioxide enrichment. Ann. Bot. 55, 835-844.

212. Cure, J.D., Israel, D.W. and Rufty, T.W., Jr. (1988) Nitrogen stress effects on growth and seed yield of nonnodulated soybean exposed to elevated carbon dioxide. Crop Sci. 28, 671-677.

213. Kimball, B. and Mauney, J. (1988) Elevated CO_2: Modeling cotton response, effects on insect-population dynamics, and long-term effects on trees. In Koomanoff, F.A. (ed) Carbon Dioxide and Climate: Summaries of Research in FY 1988. U.S. Dept. Energy, Washington, DC, pp. 50-51.

214. Cure, J.D., Rufty, T.W., Jr. and Israel, D.W. (1988) Phosphorus stress effects on growth and seed yield responses of nonnodulated soybean to elevated carbon dioxide. Agron. J. 80, 897-902.

215. Peet, M.M. and Willits, D.H. (1984) CO_2 enrichment of greenhouse tomatoes using a closed-loop heat storage: Effects of cultivar and nitrogen. Sci. Hort. 24, 21-32.

216. Johnston, M., Grof, C.P.L. and Brownell, P.F. (1986) Sodium deficiency in the C_4 species Amaranthus tricolor L. is not completely alleviated by high CO_2 concentrations. Photosynthetica. 20, 476-479.

217. Brooks, A., Woo, K.C. and Wong, S.C. (1988) Effects of phosphorus nutrition on the response of photosynthesis to CO_2 and O_2, activation of ribulose bisphosphate carboxylase and amounts of ribulose bisphosphate and 3-phosphoglycerate in spinach leaves. Photosyn. Res. 15, 133-141.

218. Marks, G.C. and Kozlowski, T.T., Eds. (1973) Ectomycorrhizae. Academic Press, New York, NY.

219. Sanders, F.E., Mosse, B. and Tinker, P.B., Eds. (1975) Endomycorrhizas. Academic Press, New York, NY.

220. Harley, J.L. and Smith, S.E. (1983) Mycorrhizal Symbiosis. Academic Press, New York, NY.

REFERENCES   SECTION 8

221. Norby, R.J., Luxmoore, R.J., O'Neill, E.G. and Weller, D.G. (1984) Plant
     Responses to Elevated Atmospheric CO$_2$ with Emphasis on Belowground
     Processes. Oak Ridge Nat. Lab., Oak Ridge, TN.
222. Higginbotham, K.O., Mayo, J.M., l'Hirondelle, S. and Krystofiak, D.K.
     (1985) Physiological ecology of lodgepole pine (Pinus contorta) in an
     enriched CO$_2$ environment. Can. J. For. Res. 15, 417-421.
223. Luxmoore, R.J., O'Neill, E.G., Ells, J.M. and Rogers, H.H. (1986)
     Nutrient-uptake and growth responses of Virginia pine to elevated
     atmospheric CO$_2$. J. Environ. Qual. 15, 244-251.
224. Norby, R.J., O'Neill, E.G. and Luxmoore, R.J. (1986) Effects of
     atmospheric CO$_2$ enrichment on the growth and mineral nutrition of
     Quercus alba seedlings in nutrient-poor soil. Plant Physiol. 82, 83-
     89.
225. Todd, R.L., Giddens, J.E., Kral, D.M. and Hawkins, S.L., Eds. (1984)
     Microbial-Plant Interactions. Amer. Soc. Agron., Madison, WI.
226. O'Neill, E.G., Luxmoore, R.J. and Norby, R.J. (1987) Elevated
     atmospheric CO$_2$ effects on seedling growth, nutrient uptake, and
     rhizosphere bacterial populations of Liriodendron tulipifera L. Plant
     Soil 104, 3-11.
227. O'Neill, E.G., Luxmoore, R.J. and Norby, R.J. (1987) Increases in
     mycorrhizal colonization and seedling growth in Pinus echinata and
     Quercus alba in an enriched CO$_2$ atmosphere. Can. J. For. Res. 17, 878-
     883.
228. Mikola, P. (1948) On the physiology and ecology of Cenococcum
     graniforme, especially as a mycorrhizal fungus of birch. Connum. Inst.
     For. Fenn. 36, 1-101.
229. Theodorou, C. and Bowen, G.D. (1971) Influence of temperature on the
     mycorrhizal associations of Pinus radiata D. Don. Austr. J. Bot. 19,
     13-20.
230. Hayman, D.S. (1974) Plant growth responses to vesicular-arbuscular
     mycorrhiza. VI. Effect of light and temperature. New Phytol. 73, 71-
     80.
231. Rouatt, J.W. and Katznelson, H. (1960) Influence of light on bacterial
     flora of roots. Nature 186, 659-660.
232. Rouatt, J.W., Peterson, E.A., Katznelson, H. and Henderson, V.E. (1963)
     Microorganisms in the root zone in relation to temperature. Can. J.
     Microbiol. 9, 227-236.
233. Vrany, J. (1963) Effect of foliar application of urea on the root
     microflora. Folia Microbiol. 8, 351-355.
234. Fenwick, L. (1973) Studies on the rhizosphere microflora of onion plants
     in relation to temperature changes. Soil Biol. Biochem. 5, 315-320.
235. Van Vuurde, J.W.L. and DeLange, A. (1978) The rhizosphere microflora of
     wheat grown under controlled conditions. II. Influence of the stage of
     growth of the plant, soil fertility, and leaf treatment with urea on
     the rhizosphere soil microflora. Plant Soil 50, 461-472.
236. Lamborg, M.R., Hardy, R.W. and Paul, E.A. (1983) Microbial effects. In
     Lemon, E.R. (ed) CO$_2$ and Plants: The Response of Plants to Rising
     Levels of Atmospheric Carbon Dioxide. Westview Press, Boulder, CO,
     pp. 131-176.
237. Greaves, M.P. and Darbyshire, J.F. (1972) The ultra structure of the
     mucilaginous layer on plant roots. Soil Biol. Biochem. 4, 443-449.
238. Bowen, G.D. and Theodorou, C. (1973) Growth of ectomycorrhizal fungi
     around seeds and roots. In Marks, G.C. and Kozlowski, T.T. (eds)
     Ectomycorrhizae. Academic Press, New York, NY, pp. 107-150.
239. Feldman, L.J. (1988) The habits of roots: What's up down under?
     BioScience 38, 612-618.
240. Gardner, W.K., Parbevy, D.G. and Barber, D.A. (1982) The acquisition of
     phosphorous by Lupinus albus L. I. Some characteristics of the
     soil/root interface. Plant Soil 68, 19-32.

REFERENCES

241. Lambers, H. (1987) Growth, respiration, and exudation and symbiotic associations: The fate of carbon translocated in the root. In Gregory, P.J., Lake, J.V. and Rose, D.A. (eds) Root Development and Function. Cambridge Univ. Press, Cambridge, UK, pp. 125-146.

242. Dart, P.J. and Mercer, F.V. (1964) The legume rhizosphere. Arch. Mikrobiol. 47. 344-378.

243. Routien, J.B. and Dawson, R.F. (1943) Some interrelationships of growth, salt absorption, respiration, and mycorrhizal development in Pinus echinata Mill. Amer. J. Bot. 30, 440-451.

244. Gray, L.E. and Gerdemann, J.W. (1967) Influence of vesicular-arbuscular mycorrhizas on the uptake of Phosphorus-32 by Liriodendron tulipifera and Liquidambar styraciflua. Nature 213, 106-107.

245. Zukovskaya, P.W. (1941) Changes in bacteriorhiza of cultivated plants. Microbiol. (USSR) 10, 919-923.

246. Meyer, F.H. (1987) Das Wurzelsystem geschadister Waldbestande. Allg. Forst. Zeitschr. 42, 754-757.

247. Ratnayaka, M., Leonard, R.T. and Menge, J.A. (1978) Root exudation in relation to supply of phosphorus and its possible relevance to mycorrizal formation. New Phytol. 81, 543-552.

248. Fogel, R. (1983) Root turnover and productivity of forests. Plant Soil 71, 75-85.

249. Harley, J.L. (1975) Problems of mycotrophy. In Sanders, F.E., Mosse, B. and Tinker, P.B. (eds) Endomycorrhizas. Academic Press, New York, NY, pp. 1-24.

250. Tinker, P.B. (1984) The role of microorganisms in mediating and facilitating the uptake of plant nutrients from the soil. Plant Soil 76, 77-91.

251. Clarkson, D.T. (1985) Factors affecting mineral nutrient acquisition by plants. Ann. Rev. Plant Physiol. 36, 77-115.

252. Slankis, V. (1973) Hormonal relationships in mycorrhizal development. In Marks, G.C. and Kozlowski, T.T. (eds) Ectomycorrhizae. Academic Press, New York, NY, pp. 231-298.

253. Simmons, G.L. and Pope, P.E. (1987) Influence of soil compaction and vesicular-arbuscular mycorrhizae on root growth of yellow poplar and sweet gum seedlings. Can. J. For. Res. 17, 970-975.

254. Simmons, G.L. and Pope, P.E. (1988) Influence of soil water potential and mycorrhizal colonization on root growth of yellow-poplar and sweet gum seedlings grown in compacted soil. Can. J. For. Res. 18, 1392-1396.

255. Tien, T.M., Gaskin, M.H. and Hubbel, D.H. (1979) Plant growth substances produced by Azospirillum brasilense and their effect on the growth of pearl millet (Pennisetum americanum L.). Appl. Environ. Microbiol. 37, 1016-1024.

256. Zimmer, W., Roeben, K. and Bothe, H. (1988) An alternative explanation for plant growth promotion by bacteria of the genus Azospirillum. Planta 176, 333-342.

257. Umali-Garcia, M., Hubbel, D.H., Gaskin, M.H. and Dazzo, F.B. (1980) Association of Azospirillum with grass roots. Appl. Environ. Microbiol. 39, 219-226.

258. Kapulnik, Y., Gafny, R. and Okon, Y. (1983) Effect of Azospirillum spp. inoculation development and NO_3^- uptake in wheat (Triticum aestivum cv. Miriam) in hydroponic system. Can. J. Bot. 63, 627-631.

259· Luxmoore, R.J., Norby, R.J. and Weller, D.G. (1983) Forest response to elevated atmospheric CO_2. Agron. Abstr. 75, 209.

260. Pitman, M. (1977) Ion transport into the xylem. Ann. Rev. Plant Physiol. 28, 71-88.

261. Jackson, W.A., Volk, R.J. and Israel, D.W. (1980) Energy supply and nitrate assimilation in root systems. In Tanaka, A. (ed) Carbon-Nitrogen Interaction in Crop Production. Jap. Soc. for Promotion of

REFERENCES   SECTION 8
Science, Tokyo, Japan, pp. 25-40.

262. Touraine, B. and Grignon, C. (1982) Energetic coupling of nitrate
secretion into the xylem of corn roots. Physiol. Veg. 20, 33-39.

263. Clement, C.R., Hopper, M.J., Jones, L.H.P. and Leafe, E.L. (1978) The
uptake of nitrate by Lolium perenne from flowering nutrient solution
II. Effect of light, defoliation, and relationship to CO_2 flux. J.
Exp. Bot. 29, 1173-1183.

264. Pearson, C.J., Volk, R.J. and Jackson, W.A. (1981) Daily changes in
nitrate influx, efflux and metabolism in maize and pearl millet.
Planta 152, 319-324.

265. Rufty, T.W., Jr., MacKown, C.T. and Volk, R.M. (1989) Effects of altered
carbohydrate availability on whole-plant assimilation of $^{15}NO_3^{-1}$.
Plant Physiol. 89, 457-463.

266. Jakucs, P. (1988) Ecological approach to forest decay in Hungary. Ambio
17, 267-274.

267. Meentemeyer, V. (1978) Macroclimate and lignin control of litter
decomposition rates. Ecology 59, 465-472.

268. Aber, J.D. and Melillo, J.M. (1980) Litter decomposition: Measuring
state of decay and percent transfer into forest soils. Can. J. Bot.
58, 416-421.

269. Bosatta, E. and Staaf, H. (1982) The control of nitrogen turnover in
forest litter. Oikos 39, 143-151.

270. Melillo, J.M., Aber, J.D. and Muratore, J.F. (1982) Nitrogen and lignin
control of hardwood leaf litter decomposition dynamics. Ecology 63,
621-633.

271. Melillo, J.M. (1983) Will increases in atmospheric CO_2 concentrations
affect decay processes? Annual Rept. Marine Biol. Lab., Woods Hole,
MA, pp. 10-11.

272. Curtis, P.S., Drake, B.G. and Whigham, D.F. (1989) Nitrogen and carbon
dynamics in C_3 and C_4 estuarine marsh plants grown under elevated CO_2
in situ. Oecologia 78, 297-301.

273. Boyle, J.R. and Voigt, G.K. (1973) Biological weathering of silicate
minerals. Implications for tree nutrition and soil genesis. Plant Soil
38, 191-201.

274. Boyle, J.R., Voight, G.K. and Sawhney, B.L. (1974) Chemical weathering
of biotite by organic acids. Soil Sci. 117, 42-45.

275. Graustein, W.C., Cromack, K., Jr. and Sollins, P. (1977) Calcium
oxalate: Occurrence in soils and effect on nutrient and geochemical
cycles. Science 198, 1252-1254.

276. Luxmoore, R.J. (1981) CO_2 and phytomass. BioScience 31, 626.

277. Robbins, C.W. (1986) Carbon dioxide partial pressure in lysimeter soils.
Agron. J. 78, 151-158.

278. Tiwari, V.N., Lehri, L.K. and Pathak, A.N. (1989) Effect of inoculating
crops with phospho-microbes. Exp. Agric. 25, 47-50.

279. Ortuno, A., Hermansarz, A., Noguera, J., Morales, V. and Armero, T.
(1978) Phosphorus solubilizing effect of A. niger and Pseudomonas
fluorescens. Microbiol. Esp. 30, 113-120.

280. Molla, M.A.Z., Chowdhary, A.A. and Islam, A.H. (1984) Microbial
mineralization of organic phosphate in soil. Plant Soil 78, 393-399.

281. Babenko, Y.S., Tyrugina, G.I., Origoreev, E.F., Dalgikh, L.M. and
Borisova, T.I. (1985) Biological activity and physiological
biochemical properties of phosphate dissolving bacteria. Microbiology
53, 427-433.

282. Sethi, R.P. and Subba Rao, N.S. (1968) Solubilization of tricalcium
phosphate and calcium phytate by soil fungi. J. Gen. Appl. Microbiol.
14, 329-331.

283. Bajpai, P.D. and Sundara Rao, W.V.B. (1971) Phosphate solubilizing
bacteria (1) Solubilization of phosphate in liquid culture by selected
bacteria as affected by different pH value. Soil Sci. Plant Nutr.
17(2), 41-43.

REFERENCES

284. Khalafallah, M.A., Saber, M.S.M. and Abd-el-Maksoud, H.K. (1982) Influence of phosphorus solubilising bacteria on efficiency of super phosphate in calcareous soil cultivated in Vicia faba. Zeitschr. Pflanz. nahr. Bodenk. 145, 455-459.
285. Badr, S.M.S., El-Din and Sabar, M.S.M. (1986) Response of soybean to dual innoculation with R. japonicum and phosphorus solubilising bacteria. Zeitschr. Pflanz. nahr. Bodenk. 149, 130-135.
286. Rosenberg, N.J., Blad, B.L. and Verma, S.B. (1983) Microclimate: The Biological Environment. Wiley Interscience, New York, NY.
287. Goldsmith, E. (1988) Gaia: Some implications for theoretical ecology. The Ecologist 18(2/3), 64-74.
288. Felker, P. and Clark, P.R. (1982) Position of mesquite (Prosopis spp.) nodulation and nitrogen fixation (acetylene reduction) in 3-m long phreatophytically simulated soil columns. Plant Soil 64, 297-305.
289. Virginia, R.A. and Jarrell, W.M. (1983) Soil properties in a mesquite-dominated Sonoran Desert ecosystem. Soil Sci. Soc. Amer. J. 47, 138-144.
290. Tiedemann, A.R. and Klemmedson, J.O. (1977) Effect of mesquite trees on vegetation and soils in the desert grassland. J. Range Manage. 30, 361-367.
291. Barth, R.C. and Klemmedson, J.O. (1978) Shrub-induced spatial patterns of dry matter, nitrogen, and organic carbon. Soil Sci. Soc. Amer. J. 42, 804-809.
292. Martyniuk, S. and Wagner, G.M. (1978) Quantitative and qualitative examination of soil microflora associated with different management systems. Soil Sci. 125, 343-350.
293. Bolton, H., Jr., Elliott, L.F., Papendick, R.I. and Bezdicek, D.F. (1985) Soil microbial biomass and selected soil enzyme activities: Effect of fertilization and cropping practices. Soil Biol. Biochem. 17, 297-302.
294. Fraser, D.G., Doran, J.W., Sahs, W.W. and Lesoing, G.W. (1988) Soil microbial populations and activities under conventional and organic management. J. Environ. Qual. 17, 585-590.
295. Jenkinson, D.S. and Ladd, J.N. (1981) Microbial biomass in soil: Measurement and turnover. In Paul, E.A. and Ladd, J.N. (eds) Soil Biochemistry. Vol. 5. Marcel Dekker, New York, NY, pp. 415-471.
296. Smith, G.E. (1942) Sanborn Field: Fifty Years of Field Experiments with Crop Rotations, Manure, and Fertilizers. Missouri Agric. Exp. Stn. Bull. No. 458.
297. Odell, R.T., Walker, W.M., Boone, L.V. and Oldham, M.G. (1982) The Morrow Plots: A Century of Learning. Univ. Illinois Agric. Exp. Stn. Bull. No. 775.
298. Nicholas, W.L. (1984) The Biology of Free-Living Nematodes. Clarendon Press, Oxford, UK.
299. Santos, P.F., Phillips, J. and Whitford, W.G. (1981) The role of mites and nematodes in early stages of buried litter decomposition in a desert. Ecology 62, 664-669.
300. Whitford, W.G., Freckman, D.W., Santos, P.F., Elkins, N.Z. and Parker, L.W. (1982) The role of nematodes in decomposition in desert ecosystems. In Freckman, D.W. (ed) Nematodes in Soil Ecosystems. Univ. Texas Press, Austin, TX, pp. 98-116.
301. Wasilewska, L. (1979) The structure and function of soil nematode communities in natural ecosystems and agrocenoses. Pol. Ecol. Stud. 5, 97-145.
302. Sohlenius, B., Bostrom, S. and Sander, A. (1987) Long-term dynamics of nematodes in arable soil under four cropping systems. J. Appl. Ecol. 24, 131-144.
303. Sohlenius, B., Bostrom, S. and Sander, A. (1988) Carbon and nitrogen budgets of nematodes in arable soil. Biol. Fertil. Soils 6, 1-8.

REFERENCES   SECTION 8

304. Anderson, R.V., Coleman, D.C. and Cole, C.V. (1981) Effects of saprotrophic grazing on net mineralization. In Clark, F.E. and Rosswall, T. (eds) Terrestrial Nitrogen Cycles. Ecol. Bull. (Stockholm) 33, 201-216.

305. Gerlach, S.A. (1978) Food-chain relationships in subtidal silty sand marine sediments and the role of meiofauna in stimulating bacterial productivity. Oecologia 33, 55-69.

306. Platt, H.M. and Warwick, R.M. (1980) The significance of free-living nematodes to the littoral ecosystem. In Price, J.H., Irvine, D.E.G. and Farnham, W.F. (eds) The Shore Environment, Vol. 2, Ecosystems. Academic Press, New York, NY, pp. 729-759.

307. Freckman, D.W. (1988) Bacterivorous nematodes and organic-matter decomposition. Agric. Ecosys. Environ. 24, 195-217.

308. Gaugler, R. (1988) Ecological considerations in the biological control of soil-inhabiting insects with entomopathogenic nematodes. Agric. Ecosys. Environ. 24, 351-360.

309. Deverel, S.J., Gillion, R.J., Fujii, R., Izbicki, J.A. and Fields, J.C. (1984) Water Resources Investigations Report 84-4319: Areal Distribution of Selenium and Other Inorganic Constituents in Shallow Groundwater of the San Luis Drain Service Area, San Joaquin Valley, California: A Preliminary Study. U.S. Geological Survey, Sacramento, CA.

310. Letey, J., Roberts, C., Penberth, M. and Vasek, C. (1986) Pub. No. 3319: An Agricultural Dilemma: Drainage Water and Toxics Disposal in the San Joaquin Valley. Div. Agric. Nat. Res., Univ. Calif., Riverside, CA.

311. Franke, K.W. and Moxon, A.L. (1936) A comparison of the minimum fatal doses of selenium, tellurium, arsenic and vanadium. J. Pharmacol. Exp. Ther. 58, 454-459.

312. McConnell, K.P. and Portman, O.W. (1952) Toxicity of dimethyl selenide in the rat and mouse. Proc. Soc. Exp. Biol. Med. 79, 230-231.

313. Vallee, B.L. (1986) A synopsis of zinc biology and pathology. In Bertini et al. (eds) Zinc Enzymes. Progress in Inorganic Biochemistry and Biophysics, Vol. 1. Birkhauser, Boston, MA, pp. 1-15.

314. Tuve, T. and Williams, H.H. (1961) Metabolism of selenium by Escherichia coli: Biosynthesis of selenomethionine. J. Biol. Chem. 236, 597-601.

315. Wood, J.M., Kennedy, F.S. and Rosen, C.G. (1968) Synthesis of methyl-mercury compounds by extracts of a methanogenic bacterium. Nature 220, 173-174.

316. Dovan, J.W. and Alexander, M. (1977) Microbial formation of volatile selenium compounds in soil. Soil Sci. Soc. Amer. J. 41, 70-73.

317. Craig, P.J. (1986) Occurrence and pathways of organometallic compounds in the environment -- General considerations. In Craig, P.J. (ed) Organometallic Compounds in the Environment. Longman Group Ltd., Harlow, Essex, UK, pp. 1-64.

318. Koch, J.T., Rachar, D.B. and Kay, B.D. (1989) Microbial participation in iodide removal from solution by organic soils. Can. J. Soil Sci. 69, 127-135.

319. Francis, A.J., Duxbury, J.M. and Alexander, M. (1974) Evolution of dimethylselenide from soils. Appl. Microbiol. 28, 248-250.

320. Karlson, U. and Frankenberger, W.T., Jr. (1988) Effects of carbon and trace element addition on alkylselenide production by soil. Soil Sci. Soc. Amer. J. 52, 1640-1644.

321. Abu-Erreish, G.M., Whitehead, E.I. and Olson, O.E. (1968) Evolution of volatile selenium from soils. Soil Sci. 106, 415-420.

322. Teramura, A.H. (1983) Effects of ultraviolet-B radiation on the growth and yield of crop plants. Physiol. Plant. 58, 415-427.

323. Sisson, W.B. (1986) Effects of UV-B radiation on photosynthesis. In Worrest, R.C. and Caldwell, M.M. (eds) Stratospheric Ozone Reduction, Solar Ultraviolet Radiation, and Plant Life. Springer-Verlag, New

REFERENCES   SECTION 8

    York, NY, pp. 161-169.

324. Flint, S.D., Jordan, P.W. and Caldwell, M.M. (1985) Plant protective
     response to enhanced UV-B radiation under field conditions: Leaf
     optical properties and photosynthesis. Photochem. Photobiol. 41, 95-
     99.

325. Barnes, P.W., Flint, S.D. and Caldwell, M.M. (1987) Photosynthesis
     damage and protective pigments in plants from a latitudinal
     arctic/alpine gradient exposed to supplemental UV-B radiation in the
     field. Arctic Alpine Res. 19, 21-27.

326. Lovelock, J. (1988) The Ages of Gaia: A Biography of Our Living Earth.
     W.W. Norton & Co., New York, NY.

327. Beyschlag, W., Barnes, P.W., Flint, S.D. and Caldwell, M.M. (1989)
     Enhanced UV-B irradiation has no effect on photosynthetic
     characteristics of wheat (Triticum aestivum L.) and wild oat (Avena
     fatua L.) under greenhouse and field conditions. Photosynthetica 22,
     516-525.

328. Billings, W.D., Luken, J.O., Mortensen, D.A. and Peterson, K.M. (1983)
     Increasing atmospheric carbon dioxide: Possible effects on arctic
     tundra. Oecologia 58, 286-289.

329. Billings, W.D., Peterson, K.M., Luken, J.O. and Mortensen, D.A. (1984)
     Interaction of increasing atmospheric carbon dioxide and soil nitrogen
     on the carbon balance of tundra microcosms. Oecologia 65, 26-29.

330. Oberbauer, S.F., Sionit, N., Hastings, S.J. and Oechel, W.C. (1986)
     Effects of CO_2 enrichment and nutrition on growth, photosynthesis, and
     nutrient concentration of Alaskan tundra plant species. Can. J. Bot.
     64, 2993-2998.

331. Williams, W.E., Garbutt, K., Bazzaz, F.A. and Vitousek, P.M. (1986) The
     response of plants to elevated CO_2 -- IV. Two deciduous-forest tree
     communities. Oecologia 69, 454-459.

332. Williams, W.E., Garbutt, K. and Bazzaz, F.A. (1988) The response of
     plants to elevated CO_2 -- V. Performance of an assemblage of
     serpentine grassland herbs. Environ. Exp. Bot. 28, 123-130.

333. Idso, S.B. (1989) Three stages of plant response to atmospheric CO_2
     enrichment. Plant Physiol. Biochem. 27, 131-134.

334. Hughes, P.R., Potter, J.E. and Weinstein, L.H. (1982) Effects of air
     pollution on plant-insect interactions: Increased susceptibility of
     greenhouse-grown soybeans to the Mexican bean beetle after plant
     exposure to SO_2. Environ. Entomol. 11, 173-176.

335. Bolsinger, M. and Fluckiger, W. (1984) Effect of air pollution at a
     motorway on the infestation of Viburnum opulus by Aphis fabae. Eur. J.
     For. Pathol. 14, 256-260.

336. Braun, S. and Fluckiger, W. (1985) Increased population of the aphid
     Aphis pomi at a motorway: Part 3 -- The effect of exhaust gases.
     Environ. Poll. Ser. A 39, 183-192.

337. Fluckiger, W., Braum, S. and Bolsinger, M. (1988) Air pollution: Effect
     on host plant-insect relationship. In Schulte-Hostede, S., Darrall,
     N.M., Blank, L.W. and Wellburn, A.R. (eds) Air Pollution and Plant
     Metabolism. Elsevier Applied Science Ltd., London, UK, pp. 360-380.

338. Dohmen, G.P. (1988) Indirect effects of air pollutants: Changes in
     plant/parasite interactions. Environ. Poll. 53, 197-207.

REFERENCES   SECTION 9

1. Gifford, R.M. (1979) Growth and yield of CO_2-enriched wheat under water-
   limited conditions. Austr. J. Plant Physiol. 6, 367-378.

2. Wittwer, S.H. (1982) Carbon dioxide and crop productivity. New Sci. 95,
   233-234.

3. Allen, L.H., Jr., Boote, K.J., Jones, J.W., Jones, P.H., Valle, R.R.,

REFERENCES   SECTION 9

   Acock, B., Rogers, H.H. and Dahlman, R.C. (1987) Response of
   vegetation to rising carbon dioxide: Photosynthesis, biomass, and seed
   yield of soybean. Global Biogeochem. Cycles 1, 1-14.
4. Warrick, R.A. (1986) Photosynthesis seen from above. Nature 319, 181.
5. Matson, P.A. and Harriss, R.C. (1988) Prospects for aircraft-based gas
   exchange measurements in ecosystem studies. Ecology 69, 1318-1325.
6. Goward, S.N., Tucker, C.J. and Dye, D.G. (1985) North American
   vegetation patterns observed with the NOAA-7 advanced very high
   resolution radiometer. Vegetatio 64, 3-14.
7. Justice, C.O., Townshend, J.R.G., Holben, B.N. and Tucker, C.J. (1985)
   Phenology of global vegetation using meteorological satellite data.
   Internat. J. Remote Sens. 6, 1278-1318.
8. Tucker, C.M., Fung, I.Y., Keeling, C.D. and Gammon, R.H. (1986)
   Relationship between atmospheric CO_2 variations and a satellite-
   derived vegetation index. Nature 319, 195-199.
9. Choudhury, B.J. and Tucker, C.J. (1987) Monitoring global vegetation
   using Nimbus-7 37GHz data: Some empirical relations. Internat. J.
   Remote Sens. 8, 1085-1090.
10. Fung, I.Y., Tucker, C.J. and Prentice, K.C. (1987) Application of
    advanced very high resolution radiometer vegetation index to study
    atmosphere-biosphere exchange of CO_2. J. Geophys. Res. 92, 2999-3015.
11. Choudhury, B.J. (1988) Relating Nimbus-7 37GHz data to global land-
    surface evaporation, primary productivity and the atmospheric CO_2
    concentration. Internat. J. Remote Sens. 9, 169-176.
12. Wessman, C.A., Aber, J.D., Peterson, D.L. and Melillo, J.M. (1988)
    Remote sensing of canopy chemistry and nitrogen cycling in temperate
    forest ecosystems. Nature 335, 154-156.
13. Card, D.H., Peterson, D.L., Matson, P.A. and Aber, J.D. (1988)
    Prediction of leaf chemistry by the use of visible and near infrared
    reflectance spectroscopy. Remote Sens. Environ. 26, 123-147.
14. Falkowski, P. (1988) Ocean productivity from space. Nature 335, 205.
15. Lohrenz, S.E., Arnone, R.A., Wiesenburg, D.A. and DePalma, I.P. (1988)
    Satellite detection of transient enhanced primary production in the
    western Mediterranean Sea. Nature 335, 245-247.
16. Balch, W.M (1988) An algorithm for the remote sensing of primary
    production based on pigments, temperature and light. Trans. Amer.
    Geophys. Union 69, 1081.
17. LaMarche, V.C., Jr., Graybill, D.A., Fritts, H.C. and Rose, M.R. (1984)
    Increasing atmospheric carbon dioxide: Tree ring evidence for growth
    enhancement in natural vegetation. Science 223, 1019-1021.
18. Wigley, T.M.L., Briffa, K.R. and Jones, P.D. (1984) Atmospheric carbon
    dioxide: Predicting plant productivity and water resources. Nature
    312, 102-103.
19. Cooper, C.F. (1986) Carbon dioxide enhancement of tree growth at high
    elevations. Science 231, 859.
20. Gale, J. (1986) Carbon dioxide enhancement of tree growth at high
    elevations. Science 231, 859-860.
21. LaMarche, V.C., Jr., Graybill, D.A., Fritts, H.C. and Rose, M.R. (1986)
    Carbon dioxide enhancement of tree growth at high elevations. Science
    231, 860.
22. Payette, S., Filion, L., Gauthier, L. and Boutin, Y. (1985) Secular
    climate change in old-growth tree-line vegetation in northern Quebec.
    Nature 315, 135-138.
23. Oikawa, T. (1986) Simulation of forest carbon dynamics based on a dry-
    matter production model. III. Effects of increasing CO_2 upon a
    tropical rainforest ecosystem. Bot. Mag. Tokyo 99, 419-430.
24. Downton, W.J.S., Grant, W.J.R. and Loveys, B.R. (1987) Carbon dioxide
    enrichment increases yield of Valencia orange. Austr. J. Plant
    Physiol. 14, 493-501.

25. Koch, K.E., Allen, L.H., Jr., Jones, P. and Avigne, W.T. (1987) Growth of citrus rootstock (Carrizo Citrange) seedlings during and after long-term CO_2 enrichment. J. Amer. Soc. Hort. Sci. 112, 77-82.
26. O'Neill, E.G., Luxmoore, R.J. and Norby, R.J. (1987) Increases in mycorrhizal colonization and seedling growth in Pinus echinata and Quercus alba in an enriched CO_2 atmosphere. Can. J. For. Res. 17, 878-883.
27. Wong, S.C. and Dunin, F.X. (1987) Photosynthesis and transpiration of trees in a Eucalypt forest stand: CO_2, light and humidity responses. Austr. J. Plant Physiol. 14, 619-632.
28. Green, K. and Wright, R. (1977) Field response of photosynthesis to CO_2 enhancement in ponderosa pine. Ecology 58, 687-692.
29. Koch, K.E., Jones, P.H., Avigne, W.T. and Allen, L.H., Jr. (1986) Growth, dry matter partitioning, and diurnal activities of RuBP carboxylase in citrus seedlings maintained at two levels of CO_2. Physiol. Plant. 67, 477-484.
30. O'Neill, E.G., Luxmoore, R.J. and Norby, R.J. (1987) Elevated atmospheric CO_2 effects on seedling growth, nutrient uptake, and rhizosphere bacterial populations of Liriodendron tulipifera L. Plant Soil 104, 3-11.
31. Graybill, D.A. (1987) A network of high elevation conifers in the western U.S. for detection of tree-ring growth response to increasing atmospheric carbon dioxide. In Jacoby, G.C., Jr. and Hornbeck, J.W. (eds) Proc. Internat. Symp. Ecological Aspects Tree-Ring Analysis. U.S. Dept. Energy, Washington, DC, pp. 463-474.
32. West, D.C. (1988) Detection of forest response to increased atmospheric carbon dioxide. In Koomanoff, F.A. (ed) Carbon Dioxide and Climate: Summaries of Research in FY 1988. U.S. Dept. Energy, Washington, DC, p. 57.
33. Parker, M.L. (1987) Recent abnormal increase in tree-ring widths: A possible effect of elevated atmospheric carbon dioxide. In Jacoby, G.C., Jr. and Hornbeck, J.W. (eds) Proc. Internat. Symp. Ecological Aspects of Tree-Ring Analysis. U.S. Dept. Energy, Washington, DC, pp. 511-521.
34. Hari, P. and Arovaara, H. (1988) Detecting CO_2 induced enhancement in the radial increment of trees. Evidence from the northern timber line. Scand. J. For. Res. 3, 67-74.
35. Mikola, P. (1950) On variations in tree growth and their significance to growth studies. Commun. Inst. For. Fenn. 38(5).
36. Kienast, F. and Luxmoore, R.J. (1988) Tree-ring analysis and conifer growth responses to increased atmospheric CO_2 levels. Oecologia 76, 487-495.
37. Haines, B., Stefani, M. and Hendrix, F. (1980) Acid rain: Threshold of leaf damage in eight plant species from a southern Appalachian forest succession. Water Air Soil Poll. 14, 403-407.
38. Evans, L.S. (1982) Biological effects of acidity in precipitation on vegetation: A review. Environ. Exp. Bot. 22, 155-169.
39. Scherbatskoy, T. and Klein, R.M. (1983) Response of spruce and birch foliage to leaching by acidic mists. J. Environ. Qual. 12, 189-195.
40. Waldman, J.D., Munger, J.W., Jacob, D.J. and Hoffmann, M.R. (1985) Chemical characterization of stratus cloudwater and its role as a vector for pollutant deposition in a Los Angeles pine forest. Tellus 37B, 91-108.
41. Saxena, V.K., Stogner, R.E., Hendler, A.H., De Felice, T.P., Yeh, R.J.-Y. and Lin, N.-H. (1989) Monitoring the chemical climate of the Mt. Mitchell State Park for evaluation of its impact on forest decline. Tellus 41B, 92-109.
42. Billings, W.D., Clebsch, E.E.C. and Mooney, H.A. (1961) Effect of low concentration of carbon-dioxide on photosynthesis rates of two races

REFERENCES   SECTION

   of Oxyria. Science 133, 1834.
43. Mooney, H.A., Wright, R.D. and Strain, B.R. (1964) The gas exchange
    capacity of plants in relation to vegetation zonation in the White
    Mountains of California. Amer. Midl. Nat. 72, 281-297.
44. Mooney, H.A., Strain, B.R. and West, M. (1966) Photosynthetic efficiency
    at reduced carbon dioxide tensions. Ecology 47, 490-491.
45. Tranquillini, W. (1979) Physiological Ecology of the Alpine
    Timberline. Springer, New York, NY.
46. Decker, J.P. (1959) Some effects of temperature and carbon dioxide
    concentration on photosynthesis of mimulus. Plant Physiol. 34, 103-
    106.
47. Gale, J. (1972) Availability of carbon dioxide for photosynthesis at
    high altitudes. Ecology 53, 494-497.
48. Korner, C.H., Scheel, J.A. and Bauer, H. (1979) Maximum leaf diffusive
    conductance in vascular plants. Photosynthetica 13, 45-82.
49. Korner, C. and Mayr, R. (1980) Stomatal behavior in alpine plant
    communities between 600 and 2,600 metres above sea level. In Grace,
    J., Ford, E.D. and Jarvis, P.G. (eds) Plants and their Atmospheric
    Environment. Blackwell, Oxford, UK, pp. 205-218.
50. Woodward, F.I. (1986) Ecophysiological studies on the shrub Vaccinium
    myrtillus L. taken from a wide altitudinal range. Oecologia 70, 580-
    586.
51. Woodward, F.I. (1987) Stomatal numbers are sensitive to increases in CO_2
    from pre-industrial levels. Nature 327, 617-618.
52. Morison, J.I.L. (1987) Plant growth and CO_2 history. Nature 327, 560.
53. Woodward, F.I. (1988) The responses of stomata to changes in
    atmospheric levels of CO_2. Plants Today 1, 132-135.
54. Allen, L.H. (1988) Assessment of crop response to increased atmospheric
    carbon dioxide: Rising CO_2 effects on rice. In Koomanoff, F.A. (ed)
    Carbon Dioxide and Climate: Summaries of Research in FY 1988. U.S.
    Dept. Energy, Washington, DC, pp. 45-46.
55. Korner, C. (1988) Does global increase of CO_2 alter stomatal density.
    Flora 181, 253-257.
56. O'Leary, J.W. and Knecht, G.N. (1981) Elevated CO_2 concentration
    increases stomate numbers in Phaseolus vulgaris leaves. Bot. Gaz. 142,
    438-441.
57. Bristow, J.M. and Looi, A. (1968) Effects of carbon dioxide on the
    growth and morphogenesis of Marsilea. Amer. J. Bot. 55, 884-889.
58. Mayeux, H.S., Jr. and Johnson, H.B. (1986) Causes and consequences of
    vegetation change on rangeland. ARS-Research Project Statement. CRIS
    Work Unit No. 6206-20110-004.
59. Smeins, F.E. (1983) Origin of the brush problem -- A geological and
    ecological perspective of contemporary distributions. Proc. Brush
    Manage. Symp. Soc. Range Manage., Albuquerque, NM, pp. 5-16.
60. Malin, J.C. (1953) Soil, animal, and plant relations of the grassland,
    historically reconsidered. Sci. Mon. 76, 207-220.
61. Davis, O.K. and Turner, R.M. (1986) Palynological evidence for the
    historic expansion of juniper and desert shrubs in Arizona, U.S.A.
    Rev. Palaeobot. Palynol. 49, 177-193.
62. Gregg, J. (1844) Commerce of the Prairies: or, the Journal of a Santa
    Fe Trader, During Eight Expeditions Across the Great Western Prairies,
    and a Residence of Nearly Nine Years in Northern Mexico. Henry G.
    Langley, New York, NY.
63. Bray, W.L. (1906) Bull. No. 82, Sci. Ser. 10: Distribution and
    Adaptation of the Vegetation of Texas. Univ. Texas, Austin, TX.
64. Cook, O.F. (1908) Circ. No. 14: Change of Vegetation on the South Texas
    Prairies. USDA Bur. Plant Industry, Washington, DC.
65. Bogush, E.R. (1952) Brush invasion of the Rio Grande Plain of Texas.
    Texas J. Sci. 4, 85-91.

66. Inglis, J. (1962) Bull. No. 45: A History of Vegetation on the Rio
     Grande Plain. Texas Parks and Wildlife Dept., Austin, TX.
67. Humphrey, R.R. (1958) The desert grassland: A history of vegetational
     change and an analysis of causes. Bot. Rev. 24, 193-252.
68. Johnston, M.C. (1963) Past and present grasslands of southern Texas and
     northern Mexico. Ecology 44, 456-466.
69. York, J.C. and Dick-Peddie, W.A. (1969) Vegetation changes in southern
     New Mexico during the past hundred years. In McGinnies, W.G. and
     Goldman, B.J. (eds) Arid Lands in Perspective. Univ. Ariz. Press,
     Tucson, AZ, pp. 155-166.
70. Buffington, L.C. and Herbel, C.H. (1965) Vegetation changes on a semi-
     desert grassland range from 1858 to 1963. Ecol. Monogr. 35, 139-164.
71. Humphrey, R.R. and Mehrhoff, L.A. (1958) Vegetation changes on Arizona
     grassland ranges. Ecology 39, 720-726.
72. Smith, A.N. and Rechenthin, C.A. (1964) Rept. No. 4-19114: Grassland
     Restoration. Part I. The Texas Brush Problem. USDA Soil Conserv.
     Serv., Ft. Worth, TX.
73. Anonymous. (1974) Acres of grasslands infested with woody plants. Range
     Newsletter. Texas A&M Univ., College Station, TX.
74. Soil Conserv. Serv. (1983) Texas Brush Inventory. U.S. Dept. Agric.,
     Washington, DC.
75. Markgraf, V. (1983) Late and postglacial vegetational paleoclimatic
     changes in subantarctic, temperate, and arid environments in Argentina
     Palynology 7, 43-70.
76. Markgraf, V. (1984) Late Pleistocene and Holocene vegetation history of
     temperate Argentina: Lago Morenito, Bariloche. Dissert. Bot. 72, 235-
     254.
77. Markgraf, V. (1987) Paleoclimates of the Southern Argentine Andes. Cur.
     Res. Pleist. 4, 150-157.
78. Veblen, T.T. and Markgraf, V. (1988) Steppe expansion in Patagonia?
     Quat. Res. 30, 331-338.
79. Willis, B. (1914) Northern Patagonia: Character and Resources. Ministry
     of Public Works, Buenos Aires, Argentina.
80. Rothkugel, M. (1916) Los Bosques Patagonicos. Ministry of Agriculture,
     Buenos Aires, Argentina.
81. Weber, T.F. (1951) Tendencias de las lluvias en la Argentina en lo que
     va del siglo. Idia 48, 6-12.
82. Barros, V. (1978) Algunos aspectos de las fluctuaciones climaticas de
     los ultimos 50 anos en la provincia de Chubut. Ciencia Interamer. 19,
     18-21.
83. Holmes, R.L., Stockton, C.W. and LaMarche, V.C. (1979) Extension of
     river flow records in Argentina from long tree-ring chronologies.
     Water Res. Bull. 15, 1081-1085.
84. Donaldson, C.H. (1969) Brush Encroachment with Special Reference to the
     Blackthorn Problem of the Molopo Area. Dept. Agric. Tech. Services,
     Pretoria, South Africa.
85. Harris, D.R. (1966) Recent plant invasions in the arid and semi-arid
     southwest of the United States. Ann. Assoc. Amer. Geogr. 56, 408-422.
86. Forse, W. (1989) The myth of the marching desert. New Sci. 121(1650),
     31-32.
87. Idso, S.B. (1985) Industrial age leading to the greening of the Earth?
     Nature 320, 22.
88. Idso, S.B. (1987) Detection of global carbon dioxide effects. Nature
     329, 293.
89. D'Arrigo, R., Jacoby, G.C. and Fung, I.Y. (1987) Boreal forests and
     atmosphere-biosphere exchange of carbon dioxide. Nature 329, 321-323.
90. Hall, C.A., Edkahl, C.A. and Wartenberg, D.C. (1975) A fifteen-year
     record of biotic metabolism in the Northern Hemisphere. Nature 255,
     136-138.

228

REFERENCES   SECTION §

91. Pearman, G.I. and Hyson, P. (1981) The annual variation of atmospheric
     CO_2 concentration observed in the northern hemisphere. J. Geophys.
     Res. 86, 9839-9847.
92. Cleveland, W.S., Frenny, A.E. and Graedel, T.E. (1983) The seasonal
     component of atmospheric CO_2: Information from new approaches to the
     decomposition of seasonal time-series. J. Geophys. Res. 88, 10,934-
     10,940.
93. Keeling, C.D., Whorf, T.P., Wong, C.S. and Bellagay, R.D. (1985) The
     concentration of atmospheric carbon dioxide at ocean weather station P
     from 1969-1981. J. Geophys. Res. 90, 10,511-10,528.
94. Bacastow, R.B., Keeling, C.D. and Whorf, T.P. (1985) Seasonal amplitude
     increase in atmospheric CO_2 concentration at Mauna Loa, Hawaii, 1959-
     1982. J. Geophys. Res. 90, 10,529-10,540.
95. Enting, I.G. (1987) The interannual variation in the seasonal cycle of
     carbon dioxide concentration at Mauna Loa. J. Geophys. Res. 92, 5497-
     5504.
96. Houghton, R.A. (1987) Biotic changes consistent with the increased
     seasonal amplitude of atmospheric CO_2 concentrations. J. Geophys. Res.
     92, 4223-4230.
97. Idso, S.B. (1988) Comment on "Biotic changes consistent with the
     increased seasonal amplitude of atmospheric CO_2 concentrations. J.
     Geophys. Res. 93, 1745-1746.
98. Blank, L.W., Roberts, T.M. and Skeffington, R.A. (1988) New perspectives
     on forest decline. Nature 336, 27-30.
99. Ryther, J.H. (1969) Photosynthesis and fish production in the sea.
     Science 166, 72-76.
100. Shulenberger, E. and Reid, L. (1981) The Pacific shallow oxygen maximum,
     deep chlorophyll maximum, and primary production, reconsidered. Deep-
     Sea Res. 28, 901-919.
101. Jenkins, W.J. (1982) Oxygen utilization rates in North Atlantic
     subtropical gyre and primary production in oligotrophic systems.
     Nature 300, 246-248.
102. Kerr, R.A. (1983) Are the ocean's deserts blooming? Science 220, 397-
     398.
103. Jenkins, W.J. and Goldman, J.C. (1985) Seasonal oxygen cycling and
     primary production in the Sargasso Sea. J. Mar. Res. 43, 465-491.
104. Reid, J.L. and Shulenberger, E. (1986) Oxygen saturation and carbon
     uptake near 28°N, 155°W. Deep-Sea Res. 33, 267-271.
105. Marra, J. and Heinemann, K.R. (1987) Primary production in the North
     Pacific central gyre: Some new measurements based on [14]C. Deep-Sea
     Res. 34, 1821-1829.
106. Laws, E.A., DiTullio, G.R. and Redalje, D.G. (1987) High phytoplankton
     growth and production rates in the North Pacific subtropical gyre.
     Limnol. Oceanogr. 32, 905-918.
107. Packard, T.T., Denis, M., Rodier, M. and Garfield, P. (1988) Deep-ocean
     metabolic CO_2 production: Calculations from ETS activity. Deep-Sea
     Res. 35, 371-382.
108. Kerr, R.A. (1986) The ocean's deserts are blooming. Science 232, 1345.
109. Gieskes, W.C.C., Kraay, G.W. and Baars, M.A. (1979) Current ^{14}C methods
     for measuring primary production: Gross underestimates in oceanic
     waters. Neth. J. Sea Res. 13, 58-78.
110. Fitzwater, S.E., Knauer, G.A. and Martin, J.H. (1982) Metal
     contamination and primary production: Field and laboratory methods of
     control. Limnol. Oceanogr. 27, 544-551.
111. Rivkin, R.B. and Putt, M. (1987) Diel periodicity of photosynthesis in
     polar phytoplankton: Influence on primary production. Science 238,
     1285-1288.
112. Sharp, J.H., Perry, M.J., Renger, E.H. and Eppley, R.W. (1980)
     Phytoplankton rate processes in the oligotrophic waters of the central

North Pacific Ocean. J. Plankton Res. 2, 335-353.
13. Marra, J. and Heinemann, K.R. (1984) A comparison between noncontaminating and conventional procedures in primary productivity measurements. Limnol. Oceanogr. 29, 389-392.
14. Venrick, E.L., McGowan, J.A., Cayan, D.R. and Hayward, T.L. (1987) Climate and chlorophyll a: Long-term trends in the central North Pacific ocean. Science 238, 70-72.
15. Platt, T. and Harrison, W.G. (1985) Biogenic fluxes of carbon and nitrogen in the ocean. Nature 318, 55-58.
16. Glover, H.E., Prezelin, B.B., Campbell, L., Wymans, M. and Garside, C. (1988) A nitrate-dependent Synechococcus bloom in Sargasso Sea water. Nature 331, 161-163.
17. Broecker, W.S. and Peng, T.H. (1982) Tracers in the Sea. Eldigio, Palisades, NY.
18. Kasting, J.F., Toon, O.B. and Pollack, J.B. (1988) How climate evolved on the terrestrial planets. Sci. Amer. 258(2), 90-97.
19. Ittekkot, V. (1988) Global trends in the nature of organic matter in river suspensions. Nature 332, 436-438.
20. Garside, C. (1985) The vertical distribution of nitrate in open ocean surface water. Deep Sea Res. 32, 723-732.
21. Klein, P. and Coste, B. (1984) Effects of wind-stress variability on nutrient transport into the mixed layer. Deep Sea Res. 31, 21-37.

1. Nahas, G., Poyart, C. and Triner, L. (1968) Acid base equilibrium changes and metabolic alterations. Ann. N. Y. Acad. Sci. 150, 562-576.
2. Brackett, N.C., Jr., Wingo, C.F., Muren, O. and Solano, J.T. (1969) Acid-base response to chronic hypercapnia in man. New Engl. J. Med. 280, 124-130.
3. van Ypersele de Strihou, C. (1974) Acid-base equilibrium in chronic hypercapnia. In Nahas, G. and Schaefer, K.E. (eds) Carbon Dioxide and Metabolic Regulations. Springer-Verlag, New York, NY, pp. 266-272.
4. Poyart, C.F. and Nahas, G. (1968) Inhibition of activated lipolysis by acidosis. Molecular Pharmacol. 4, 389-401.
5. Turino, G.M., Goldring, R.M. and Heinemann, H.O. (1974) The extracellular bicarbonate concentration and the regulation of ventilation in chronic hypercapnia in man. In Nahas, G. and Schaefer, K.E. (eds) Carbon Dioxide and Metabolic Regulations. Springer-Verlag, New York, NY, pp. 273-281.
6. Luft, U.C., Finkelstein, S. and Elliott, J.C. (1974) Respiratory gas exchange, acid-base balance, and electrolytes during and after maximal work breathing 15 mm Hg P_{ICO2}. In Nahas, G. and Schaefer, K.E. (eds) Carbon Dioxide and Metabolic Regulations. Springer-Verlag, New York, NY, pp. 282-293.
7. Schaefer, K.E. (1982) Effects of increased ambient CO_2 levels on human and animal health. Experientia 38, 1163-1168.
8. Bown, A.W. (1985) CO_2 and intracellular pH. Plant Cell Environ. 8, 459-465.
9. McAleavy, J.C., Way, W.L., Altotatt, A.H., Guadagni, N.P. and Severinghaus, J.W. (1961) The effect of pCO_2 on the depth of anesthesia. Anesthesiology 22, 260-264.
10. Dundee, J.W., Black, G.W. and Nicholl, R.M. (1962) Alterations in response to somatic pain associated with anesthesia. Brit. J. Anesth. 34, 24-30.
11. Eisele, J.H., Eger, E.I., II and Muallem, M. (1967) Narcotic properties of carbon dioxide in the dog. Anesthesiology 28, 856-865.
12. Fenn, W.O. (1974) Carbon dioxide and the sea. In Nahas, G. and Schaefer

K.E. (eds) <u>Carbon Dioxide</u> and <u>Metabolic Regulations</u>. Springer-Verlag,
New York, NY, pp. xix-xxv.

13. Albers, C. (1974) Carbon dioxide and the utilization of oxygen. Ir
    Nahas, G. and Schaefer, K.E. (eds) <u>Carbon Dioxide</u> and <u>Metabolic</u>
    <u>Regulations</u>. Springer-Verlag, New York, NY, pp. 144-151.

14. Kammermeier, H., Rudroff, W. and Gerlach, E. (1968) Beeinflussung vor
    kontraktilitat, kroonarfluss und intrazellularen metaboliten des
    isolierten herzens bei variation von pH, P_{CO2} und bikarbonat. In
    Reindell, H., Keul, J. and Doll, E. (eds) <u>Herzinsuffizienz</u>. Thieme,
    Stuttgart, W. Germany, pp. 242-247.

15. Demers, H.G., Spaich, P. and Usinger, W. (1969) Der hirnkreislauf bei
    erhkohter korpertemperatur. <u>Verh</u>. <u>Dtsch</u>. <u>Ges</u>. <u>Kreislaufforschg</u>. 35,
    340-343.

16. Lenhoff, H.M. and Loomis, W.F. (1957) Environmental factors controlling
    respiration in hydra. <u>J</u>. <u>Exp</u>. <u>Zool</u>. 134, 171-182.

17. Loomis, W.F. (1957) Sexual differentiation in hydra: Control by carbon
    dioxide tension. <u>Science</u> 126, 735-739.

18. Braverman, M.H. (1962) Studies in hydroid differentiation: I. <u>Podocoryne</u>
    <u>carnea</u>, culture methods and carbon dioxide induced sexuality. <u>Exp</u>.
    <u>Cell</u> <u>Res</u>. 27, 301-306.

19. Valley, G. and Rettger, L.F. (1927) Influence of carbon dioxide on
    bacteria. <u>J</u>. <u>Bacteriol</u>. 14, 101-123.

20. Geyer, R.P. and Chang, R.S. (1958) Bicarbonate as an essential for human
    cells in vitro. <u>Arch</u>. <u>Biochem</u>. <u>Biophys</u>. 73, 500-506.

21. Runyan, W.S. and Geyer, R.P. (1967) Partial replacement of CO_2 in strain
    L and Hela cultures. <u>Proc</u>. <u>Soc</u>. <u>Exp</u>. <u>Biol</u>. <u>Med</u>. 125, 1301-1304.

22. Nesius, K.K. and Fletcher, J.S. (1973) Carbon dioxide and pH
    requirements of non-photosynthetic tissue culture cells. <u>Physiol</u>.
    <u>Plant</u>. 28, 259-263.

23. Burk, D. (1961) On the use of carbonic anhydrase in carbonate and amine
    buffers for CO_2 exchange in manometric vessels, atomic submarines, and
    industrial CO_2 scrubbers. <u>Ann</u>. <u>New</u> <u>York</u> <u>Acad</u>. <u>Sci</u>. 92, 372-400.

24. Tozer, B.T., Cammack, K.A. and Smith, H. (1962) Separation of antigens
    by immunological specificity. 2. Release of antigen and antibody from
    their complexes by aqueous carbon dioxide. <u>Biochem</u>. <u>J</u>. 84, 80-93.

25. Hetherington, C.M (1968) Induction of deciduomata in the mouse by carbon
    dioxide. <u>Nature</u> 219, 863-864.

26. McLaren, A. (1969) Stimulus and response during early pregnancy in the
    mouse. <u>Nature</u> 221, 739-741.

27. Hastings, A.B. (1970) A biochemist's Anabasis. <u>Ann</u>. <u>Rev</u>. <u>Biochem</u>. 39,
    1-24.

28. Roughton, F.J.W. (1970) Some recent work on the interactions of oxygen,
    carbon dioxide and haemoglobin. <u>Biochem</u>. <u>J</u>. 117, 801-812.

29. Lorimer, G.H. (1983) Carbon dioxide and carbamate formation: The makings
    of a biochemical control system. <u>Trends</u> <u>Biochem</u>. <u>Sci</u>. 8, 65-68.

30. Mitz, M.A. (1979) CO_2 biodynamics: A new concept of cellular control. <u>J</u>.
    <u>Theoret</u>. <u>Biol</u>. 80, 537-551.

31. Forster, R.E., Edsall, J.T., Otis, A.B. and Roughton, F.J.W., Eds.
    (1968) <u>CO_2</u>: <u>Chemical</u>, <u>Biochemical</u> <u>and</u> <u>Biophysical</u> <u>Aspects</u>. NASA,
    Washington, DC.

32. Nahas, G. and Schaefer, K.E., Eds. (1974) <u>Carbon Dioxide</u> and <u>Metabolic</u>
    <u>Regulations</u>. Springer-Verlag, New York, NY.

33. Bauer, C., Gros, G. and Bartels, H. (1980) <u>Biophysics</u> <u>and</u> <u>Physiology</u> <u>of</u>
    <u>Carbon Dioxide</u>. Springer-Verlag, New York, NY.

34. Carpenter, D.O., Hubbard, J.H., Humphrey, D.R., Thompson, H.K. and
    Marshall, W.H. (1974) Carbon dioxide effects on nerve cell function.
    <u>In</u> Nahas, G. and Schaefer, K.E. (eds) <u>Carbon Dioxide</u> and <u>Metabolic</u>
    <u>Regulations</u>. Springer-Verlag, New York, NY, pp. 49-62.

35. Longmore, W.J., Liang, A.M. and McDaniel, M.L. (1974) The regulation by

carbon dioxide of carbohydrate and lipid metabolism in isolated perfused tissue. In Nahas, G. and Schaefer, K.E. (eds) Carbon Dioxide and Metabolic Regulations. Springer-Verlag, New York, NY, pp. 15-23.

36. Chalazonitis, N. (1974) Simultaneous recordings of pH, P_{CO_2}, and neuronal activity during hypercapnic transients (identifiable neurons of Aplysia). In Nahas, G. and Schaefer, K.E. (eds) Carbon Dioxide and Metabolic Regulations. Springer-Verlag, New York, NY, pp. 63-80.

37. Mitz, M.A. (1957) The solubility of proteins in the presence of carbon dioxide. Biochim. Biophys. Acta 25, 425-426.

38. Moyse, A. (1974) Carbon dioxide and metabolic regulations in plant photosynthesis. In Nahas, G. and Schaefer, K.E. (eds) Carbon Dioxide and Metabolic Regulations. Springer-Verlag, New York, NY, pp. 3-14.

39. Chance, B. and Park, J.H. (1967) The properties and enzymatic significance of the enzyme-diphosphopyridine nucleotide compound of 3-phosphoglyceraldehyde dehydrogenase. J. Biol. Chem. 242, 5093-5105.

40. Morgan, T.H. (1927) Experimental Embryology. Columbia Univ. Press, New York, NY.

41. Lorimer, G.H. and Miziorko, H.M. (1980) Carbamate formation on the ϵ-amino group of a lysyl residue as the basis for the activation of ribulosebisphosphate carboxylase by CO_2 and Mg^{2+}. Biochemistry 19, 5321-5328.

42. Lorimer, G.H. (1981) The carboxylation and oxygenation of ribulose 1,5-bisphosphate: The primary events in photosynthesis and photorespiration. Ann. Rev. Plant Physiol. 32, 349-383.

43. Appleby, C.A. (1985) Plant hemoglobin properties, function and genetic origin. In Ludden, P.W. and Burris, J.E. (eds) Nitrogen Fixation and CO_2 Metabolism. Elsevier, New York, NY, pp. 41-51.

44. Landsmann, J., Dennis, E.S., Higgins, T.J.V., Appleby, C.A., Kortt, A.A. and Peacock, W.J. (1986) Common evolutionary orgin of legume and non-legume plant haemoglobins. Nature 324, 166-168.

45. Appleby, C.A., Bogusz, D., Dennis, E.S. and Peacock, W.J. (1988) A role for haemoglobin in all plant roots? Plant Cell Environ. 11, 359-367.

46. Stupfel, M. (1974) Carbon dioxide and temperature regulation of homeothermic mammals. In Nahas, G. and Schaefer, K.E. (eds) Carbon Dioxide and Metabolic Regulations. Springer-Verlag, New York, NY, pp. 163-186.

47. Rahn, H. (1974) P_{CO_2}, pH and body temperature. In Nahas, G. and Schaefer, K.E. (eds) Carbon Dioxide and Metabolic Regulations. Springer-Verlag, New York, NY, PP. 151-162.

48. Siegfried, M. (1907) Bemerkung zur methode der bestimmung des quotienten CO_2/N bei der carbaminoreaktion. Z. Physiol. Chem. 52, 506.

49. Mitz, M.A. (1956) New insoluble active derivative of an enzyme as a model for study of cellular metabolism. Science 123, 1076-1077.

50. Wittebort, R.J., Hayes, D.F., Rothgeb, M. and Gurd, R.S. (1978) The quantitation of carbamino adduct formation of angiotension II and bradykinin. Biophys. J. 24, 765-778.

51. Salisbury, G.W. and VanDemark, N.L. (1957) Sulfa compounds in reversible inhibition of sperm metabolism by carbon dioxide. Science 126, 1118-1119.

52. Eckhardt, R.B., Eckhardt, D.A. and Eckhardt, J.T. (1988) Are racehorses becoming faster? Nature 335, 773.

53. Hill, W.G. (1988) Why aren't horses faster? Nature 332, 678.

54. Gaffney, B. and Cunningham, E.P. (1988) Estimation of genetic trend in racing performance of thoroughbred horses. Nature 332, 722-724.

55. Bryant, J.P., Wieland, G.D., Reichardt, P.B., Lewis, V.E. and McCarthy, M.C. (1983) Pinosylvin methyl ether deters snowshoe hare feeding on green alder. Science 222, 1023-1025.

56. Bryant, J.P. (1987) Feltleaf willow-snowshoe hare interactions: Plant

carbon/nutrient balance and floodplain succession. Ecology 68, 1319-1327.

57. Rosenthal, G.A. and Janzen, D.J., Eds. (1979) Herbivores: Their Interaction with Secondary Plant Metabolites. Academic Press, New York, NY.

58. Crawley, M.J. (1983) Herbivory: The Dynamics of Animal-Plant Interactions. Univ. Calif. Press, Berkeley, CA.

59. Hassell, M.P. (1976) The Dynamics of Competition and Predation. Edward Arnold, London, UK.

60. Taylor, R.J. (1984) Predation: Population and Community Biology. Chapman and Hall, New York, NY.

61. Fraenkel, G.S. (1959) The raison d'etre of secondary plant substances. Science 129, 1466-1470.

62. Abrahamson, W.G., Ed. (1989) Plant-Animal Interactions. McGraw-Hill, New York, NY.

1. Brown, L.R. and others. (1984, 1985, 1986, 1987, 1988, 1989) State of the World. W.W. Norton & Co., New York, NY.

2. Maggs, W.W. (1989) Congress tackles human-induced climate change. Trans. Amer. Geophys. Union 70, 97 & 105.

3. McElroy, M. (1988) The challenge of global change. New Sci. 119(1623), 34-36.

4. Miller, R.L. (1989) From the publisher. Time 133(1), 3.

5. Stoel, T.B., Jr. (1988) Global tomorrow. Amicus J. 10(4), 21-27.

6. Sancton, T.A. (1989) Hands across the sea. Time 133(1), 54, 63.

7. Thompson, D. (1989) The greening of the U.S.S.R. Time 133(1), 68-69.

8. Revkin, A.C. (1989) Cooling off the greenhouse. Discover 10(1), 30-32.

9. Hecht, A.D. (1989) Climate of change. Nature 337, 128.

10. Wirth, T.E. (1989) The ripple of the greenhouse effect. Discover 10(2), 84.

11. McIntyre, A. (1988) Once and future climates: Modeling with paleoclimatology. Lamont-Doherty Geological Observatory Yearbook 1987, 6-10.

12. Thiele, O. and Schiffer, R.A., Eds. (1985) Understanding Climate: A Strategy for Climate Modeling and Predictability Research. U.S. Nat. Aeronaut. Space Admin., Washington, DC.

13. Bryson, R.A. (1988) Civilization and rapid climatic change. Environ. Conserv. 15, 7-15.

14. Tangley, L. (1988) Preparing for climate change. BioScience 38, 14-18.

15. McGourty, C. (1988) Global warming becomes an international political issue. Nature 336, 194.

16. Michaels, P. (1989) The greenhouse climate of fear. The Washington Post 112(34), C3.

17. Broecker, W.S. (1988) Greenhouse surprises. Lamont-Doherty Geological Observatory Yearbook 1987, 11-15.

18. Marshall, E. (1989) EPA's plan for cooling the global greenhouse. Science 243, 1544-1545.

19. Lovelock, J.E. (1988) The Ages of Gaia: A Biography of Our Living Earth. W.W. Norton & Co., New York, NY.

20. Revelle, R. (1974) Food and population. Sci. Amer. 231(3), 161-170.

21. Roberts, L. (1988) Is there life after climate change? Science 242, 1010-1012.

22. Morris, S.C. (1985) Polar forests of the past. Nature 313, 739.

23. Saini, H.S., Waraich, R.S., Malhotra, N.K., Nagpaul, K.K. and Sharma, K.K. (1979) Cooling and uplift rates and dating of thrusts from Kinnaur Himachal, Himalaya. Himal. Geol. 9, 549-567.

24. Burbank, D.W. and Johnson, G.D. (1983) The late Cenozoic chronologic and stratigraphic development of the Kashmir Intermontane Basin. Palaeogeog. Palaeoclimatol. Palaeoecol. 43, 205-235.
25. McCourt, W.J., Aspden, J.A. and Brook, M. (1984) New geological and geochronological data from the Colombian Andes: Continental growth by multiple accretion. J. Geol. Soc. Lond. 141, 831-845.
26. Allmendinger, R.W. (1986) Tectonic development, southeastern border of the Puna Plateau, northwestern Argentine Andes. Bull. Geol. Soc. Amer. 97, 1070-1082.
27. Benjamin, M.T., Johnson, N.M. and Naeser, C.W. (1987) Recent rapid uplift in the Bolivian Andes: Evidence from fission-track dating. Geology 15, 680-683.
28. Ruddiman, W.F. and Raymo, M.E. (1989) Northern Hemisphere climate regimes during the last 3 Myr: Possible tectonic connections. J. Roy. Soc. Lond., in press.
29. Raymo, M.E., Ruddiman, W.F. and Froelich, P.N. (1988) Influence of late Cenozoic mountain building on ocean geochemical cycles. Geology 16, 649-653.
30. Arthur, M.A., Dean, W.E. and Schlanger, S.O. (1985) Variations in the global carbon cycle during the Cretaceous related to climate, volcanism, and changes in atmospheric CO_2. In Sundquist, E.T. and Broecker, W.S. (eds) The Carbon Cycle and Atmospheric CO_2: Natural Variations Archean to Present. Amer. Geophys. Union, Washington, DC, pp. 504-529.
31. Dean. W.E., Arthur, M.A. and Claypool, G.E. (1986) Depletion of ^{13}C in Cretaceous marine organic matter: Source, diagenetic, or environmental signal. Mar. Geol. 70, 119-157.
32. Arthur, M.A., Dean, W.E. and Pratt, L.M. (1988) Geochemical and climatic effects of increased marine organic carbon burial at the Cenomanian/Turonian boundary. Nature 335, 714-717.
33. Bjorkman, O. (1973) Comparative studies on photosynthesis in higher plants. Photophysiol. 8, 1-63.
34. Ehrleringer, J. and Bjorkman, O. (1977) Quantum yields for CO_2 uptake in C_3 and C_4 plants. Plant Physiol. 59, 86-90.
35. Cantrell, A. and Bryant, D.A. (1987) Molecular cloning and nucleotide sequence of the psaA and psaB genes of the cyanobacterium Synechococcus sp. PCC 7002. Plant Molec. Biol. 9, 453-468.
36. Mauney, J.R., Guinn, G., Fry, K.E. and Hesketh, J.D. (1979) Correlation of photosynthetic carbon dioxide uptake and carbohydrate accumulation in cotton, soybean, sunflower and sorghum. Photosynthetica 13, 260-266.
37. Marc, J. and Gifford, R.M. (1983) Flower initiation in wheat, sunflower and sorghum under carbon dioxide enrichment. Can. J. Bot. 62, 9-14.
38. Sionit, N. and Patterson, D.T. (1984) Responses of C_4 grasses to atmospheric CO_2 enrichment. I. Effect of irradiance. Oecologia 65, 30-34.
39. Bonner, W. and Bonner, J. (1948) The role of carbon dioxide in acid formation by succulent plants. Amer. J. Bot. 35, 113-117.
40. Walker, D.A. and Brown, J.M.A. (1957) Physiological studies on acid metabolism. 5. Effects of carbon dioxide concentration on phosphoenolpyruvate carboxylase activity. Biochem. J. 67, 79-83.
41. Allaway, W.G., Austin, B. and Slatyer, R.O. (1974) Carbon dioxide and water vapor exchange parameters of photosynthesis in a Crassulacean plant, Kalanchoe diagremontiana. Aust. J. Plant Physiol. 1, 397-405.
42. Moradshahi, A., Vines, H.M. and Black, C.C. (1977) CO_2 exchange and acidity levels in detached pineapple, Ananas comosus (L.) Merr., leaves during the day at various temperatures, O_2 and CO_2 concentrations. Plant Physiol. 59, 274-278.
43. Nobel, P.S. and Hartsock, T.L. (1986) Short-term and long-term responses

of Crassulacean acid metabolism plants to elevated CO_2. Plant Physiol. 82, 604-606.

44. Linden, E. (1989) Biodiversity: The death of birth. Time 133(1), 32-35.
45. Ellis, W.S. (1988) Brazil's imperiled rain forest. Nat. Geogr. 174, 772-799.
46. Johns, A.D. (1988) Economic development and wildlife conservation in Brazilian Amazonia. Ambio 17, 302-306.
47. Sancton, T.A. (1989) Planet of the year. Time 133(1), 26-30.
48. Gradwohl, J. and Greenberg, R. (1988) Saving the Tropical Forests. Island Press, Covelo, CA.
49. Myers, N. (1988) Threatened biotas: "Hotspots" in tropical forests. The Environmentalist 8, 1-200.
50. Jablonski, D. (1986) Background and mass extinctions: The alternation of macroevolutionary regimes. Science 231, 129-133.
51. Raup, D.M. (1986) Biological extinction in earth history. Science 231, 1528-1533.
52. Myers, N. (1989) Extinction rates past and present. BioScience 39, 39-41.
53. Ehrlich, P.R. and Ehrlich, A.H. (1981) Extinction: The Causes and Consequences of the Disappearance of Species. Random House, New York, NY.
54. Norton, B.G., Ed. (1986) The Preservation of Species: The Value of Biological Diversity. Princeton Univ. Press, Princeton, NJ.
55. Soule, M.E., Ed. (1986) Conservation Biology: The Science of Scarcity and Diversity. Sinauer Associates, Sunderland, MA.
56. Wilson, E.O. (1988) The current slate of biological diversity. In Wilson, E.O. (ed) Biodiversity. National Academy Press, Washington, DC, pp. 3-18.
57. Langone, J. (1989) Waste: A stinking mess. Time 133(1), 44-47.
58. Lemonick, M.D. (1989) Deadly danger in a spray can. Time 133(1), 42.
59. Elmer-DeWitt, P. (1989) Nuclear power plots a comeback. Time 133(1), 41.
60. Toufexis, A. (1989) Overpopulation: Too many mouths. Time 133(1), 48-50.
61. Vitousek, P.M., Ehrlich, P.R., Ehrlich, A.H. and Matson, P.A. (1986) Human appropriation of the products of photosynthesis. BioScience 36, 368-373.
62. Wright, D.H. (1987) Estimating human effects on global extinction. Internat. J. Biometeorol. 31, 293-299.
63. Brown, J.H. and Maurer, B.A. (1989) Macroecology: The division of food and space among species on continents. Science 243, 1145-1150.
64. Lemonick, M.D. (1989) Global warming: Feeling the heat. Time 133(1), 36-39.
65. Sagan, C. and Druyan, A. (1988) Give us hope. Parade. 27 Nov. 1988, pp. 5-9.
66. Roberts, L. (1988) Is there life after climate change? Science 242, 1010-1012.
67. Elmer-DeWitt, P. (1989) Preparing for the worst. Time 133(1), 70-71.
68. Lal, R. (1988) Soil degradation and the future of agriculture in sub-Saharan Africa. J. Soil Water Conserv. 43, 444-451.
69. Gore, A. (1989) What is wrong with us? Time 133(1), 66.
70. Price, P.W. (1988) An overview of organismal interactions in ecosystems in evolutionary and ecological time. Agirc. Ecosys. Environ. 24, 369-377.
71. Mallock, D.W., Pirozynski, K.A. and Raven, P.H. (1980) Ecological and evolutionary significance of mycorrhizal symbioses in vascular plants (a review). Proc. Nat. Acad. Sci. U.S.A. 77, 2113-2118.
72. Margulis, L. (1981) Symbiosis in Cell Evolution: Life and its Environment on the Early Earth. W.H. Freeman, San Francisco, CA.
73. Boucher, D.H., James, S. and Keeler, K.H. (1982) The ecology of mutualism. Ann. Rev. Ecol. Syst. 13, 315-347.

74. Klir, G.J. (1985) Complexity: Some general observations. <u>Syst</u>. <u>Res</u>. 2, 131-140.
75. Wood, F.B., Jr. (1988) The need for systems research on global climate change. <u>Syst</u>. <u>Res</u>. 5, 225-240.
76. AMS Council and UCAR Board of Trustees. (1988) The changing atmosphere- - Challenges and opportunities. <u>Bull</u>. <u>Amer</u>. <u>Meteorol</u>. <u>Soc</u>. 69, 1434- 1440.
77. Committee on Earth Sciences, Federal Coordinating Council for Science, Engineering, and Technology. (1989) <u>Our</u> <u>Changing</u> <u>Planet</u>: <u>A</u> <u>U.S.</u> <u>Strategy</u> <u>for</u> <u>Global</u> <u>Change</u> <u>Research</u>. U.S. Geol. Survey, Reston, VA.

SUBJECT INDEX

240

cucumbers 90, 184
cultures, bacterial 118, 230
            cell 186
            Hela cell 230
            human cell 118, 230
            meristem 205
            plant cell 68, 185, 186
            tissue 185, 230
cumulus clouds 156
            convection 29, 157, 162
currents, air 45
            ocean 18, 58
cuticle, leaf 97, 98, 211
cuttings 82, 91, 96, 198, 199, 205,
    209
cyanobacteria 83, 91, 116, 150, 200,
    206, 233
cyclones 53, 171
cytokinins 197

dark CO_2 fixation 192, 203
    respiration 89, 184, 203
Daucus carota 88
Death Valley, CA 41
death of biosphere 4, 5, 131, 234
    of people 150
decarboxylation 118
Deccan flood basalts 175
decidual response 120, 121
deciduomata 230
deciduous forests 103, 189, 194, 223
decomposer microflora 105
decomposition 2, 113, 114, 194, 220-
    222
deep ocean 52, 53, 138, 173, 186,
    228
    water 18, 154
defense chemicals 100, 101, 215
            mechanisms 100
            strategies 100, 214
            systems 100
deficiency, nutrient 99, 101
defoliation 101, 220
deforestation 5, 18, 113, 133, 140,
    145, 154
degassing 137
degradation of herbicides 195
degree days 160
dehydration 95
dendroclimatology 45, 46, 169
Denver, Colorado 159
depletion, ozone 14, 60, 61, 105,
    106, 131, 150, 151, 179, 180
deposition, pollution 225
desert flora 73, 206, 226
        grasslands 221, 227
        legumes 83, 201
deserts 1, 72, 73, 104, 111-113,
    115, 136, 154, 201, 221, 227
desertscrub 70-73

Desmodium paniculatum 216
detection of warming 13, 148-150,
    176-178, 227
    of biospheric rejuvenation
    108-116, 132, 141, 225,
    227
detritus 78, 191, 195
development times, insect 100
Devil's Hole 174
diagenesis 233
diatoms 173
dieback, forest 1, 65, 110, 114
differentiation, sexual 230
diffusion 153
diffusion resistance, stomatal 183,
    187, 203, 211
digestion 77, 215
dimethylpropiothetin 157
dimethylselenide 222
dimethylsulfide 21, 157, 158, 174
dimethylsulfonio propionate 22, 158
dinosaurs 56, 131, 174, 175
diphosphopyridine 231
disclimax 75
diseases, plant 10, 74, 90, 96, 98,
    99, 107, 129, 142, 143, 192, 200,
    213
disposal, waste 78
dissemination of spores 75
dissolved CO_2 118
diversity, species 1, 73, 74, 131,
    133, 190, 234
DMS 21, 22, 40, 53, 54, 56, 57, 134
dogs 229
Dome C 173
doom and gloom 9, 10, 12, 127, 165
doomsday 165
dormancy 197
downdrafts 21, 155-157
drainage, soil 77, 147
        surface water 79, 222
drizzle 23
droplets, cloud 21, 23, 158, 159
drought 1, 13, 41, 45-47, 110, 136,
    147, 154, 164, 165, 168-170, 189,
    207, 208, 212, 217
drought tolerance 75, 98, 207
dry matter content 79, 80, 94-96,
        196
        partitioning 201, 202,
        225
        production 9, 69, 88,
        184, 199, 203, 207,
        216, 217, 221, 224
dryness, soil 146
dust 52, 155, 169, 174
dust bowl 169
    storms 191
dwarf spring wheat 185
dwarfing 143
dynamical heat flux 18, 153

AUTHOR INDEX

Murphy, P.M.   200
Murray, D.W.   173
Murty, T.S.   149
Musgrave, D.R.   191
Musgrave, M.E.   204
Myers, N.   140, 234

Naeser, C.W.   233
Nagpaul, K.K.   232
Nagy, M.   215
Nahas, G.   229-230
Nakanishi, T.   204
Nakayama, F.S.   193, 208
Namias, J.   146, 168, 170
Nance, R.D.   137
Naujokat, B.   149
Navato, A.R.   162
Naylor, D.   191
Neales, T.F.   198, 204
Neelin, J.D.   157
Neeman, B.U.   153, 160
Neftel, A.   139, 140, 148, 171
Negm, F.B.   197
Neiderpruem, D.J.   204
Nelson, C.D.   201
Nelson, D.E.   173
Nesius, K.K.   204, 230
Neumann, K.H.   185
Newell, K.D.   201
Newell, N.D.   141
Newell, R.D.   151, 152, 162, 163
Newman, M.J.   138
Newton, C.W.   174
Newton, J.W.   150
Newton, W.E.   200
Nguyen, B.C.   157, 158
Nguyen, H.   139
Nicholas, W.L.   221
Nicholl, R.M.   229
Nicholls, A.O.   198
Nicholls, N.   170
Nicholls, S.   159
Nichols, H.   171
Nicholson, F.   203, 215
Nicholson, J.Y., III   179
Nicolis, C.   169, 173
Nicolson, T.H.   190
Nieman, R.A.   175
Nieman, R.H.   209
Nijs, I.   183
Nilovskaya, N.T.   203
Nilsen, E.T.   201
Nisbet, E.G.   166
Nitecki, M.H.   136
Nival, P.   157
Nixon, P.E.   208
Nobel, P.S.   188, 233
Nobs, M.A.   203, 215
Noguera, J.   220

Noonkester, V.R.   159
Norby, R.J.   218, 219, 225
North, G.R.   136, 144, 149
Norton, B.G.   234
Norton-Griffiths, M.   215
Nullet, D.   155
Nutman, P.S.   200
Nyman, C.J.   200

O'Brian, M.J.   142
O'Leary, J.W.   226
O'Neill, E.G.   217, 225
O'Neill, P.   204
Oakes, D.B.   195
Oberbauer, S.F.   223
Oberhansli, H.   176
Odell, R.T.   221
Odum, E.   191
Oebker, N.F.   193
Oechel, W.C.   197, 214, 223
Oeschger, H.   139, 140, 148, 171, 181
Officer, C.B.   175
Ogawa, T.   205, 206
Ogren, J.A.   161
Ogren, W.L.   202, 206
Ogsen, J.A.   159
Ohhara, Y.   197
Ohring, G.   153, 160
Oikawa, T.   224
Oke, T.R.   166, 167
Okon, Y.   219
Okuda, M.   158
Oldham, M.G.   221
Olivieri, I.   214
Olson, J.S.   140
Olson, O.E.   222
Omata, T.   205, 206
Oort, A.H.   162, 167
Ooshima, Y.   197
Oota, H.   198
Origoreev, E.F.   220
Ormrod, D.P.   210
Orth, C.J.   175
Ortuno, A.   220
Osbrink, W.L.A.   214
Oschwald, W.R.   192
Osmond, C.B.   187, 188, 203
Otis, A.B.   230
Overdieck, D.   188, 189
Owen, T.   137, 161
Oyama, Y.I.   161

Packard, T.T.   228
Paez, A.   184, 207
Page, N.R.   191
Pain, S.   144, 188, 189